BILDER VOM MARS
NICOLAS THOMAS UND DAS CaSSIS-TEAM

WEBERVERLAG.CH

BILDER
VOM
MARS

NICOLAS THOMAS
UND DAS
CaSSIS-TEAM

IMAGES OF MARS
IMAGES DE MARS
IMMAGINI DA MARTE
ИЗОБРАЖЕНИЯ С МАРСА

WEBERVERLAG.CH

INHALT
CONTENTS
CONTENU
CONTENUTO
СОДЕРЖАНИЕ

- 7 VORWORT
- 8 PREFACE
- 9 PRÉFACE
- 10 PREFAZIONE
- 11 РПРЕДИСЛОВИЕ

13 DER EXOMARS TRACE GAS ORBITER

33 DAS COLOUR AND STEREO SURFACE IMAGING SYSTEM

55 EINE AUSWAHL VON BILDERN

- 58 A SELECTION OF IMAGES
- 60 UNE SÉLECTION D'IMAGES
- 62 UNA SELEZIONE DI IMMAGINI
- 64 ПОДБОРКА ИЗОБРАЖЕНИЙ

67 DELTAS, FLÜSSE, ERDRUTSCHE UND FÄCHER

DELTAS, FLOWS, LANDSLIDES AND FANS
DELTAS, COULÉES, GLISSEMENTS DE TERRAIN ET ÉVENTAILS
DELTA, FLUSSI, FRANE E CONI DI DEIEZIONE
ДЕЛЬТЫ, ПОТОКИ, ОПОЛЗНИ И ВЕЕРНЫЕ ОТЛОЖЕНИЯ

89 TÄLER UND RINNEN

VALLEYS AND CHANNELS
VALLÉES ET CHENAUX
VALLI E CANALI
ДОЛИНЫ И КАНАЛЫ

121 KRATERINNERES UND ERHEBUNGEN

CRATER INTERIORS AND UPLIFTS
INTÉRIEURS DE CRATÈRES ET SOULÈVEMENTS
AL'INTERNO DEI CRATERI E SOLLEVAMENTI CENTRALI
ВНУТРЕННИЕ ОБЛАСТИ И ЦЕНТРАЛЬНЫЕ ПОДНЯТИЯ КРАТЕРОВ

155 KRATERRÄNDER UND RINNEN

CRATER RIMS AND GULLIES
BORDURES DE CRATÈRES ET RAVINES
BORDI DI CRATERI E CALANCHI
ВАЛЫ КРАТЕРОВ И ОВРАГИ

197 ERDHÜGEL, CHAOS, TAFELBERGE UND KONTAKTE
    MOUNDS, CHAOS, MESAS, AND CONTACTS
    MONTICULES, CHAOS, MÉSAS ET CONTACTS
    TUMULI, CHAOS, MESAS E ZONE DI CONTATTO
    КУРГАНЫ, ХАОСЫ, СТОЛОВЫЕ ГОРЫ И КОНТАКТЫ РЕЛЬЕФОВ

217 WOLKEN, EIS, POLYGONE UND MUSCHELN
    CLOUDS, ICE, POLYGONS, AND SCALLOPS
    NUAGES, GLACE, POLYGONES ET PÉTONCLES
    NUVOLE, GHIACCIO, POLIGONI E SCALLOPS
    ОБЛАКА, ЛЕД, ПОЛИГОНЫ И ФЕСТОНЫ

255 VULKANE, LAVEN, GRÄBEN, GRUBEN UND GÄNGE
    VOLCANOES, LAVAS, GRABEN, PITS, AND DIKES
    VOLCANS, LAVES, GRABEN, FOSSES ET DYKES
    VULCANI, COLATE LAVICHE, GRABEN, FOSSE E DICCHI
    ВУЛКАНЫ, ЛАВЫ, ГРАБЕНЫ, ЯМЫ И ДАЙКИ

293 DÜNEN, STAUB UND HANGSTREIFEN
    DUNES, DUST, AND SLOPE STREAKS
    DUNES, POUSSIÈRE ET TRAÎNÉES DE TALUS
    DUNE, POLVERE E STRISCE DI PENDENZA
    ДЮНЫ, ПЫЛЬ И ПОЛОСЫ НА СКЛОНАХ

325 UNTERSCHIEDLICHE MINERALOGIEN UND SCHICHTEN
    DIVERSE MINERALOGIES AND LAYERS
    MINÉRALOGIES ET STRATES DIVERSES
    DIVERSE MINERALOGIE E STRATIFICAZIONI
    МНОГООБРАЗИЕ МИНЕРАЛОВ И СЛОЁВ

363 GRUNDGEBIRGE UND KOMPLEXE GELÄNDEFORMEN
    BEDROCKS AND COMPLEX TERRAINS
    SUBSTRATS ROCHEUX ET TERRAINS COMPLEXES
    PAVIMENTI ROCCIOSI E TERRENI COMPLESSI
    КОРЕННЫЕ ПОРОДЫ И СЛОЖНЫЕ РЕЛЬЕФЫ

GLOSSAR
GLOSSARY
GLOSSAIRE
GLOSSARIO
ГЛОССАРИЙ

430   DIE GEOLOGISCHEN PERIODEN DES MARS
430   THE GEOLOGICAL PERIODS OF MARS
430   TEMPS GÉOLOGIQUES MARTIENS
431   I PERIODI GEOLOGICI DI MARTE
431   ГЕОЛОГИЧЕСКИЕ ПЕРИОДЫ МАРСА

432   REGIONS OF MARS
434   TERMS IN PLANETARY GEOLOGY

Weitere Übersetzungen des Glossars sowie zusätzliche Bilder finden Sie online.

Further translations of the glossaries can be found online.

D'autres images sont disponibles sur notre site ainsi que la traduction en français du glossaire.

Ulteriori traduzioni dei termini del glossario possono essere trovate online.

Остальной перевод глоссария можно найти онлайн.

https://book.cassis.unibe.ch

436   ACKNOWLEDGEMENTS

439   AUTORENPORTRÄT
        AUTHOR PORTRAIT
        PORTRAIT DE L'AUTEUR
        RITRATTO D'AUTORE
        ОБ АВТОРЕ

VORWORT

Das Colour and Stereo Surface Imaging System (CaSSIS) ist eine Kamera, die von einem Team unter der Leitung der Universität Bern in der Schweiz für den ExoMars Trace Gas Orbiter der Europäischen Weltraumorganisation (ESA) gefertigt wurde. Zum Zeitpunkt der Erstellung dieses Buchs übermittelt das Instrument weiterhin über 200 Bilder der Marsoberfläche pro Woche. Bei den Bildern handelt es sich in der Regel um Farbaufnahmen. Sobald die Daten an die Erde übertragen sind, können wir sie verarbeiten und verbessern, um die Vielfalt der Farben der Oberfläche des Mars zu untersuchen. Diese Verbesserungen sind für die wissenschaftliche Auswertung der Daten nützlich, führen aber auch zu spektakulär schönen Bildern, die in einigen Fällen fast wie abstrakte Kunst wirken.

Ziel dieses Buches ist es, rund 180 ausgewählte CaSSIS-Bilder zu präsentieren. Zu jedem Bild gibt es eine kurze Bildunterschrift, die den Standort, den Kontext und die mögliche wissenschaftliche Interpretation beschreibt. Wir stellen diesen Text in fünf Sprachen zur Verfügung, um sowohl den mehrsprachigen Charakter der Schweiz als auch den multinationalen Charakter unserer wissenschaftlichen und technischen Teams widerzuspiegeln. Wir beginnen jedoch mit einer kurzen Beschreibung des Raumschiffs und des Instruments, um Ihnen bei Bedarf Hintergrundinformationen zu bieten.

PREFACE

The Colour and Stereo Surface Imaging System (CaSSIS) is a camera that was built by a team led by the University of Bern in Switzerland for the European Space Agency's (ESA) ExoMars Trace Gas Orbiter. At the time of writing, the instrument is continuing to return over 200 images of the surface of Mars per week. The images are usually colour acquisitions. Once the data are transmitted to ground, we can process and enhance these to study the diversity in the colour of the surface. This enhancement is useful for scientific evaluation of the data but it also results in spectacularly beautiful images that, in some cases, appear almost like abstract art.

The goal of this book is to present around 180 selected CaSSIS images. For each image, we provide a brief caption describing the location, the context, and possible scientific interpretation. We provide this text in 5 languages to reflect both the multi-lingual character of Switzerland and the multi-national nature of our Science and Engineering Teams. We begin however by briefly describing the spacecraft and the instrument to provide background information should you need it.

PRÉFACE

Le Système d'imagerie de surface en couleurs et en stéréo ("Color and Stereo Surface Imaging System", CaSSIS) est l'imageur construit par une équipe dirigée par l'Université de Berne en Suisse pour la sonde ExoMars Trace Gas Orbiter de l'Agence Spatiale Européenne (ESA). Au moment de la rédaction de ce livre, l'instrument continue de nous envoyer plus de 200 images de la surface de Mars par semaine. Les images sont généralement acquises en couleurs. Une fois les données transmises au sol, nous pouvons les traiter et les enrichir pour étudier la diversité de couleur de la surface. Ces opérations sont utiles pour l'analyse scientifique des données, mais elles produisent également des images d'une beauté spectaculaire qui, dans certains cas, ressemblent presque à de l'art abstrait.

L'objectif de ce livre est de présenter environ 180 images CaSSIS soigneusement sélectionnées. Pour chaque image, nous proposons une brève légende décrivant l'emplacement, le contexte et une interprétation scientifique possible. Nous fournissons ce texte en 5 langues pour refléter à la fois le caractère multilingue de la Suisse et la nature multinationale de nos équipes scientifiques et techniques. Nous commençons cependant par décrire brièvement la sonde spatiale et l'instrument pour fournir les quelques informations générales nécessaires à la compréhension du projet.

PREFAZIONE

Il Colour and Stereo Surface Imaging System (CaSSIS) è una fotocamera costruita da un team guidato dall'Università di Berna in Svizzera per l'ExoMars Trace Gas Orbiter dell'Agenzia Spaziale Europea (ESA). Al momento, lo strumento sta continuando ad acquisire oltre 200 immagini della superficie di Marte a settimana. Le immagini sono di solito acquisizioni a colori. Una volta che i dati sono trasmessi a terra, possiamo elaborarli e migliorarli per studiare la diversità del colore della superficie. Questo miglioramento dei dati è utile per la loro valutazione scientifica, ma risulta anche in immagini spettacolarmente belle che, in alcuni casi, sembrano quasi arte astratta.

Questo libro presenta circa 180 immagini selezionate di CaSSIS. Per ogni immagine, forniamo una breve descrizione della zona, il contesto e la possibile interpretazione scientifica. Questo testo è scritto in cinque lingue per rispecchiare sia il carattere multilingue della Svizzera che la natura internazionale dei nostri gruppi scientifici e ingegneristici. Iniziamo prima di tutto descrivendo brevemente il veicolo spaziale e lo strumento per fornire informazioni di base, qualora il lettore ne avesse bisogno.

ПРЕДИСЛОВИЕ

Система цветной и стерео съёмки поверхности (Colour and Stereo Surface Imaging System, CaSSIS) – это камера, созданная командой под руководством Бернского университета в Швейцарии для орбитального аппарата Европейского космического агентства (ЕКА) ExoMars Trace Gas Orbiter. На момент написания книги прибор продолжает получать более 200 изображений поверхности Марса в неделю. Изображения, как правило, цветные. После передачи данных на землю мы можем обработать и усилить цвета для изучения разнообразия оттенков поверхности. Такое усиление цветов полезно для научной оценки данных, но оно также приводит к созданию впечатляюще красивых изображений, которые в некоторых случаях выглядят почти как абстрактное искусство.

Цель этой книги – представить около 180 избранных изображений CaSSIS. Каждому из них мы даём краткое описание, с указанием местоположения, контекста и возможной научной интерпретации содержания снимка. Мы приводим этот текст на 5 языках, чтобы отразить многоязычный характер Швейцарии и многонациональный характер наших научных и инженерных групп. Однако мы начинаем с краткого описания космического аппарата и прибора, чтобы предоставить справочную информацию, если она вам понадобится.

DER EXOMARS
TRACE GAS ORBITER

DAS RAUMFAHRZEUG

SPACECRAFT

Der ExoMars Trace Gas Orbiter (TGO) wurde von einem internationalen Industriekonsortium unter der Leitung von Thales Alenia Space Italia und unter Aufsicht der Europäischen Weltraumorganisation (ESA) gebaut. Unternehmen aus mehr als 20 Ländern waren daran beteiligt. TGO hat eine rechteckige, kastenförmige Struktur (3,2 m × 2 m × 2 m) mit ausfahrbaren Solarzellen, die sich bis zu 17,5 m (Spitze zu Spitze) erstrecken und eine Leistung von 2 kW erzeugen. Ursprünglich war er als Kommunikations-Orbiter zur Unterstützung der ExoMars-Rover-Mission konzipiert. Eine Antenne mit einem Durchmesser von 2,2 m und hoher Verstärkung sorgt für die Kommunikation mit der Erde. Ein grosses Triebwerk auf dem Raumschiff, welches eine Kraft von 424 Newton produzieren kann, steht für grössere Manöver zur Verfügung.

The ExoMars Trace Gas Orbiter (TGO) was built by an international industrial consortium led by Thales Alenia Space Italia and supervised by the European Space Agency (ESA). Companies in more than 20 countries were involved. TGO has a rectangular box like structure (3.2 m × 2 m × 2 m) with extendable solar panels that extend to 17.5 m (tip-to-tip) and generate 2 kW of power. It was originally conceived as a communications orbiter to support the ExoMars Rover mission. A 2.2 metre diameter high gain antenna provides the communication to Earth. A large engine (producing a force of 424 Newtons) on the spacecraft is available for major manoeuvres.

SONDE SPATIALE

La sonde ExoMars Trace Gas Orbiter (TGO) a été construite par un consortium industriel international dirigé par Thales Alenia Space Italia et supervisé par l'Agence Spatiale Européenne (ESA). Des entreprises de plus de 20 pays ont été impliquées. TGO a une structure en forme de parallélépipède (3,2 m × 2 m × 2 m) avec des panneaux solaires qui s'étendent sur 17,5 m (d'une extrémité à l'autre) et génèrent 2 kW de puissance. Il a été conçu à l'origine comme un satellite de communication pour permettre la mission future du rover ExoMars. Une antenne à haut gain de 2,2 mètres de diamètre assure la communication avec la Terre. Un gros moteur (produisant une force de 424 Newtons) est utilisé pour les manœuvres importantes de la sonde.

SPACECRAFT

L'ExoMars Trace Gas Orbiter (TGO) è stato costruito da un consorzio industriale internazionale guidato da Thales Alenia Space Italia e supervisionato dall'Agenzia Spaziale Europea (ESA). Sono state coinvolte aziende di più di 20 paesi. TGO ha una struttura rettangolare (3,2 m × 2 m × 2 m) con pannelli solari estensibili che si estendono fino a 17,5 m (da punta a punta) e generano 2 kW di potenza. È stato originariamente concepito come un orbiter di comunicazione per supportare la missione ExoMars Rover. Un'antenna ad alto guadagno di 2,2 metri di diametro provvede alla comunicazione con la Terra. Per le manovre più importanti il veicolo spaziale utilizza un potente motore (che produce una forza di 424 Newton).

ОРБИТАЛЬНЫЙ АППАРАТ

The ExoMars Trace Gas Orbiter (TGO) был построен международным промышленным консорциумом во главе с Thales Alenia Space Italia и под руководством Европейского космического агентства (ЕКА). Участие принимали компании более чем 20 стран. TGO имеет прямоугольную коробчатую форму (3,2 м x 2 м x 2 м) с выдвижными солнечными панелями, их размах составляет 17,5 м, а мощность равна 2 кВт энергии. Первоначально он был задуман как орбитальный аппарат связи для поддержки миссии марсохода «ЭкзоМарс» (ExoMars Rover). Связь с Землей обеспечивает антенна диаметром 2,2 метра с высоким коэффициентом усиления. Для совершения основных маневров, на космическом аппарате имеется большой двигатель (создающий усилие в 424 Ньютона).

DIE NUTZLAST

Die Nutzlast umfasst vier Hauptinstrumente: ACS, NOMAD, FREND und CaSSIS.

Die Atmospheric Chemistry Suite (ACS) besteht aus drei Infrarot-Spektrometerkanälen, nämlich dem Nahinfrarotkanal (NIR), dem Mittelinfrarotkanal (MIR) und dem Ferninfrarotkanal (TIRVIM). Der NIR-Kanal arbeitet im Bereich von 0,7 bis 1,7 µm und der MIR-Kanal zwischen 2,2 und 4,4 µm, wobei beide eine hohe Wellenlängenauflösung haben, um ein Inventar und eine Karte der atmosphärischen Gasen mit geringen Anteilen (Spurengase) in der Marsatmosphäre zu erstellen. Das Ziel von TIRVIM ist die Untersuchung der 15-µm-CO_2-Bande und damit die Untersuchung der dominierenden Gase in der Marsatmosphäre. ACS wurde vom Weltraumforschungsinstitut (IKI) in Moskau mit Unterstützung des Centre national d'études spatiales (CNES) in Frankreich entwickelt.

Die Nadir and Occultation for Mars Discovery (NOMAD) Suite besteht ebenfalls aus drei Kanälen. Die Kanäle Solar Occultation (SO) und Limb Nadir and Occultation (LNO) arbeiten im Infrarotbereich (2,2 bis 4,3 µm) und ergänzen ACS, während ein dritter Kanal (UVIS) im UV-visuellen Bereich arbeitet, um Ozon und Schwefelsäure zu messen und Untersuchungen von Partikeln (Aerosolen) in der Marsatmosphäre durchzuführen. NOMAD wurde von einem multinationalen Team aus Europa und Nordamerika unter der Leitung des Belgian Institute for Space Aeronomy in Brüssel gebaut.

Der Fine-Resolution Epithermal Neutron Detector (FREND) ist ein Neutronendetektor, der die Menge an gebundenem Wasserstoff in der Oberflächenschicht des Mars messen soll. Da Wasserstoff ein Bestandteil des Wassermoleküls ist, wird erwartet, dass Gebiete mit grossen Wasserstoffmengen entweder Wassereis oder Mineralien mit einem bedeutenden Anteil an Kristallwasser enthalten. FREND enthält auch ein Dosimeter zur Überwachung der Strahlungsumgebung rund um den Mars in Vorbereitung auf zukünftige bemannte Flüge. Das Instrument wurde vom Weltraumforschungsinstitut in Moskau gebaut.

CaSSIS ist das Thema des nächsten Kapitels.

PAYLOAD

The payload comprises 4 main instruments, ACS, NOMAD, FREND, and CaSSIS.

The Atmospheric Chemistry Suite (ACS) consists of three infrared spectrometer channels namely, the near-infrared channel (NIR), the mid-infrared channel (MIR), and the far infrared channel (TIRVIM). The NIR and MIR channels operate from 0.7 to 1.7 µm and 2.2–4.4 µm, respectively, with a high wavelength resolution in order to produce an inventory and map of minor atmospheric species (trace gases) in the Martian atmosphere. The aim of TIRVIM is to study the deep CO_2 band at 15 µm wavelength and therefore study the dominant species in the Martian atmosphere. ACS was developed by the Space Research Institute (IKI) in Moscow with support from the Centre national d'études spatiales (CNES) in France.

The Nadir and Occultation for Mars Discovery (NOMAD) suite also consists of three channels. The solar occultation (SO) and limb nadir and occultation (LNO) channels operate in the infrared (2.2 to 4.3 µm) and complement ACS while a third channel (UVIS) operates in the UV-visible range in order to measure to measure ozone, sulphuric acid, and perform studies of particles (aerosols) in the Martian atmosphere. NOMAD was built by a multi-national team from Europe and North America and led by the Belgian Institute for Space Aeronomy in Brussels.

The Fine-Resolution Epithermal Neutron Detector (FREND) is a neutron detector which has been designed to map the amount of bound hydrogen in the surface layer of Mars. Hydrogen is, of course, a component of the water molecule and it is expected that areas with large amounts of hydrogen will contain either water ice or minerals with significant water of hydration. FREND also includes a dosimeter to monitor the radiation environment around Mars in preparation for future manned flights. The instrument was built by the Space Research Institute in Moscow.

CaSSIS is the subject of the next chapter.

CHARGE UTILE

La charge utile comprend 4 instruments principaux : ACS, NOMAD, FREND et CaSSIS.

L'instrument « Atmospheric Chemistry Suite » (ACS) comprend trois canaux de spectrométrie infrarouge, à savoir le canal proche infrarouge (NIR), le canal moyen infrarouge (MIR) et le canal infrarouge lointain (TIRVIM). Les canaux NIR et MIR fonctionnent respectivement de 0,7 à 1,7 µm et de 2,2 à 4,4 µm, et offrent une haute résolution spectrale afin de produire un inventaire et une carte des espèces atmosphériques mineures (gaz « traces ») dans l'atmosphère martienne. L'objectif du canal TIRVIM est d'étudier la bande intense du CO_2 à la longueur d'onde de 15 µm et d'étudier ainsi l'espèce dominante de l'atmosphère martienne. ACS a été développé par l'Institut de Recherche Spatiale (IKI) de Moscou avec le soutien du Centre National d'Études Spatiales (CNES) en France.

Le spectromètre « Nadir and Occultation for Mars Discovery » (NOMAD) comprend également trois canaux. Les canaux d'« occultation solaire » (SO) et de « limbe, nadir et occultation » (LNO) fonctionnent dans l'infrarouge (2,2 à 4,3 µm) et complètent l'instrument ACS tandis qu'un troisième canal (UVIS) fonctionne dans le domaine UV-visible afin de mesurer l'ozone, l'acide sulfurique, et d'étudier les particules solides (aérosols) de l'atmosphère martienne. NOMAD a été construit par une équipe multinationale d'Europe et d'Amérique du Nord, dirigée par l'Institut royal d'Aéronomie Spatiale de Belgique à Bruxelles.

L'instrument « Fine-Resolution Epithermal Neutron Detector » (FREND) est un détecteur de neutrons qui a été conçu pour cartographier la quantité d'hydrogène dans les couches superficielle du sol martien. L'hydrogène est un atome de la molécule d'eau et on s'attend à ce que les zones avec de grandes quantités d'hydrogène contiennent soit de la glace d'eau, soit des minéraux fortement hydratés. FREND comprend également un dosimètre pour surveiller l'environnement radiatif autour de Mars en vue des futurs vols habités. L'instrument a été construit par l'Institut de Recherche Spatiale de Moscou.

CaSSIS fait l'objet du chapitre suivant.

PAYLOAD

Il carico utile comprende 4 strumenti principali, ACS, NOMAD, FREND e CaSSIS.

L'Atmospheric Chemistry Suite (ACS) consiste in tre canali spettrometrici all'infrarosso: il canale del vicino infrarosso (NIR), il canale del medio infrarosso (MIR) e il canale del lontano infrarosso (TIRVIM). I canali NIR e MIR operano rispettivamente da 0.7 a 1.7 µm e 2.(2)–(4).4 µm, con un'alta risoluzione nelle lunghezze d'onda al fine di produrre un inventario e una mappa delle specie atmosferiche minori (tracce di gas) nell'atmosfera marziana. Lo scopo di TIRVIM è quello di studiare la profonda banda di CO_2 presente alla lunghezza d'onda di 15 µm e quindi studiare le specie dominanti nell'atmosfera marziana. ACS è stato sviluppato dall'Istituto di Ricerca Spaziale (IKI) di Mosca con il supporto del Centre national d'études spatiales (CNES) in Francia.

Anche la suite Nadir and Occultation for Mars Discovery (NOMAD) consiste di tre canali. I canali di occultazione solare (SO) e di nadir e occultazione del lembo (LNO) operano nell'infrarosso (da 2,2 a 4,3 µm) e completano ACS, mentre un terzo canale (UVIS) opera nella gamma UV-visibile per misurare l'ozono, l'acido solforico ed eseguire studi sulle particelle (aerosol) nell'atmosfera marziana. NOMAD è stato costruito da un team multinazionale europeo e nordamericano ed è guidato dall'Istituto Belga per l'Aeronomia Spaziale di Bruxelles.

Il Fine-Resolution Epithermal Neutron Detector (FREND) è un rivelatore di neutroni che è stato progettato per mappare la quantità di idrogeno nello strato superficiale di Marte. Ovviamente, l'idrogeno è un componente della molecola dell'acqua e ci si aspetta che le aree con grandi quantità di idrogeno contengano o ghiaccio d'acqua o minerali con una significativa idratazione. FREND include anche un dosimetro per monitorare le radiazioni ambientali intorno a Marte, in preparazione ai futuri voli con equipaggio. Lo strumento è stato costruito dallo Space Research Institute di Mosca.

CaSSIS è il soggetto del prossimo capitolo.

ПОЛЕЗНАЯ НАГРУЗКА

Полезную нагрузку составляют 4 основных инструмента: ACS, NOMAD, FREND и CaSSIS.

ACS (The Atmospheric Chemistry Suite) – это комплекс атмосферной химии, который использует три канала инфракрасной спектрометрии, а именно: ближний инфракрасный канал (NIR), средний инфракрасный канала (MIR) и дальний инфракрасный канал (TIRVIM). NIR и MIR каналы работают в диапазоне 0,7–1,7 мкм и 2,2–4,4 мкм, соответственно, с высоким спектральным разрешением, позволяющим записать и закартировать второстепенные составляющие марсианской атмосферы (следовые газы). Целью TIRVIM является изучение мощной полосы CO_2 на длине волны 15 мкм и, следовательно, изучение доминирующих компонентов марсианской атмосферы. ACS был разработан Институтом космических исследований (ИКИ) в Москве при поддержке Национального центра космических исследований (CNES) во Франции.

Научный инструмент Nadir and Occultation for Mars Discovery (NOMAD) также использует три спектрометра. Солнечно-затменный (SO) и лимбовый, надирный и затменный (LNO) каналы работают в инфракрасном диапазоне (от 2,2 до 4,3 мкм) и дополняют ACS, в то время как третий канал (UVIS) работает в УФ и видимом диапазонах с целью измерения содержания озона и серной кислоты, а также изучения взвешенных частиц (аэрозолей) в марсианской атмосфере. NOMAD был создан многонациональной командой из Европы и Северной Америки под руководством Бельгийского института космической аэрономии в Брюсселе.

Детектор эпитепловых нейтронов высокого разрешения (FREND) – детектор нейтронов, который был разработан для отображения количества связанного водорода в поверхностных слоях на Марсе. Водород, как известно, является компонентом молекулы воды, и предполагается, что области с большим количеством водорода содержат либо водяной лёд, либо минералы со значительной долей связанной воды. FREND также включает дозиметр для мониторинга радиационной обстановки на марсианской орбите в рамках подготовки к будущим пилотируемым полетам. Прибор был построен Институтом космических исследований в Москве.

CaSSIS – тема следующей главы.

START, ATMOSPHÄRENBREMSUNG UND ENDUMLAUFBAHN

Im Rahmen der Zusammenarbeit zwischen der ESA und der russischen Raumfahrtagentur Roscosmos wurde TGO im Januar 2016 nach Kasachstan verschifft und am 14. März 2016 um 09.31 UTC mit einer Proton-Rakete vom Weltraumbahnhof Baikonur gestartet. Nach einer siebenmonatigen Reise zum Mars wurde TGO am 19. Oktober 2016 in die Umlaufbahn gebracht. Die ursprüngliche Umlaufbahn war stark elliptisch. Das Operationsteam im Europäischen Raumfahrtkontrollzentrum in Darmstadt brachte TGO in die obersten Höhen der Marsatmosphäre, um die Umlaufbahn zu zirkularisieren – ein Prozess, der als Aerobraking bekannt ist. Nach elf Monaten war TGO nahe an seiner gewünschten Umlaufbahn. Weitere Zündungen der Triebwerke des Raumfahrzeugs wurden dann verwendet, um den Prozess abzuschliessen. Dies führte zu einer endgültigen Umlaufbahn von etwa 420 × 360 km über der Marsoberfläche und ermöglichte den Übergang zu wissenschaftlichen Aktivitäten am 21. April 2018.

Die endgültige Umlaufbahn hat eine Neigung gegenüber dem Marsäquator von etwa 74°. Dies war ein Kompromiss zwischen dem Wunsch des CaSSIS-Teams, die Polarregionen abzubilden, und den Bedürfnissen der ACS- und NOMAD-Teams, Sonnenokkultationsmessungen über einen grossen Breitenbereich durchzuführen. Diese Neigung bedeutet, dass der Breitengrad des Punkts direkt unter dem Raumschiff (Nadir) auf dem Mars (Norden oder Süden) nicht grösser als 74° sein kann. Folglich kann CaSSIS keine Breitengrade oberhalb dieses Wertes beobachten.

Ein weiterer Aspekt der endgültigen Umlaufbahn ist, dass sie nicht sonnensynchron ist. Viele Raumschiffe, die für die optische Fernerkundung eingesetzt werden, überqueren den Äquator eines Planeten (z. B. der Erde oder des Mars) auf jeder Umlaufbahn zur gleichen Ortszeit. Der Mars Reconnaissance Orbiter (MRO) der NASA ist ein gutes Beispiel dafür. Die Ortszeit am Äquator, wenn

MRO ihn auf der Tagseite überquert, ist immer rund 15.00 Uhr. Das primäre bildgebende Instrument an Bord, HiRISE, macht sich dies zunutze und liefert hervorragende morphologische Beobachtungen der Oberfläche. TGO hat keine sonnensynchrone Umlaufbahn, sodass ein Bild zu jeder beliebigen Ortszeit aufgenommen werden kann. Das bedeutet, dass ein Teil der Marsoberfläche kurz nach dem Auftreten des Sonnenlichts am Morgen oder zur Mittagszeit beobachtet werden kann. Dies kann von Vorteil sein, da beispielsweise Aufnahmen um die Mittagszeit in der Regel bessere Farbbeobachtungen ergeben. Daher kann dieser Ansatz als komplementär betrachtet werden.

LAUNCH, AEROBRAKING AND FINAL ORBIT

As part of the collaboration between ESA and the Russian space agency, Roscosmos, TGO was shipped to Kazakhstan in January 2016 and launched on a Proton rocket from the Baikonur Cosmodrome at 09:31 UTC on 14 March 2016. After a seven-month cruise to Mars, TGO was injected into orbit on 19 October 2016. The original orbit was highly elliptical. The operations team at the European Space Operations Centre in Darmstadt brought TGO into the uppermost altitudes of the Martian atmosphere to start to circularize the orbit – a process known as aerobraking. After 11 months, TGO was close to its required orbit. Additional firings of the spacecraft thrusters were then used to complete the task. This resulted in a final orbit of approximately 420 × 360 km over the surface of Mars and allowed a transition to science activities on 21 April 2018.

The final orbit has an inclination with respect to the Martian equator of about 74°. This was a trade-off between the desire of the CaSSIS team to image the polar regions against the needs of the ACS and NOMAD teams to perform solar occultation measurements over a wide range of latitude. This inclination implies that the latitude of the sub-spacecraft point on Mars (north or south) cannot be greater than 74°. Consequently, CaSSIS cannot observe latitudes above this value.

Another aspect of the final orbit is that it is not Sun-synchronous. Many orbiting spacecraft devoted to optical remote sensing cross the equator of a planet (e.g. Earth or Mars) at the same local time every orbit. NASA's Mars Reconnaissance Orbiter (MRO) is a good example of this. The local time at the equator when MRO crosses it on the dayside is always around 15:00 (equivalent to 3 pm). The main imaging instrument onboard, HiRISE, takes advantage of this to provide excellent morphological observation of the surface. On the other hand, TGO does not do this and so an image can be taken at any local time. This means that part of the Mars surface might be observed just after that surface has come into sunlight in the local morning or at local midday. This can have advantages because imaging close to midday, for example, tends to produce better colour observations. Hence, this approach can be considered complementary.

LANCEMENT, AÉROFREINAGE
ET ORBITE FINALE

Dans le cadre de la collaboration entre l'ESA et l'agence spatiale russe Roscosmos, la sonde TGO a été expédiée au Kazakhstan en Janvier 2016 et lancée sur une fusée Proton depuis le cosmodrome de Baïkonour à 09 h 31 UTC le 14 mars 2016. Après un trajet de sept mois jusqu'à Mars, TGO a été insérée sur une orbite initiale fortement elliptique le 19 octobre 2016. L'équipe d'opérations du Centre Européen d'Opérations Spatiales (ESOC) de Darmstadt ont abaissé l'altitude de TGO jusqu'à traverser les couches les plus élevées de l'atmosphère martienne pour circulariser progressivement son orbite – un processus connu sous le nom d' « aérofreinage ». Au bout de 11 mois, TGO s'est rapprochée de l'orbite requise. De courts allumages supplémentaires des propulseurs de la sonde ont alors permis de finaliser la tâche. Le résultat est une orbite finale d'environ 420 × 360 km au-dessus de la surface de Mars qui a permis la transition vers des activités scientifiques le 21 avril 2018.

L'orbite finale a une inclinaison par rapport à l'équateur martien d'environ 74°. Il s'agissait d'un compromis entre le désir de l'équipe CaSSIS d'imager les régions polaires et les besoins des équipes ACS et NOMAD d'effectuer des mesures d'occultation solaire sur une large plage de latitude. Cette inclinaison implique que la latitude (nord ou sud) du point de la surface martienne situé à la verticale sous la sonde ne peut pas être supérieure à 74°. Par conséquent, il est impossible pour CaSSIS d'observer des latitudes supérieures à cette valeur.

Un autre aspect intéressant de l'orbite finale est qu'elle n'est pas héliosynchrone. De nombreuses sondes orbitales consacrées à la télédétection optique traversent l'équateur d'une planète (par exemple, la Terre ou Mars) à la même heure locale à chaque orbite. Le « Mars Reconnaissance Orbiter » (MRO) de la NASA en est un bon exemple. L'heure locale à l'équateur lorsque MRO le traverse en journée est toujours aux alentours de 15 h 00. Le principal instrument d'imagerie embarqué, HiRISE, en profite pour fournir d'excellentes observations morphologiques de la surface. Par contre, TGO n'étant pas héliosynchrone, une image de la surface peut être prise à n'importe quelle heure locale. Cela signifie qu'une zone particulière de la surface de Mars peut par exemple être observée tôt le matin, juste après le lever du soleil ou à midi avec le soleil au plus haut dans le ciel. Cela peut présenter des avantages, car l'imagerie vers midi limite les ombres et a tendance à produire de meilleures observations des couleurs. Cette approche peut donc être considérée comme complémentaire.

LANCIO, AEROFRENAGGIO E ORBITA FINALE

Come parte della collaborazione tra l'ESA e l'agenzia spaziale russa (Roscosmos), TGO è stato spedito in Kazakistan nel gennaio 2016 e lanciato su un razzo Proton dal cosmodromo di Baikonur alle 09:31 UTC del 14 marzo 2016. Dopo una crociera di sette mesi verso Marte, TGO è stato inserito nell' orbita il 19 ottobre 2016. L'orbita originale era estremamente ellittica e il team operativo del Centro europeo per le operazioni spaziali di Darmstadt ha portato TGO alle quote più alte dell'atmosfera marziana per iniziare a circolarizzare l'orbita (un processo noto come «aerofrenaggio»). Dopo 11 mesi, TGO era vicino all'orbita richiesta. Infine, i propulsori del veicolo spaziale sono stati accesi ulteriori volte per completare il compito. Questo ha portato a un'orbita finale di circa 420 × 360 km dalla superficie di Marte e ha permesso il passaggio alle attività scientifiche il 21 aprile 2018.

L'orbita finale ha un'inclinazione rispetto all'equatore marziano di circa 74°. Questo è stato un compromesso tra il desiderio del team CaSSIS di fotografare le regioni polari contro le esigenze dei team ACS e NOMAD di eseguire misure di occultazione solare su un ampio intervallo di latitudini. Questa inclinazione implica che la latitudine del punto su Marte corrispettivo alla navicella non può essere maggiore di 74° (nord o sud). Di conseguenza, CaSSIS non può osservare regioni a latitudini superiori a questo valore.

Un'altra caratteristica dell'orbita finale è che non è sincrona con il Sole. Molti veicoli spaziali orbitanti dedicati al telerilevamento ottico attraversano l'equatore di un pianeta (ad esempio la Terra o Marte) alla stessa ora locale ad ogni orbita. Il Mars Reconnaissance Orbiter (MRO) della NASA ne è un buon esempio. L'ora locale all'equatore quando MRO lo attraversa di giorno è sempre intorno alle 15:00 (equivalente alle 3 del pomeriggio). Il principale strumento per la produzione di immagini a bordo, HiRISE, ne approfitta per fornire un'eccellente osservazione morfologica della superficie. D'altra parte, TGO non può farlo e quindi un'immagine può essere presa in qualsiasi momento. Questo significa che una parte della superficie di Marte potrebbe essere osservata appena dopo l'alba al mattino locale o al mezzogiorno locale. Questo può avere dei vantaggi perché le immagini prese intorno a mezzogiorno, per esempio, tendono a produrre le migliori osservazioni a colori. Quindi, questo approccio può essere considerato complementare.

ЗАПУСК, АЭРОДИНАМИЧЕСКОЕ ТОРМОЖЕНИЕ И КОНЕЧНАЯ ОРБИТА

В рамках сотрудничества между ЕКА и Российским космическим агентством (Роскосмос), TGO был отправлен в Казахстан в январе 2016 года, и 14 марта 2016 года, в 09:31 по Гринвичу, запущен на ракете Протон с космодрома Байконур. После семи месяцев полёта к Марсу, 19 октября 2016 года, TGO был выведен на орбиту. Первоначальная орбита была высокоэллиптической. Сотрудники Европейского центра управления космическими полётами в Дармштадте направили TGO в верхние слои атмосферы Марса, чтобы начать округлять орбиту – с помощью процесса, известного как аэродинамическое торможение. После 11 месяцев, TGO был близок к требуемой орбите. Затем, дополнительные пуски маневровых двигателей аппарата завершили эту задачу. Это привело к конечной орбите высотой приблизительно 420 x 360 км над поверхностью Марса и позволило осуществить переход к научной деятельности 21 апреля 2018 года.

Конечная орбита имеет наклонение относительно марсианского экватора около 74°. Это результат компромисса между желанием команды CaSSIS получать изображения полярных регионов и потребностями команд ACS и NOMAD выполнять солнечно-затменные наблюдения в большом диапазоне широт. Такой наклон подразумевает, что широта (северная или южная) точки на Марсе под орбитальным аппаратом не может быть больше 74°. Следовательно, CaSSIS не может наблюдать широты выше этого значения.

Еще один аспект конечной орбиты заключается в том, что она не является солнечно-синхронной. Многие орбитальные космические аппараты, предназначенные для дистанционного оптического наблюдения, пересекают экватор планеты (например, Земли или Марса) в одно и то же местное время на каждом обращении. Хорошим примером этого является орбитальный аппарат НАСА Марсианский разведывательный спутник (Mars Reconnaissance Orbiter, MRO). Местное время на экваторе, когда MRO пересекает его на дневной стороне, всегда около 15:00 (в три часа дня). Основной бортовой инструмент получения изображений, HIRISE, использует данную особенность, обеспечивая прекрасные наблюдения морфологии поверхности. С другой стороны, TGO движется иначе, и поэтому изображение может быть получено в любое местное время. Это означает, что необходимая область на поверхности может наблюдаться как после освещения солнечным светом локальным утром, так и в локальный полдень. Это может иметь свои преимущества, поскольку, например, при съемке ближе к полудню, как правило, получаются более качественные цвета. Следовательно, этот подход можно считать комплементарным.

DAS
COLOUR AND STEREO
SURFACE IMAGING SYSTEM

DAS BILDGEBENDE VERFAHREN

CaSSIS ist ein Push-Frame-Bilderfassungssystem. Es nimmt kleine Bilder (sogenannte Framelets) sehr schnell auf, während das Raumschiff über die Oberfläche fliegt. Diese Einzelbilder werden zur Erde übertragen, kalibriert und dann zu langen, breiten Streifen zusammengesetzt, die in der Regel 9,5 km × 40 km der Marsoberfläche pro Bild abdecken. Die Zeit zwischen den einzelnen Einzelbildern liegt im Bereich von 350 bis 400 Millisekunden und ergibt sich aus den Eigenschaften des Teleskops und des Detektors.

CaSSIS ist auch für die Aufnahme von Stereobildern ausgelegt. Zu diesem Zweck wird das Teleskop um 10° aus der Nadir-Richtung verschoben. Mit Hilfe eines Motors wird das Teleskop so gedreht, dass es auf die Bodenspur ausgerichtet ist, aber nach vorne zeigt. Sobald das Bild aufgenommen ist, wird das Teleskop um 180° um die Nadir-Ausrichtungsachse gedreht, sodass das Teleskop nach hinten zeigt. Für diese Drehung hat es 45 Sekunden Zeit, bevor das nächste Bild aufgenommen werden muss, damit sich die beiden Bilder überlappen. Dies führt zu einem Stereokonvergenzwinkel von 22,4° (im Durchschnitt und unter Berücksichtigung der Krümmung des Mars) und liefert hervorragende Stereobilder der Oberfläche. Der Rotationsmotor kann für die Ausrichtung aller Bilder (einschliesslich der Nicht-Stereoaufnahmen) verwendet werden. Das Operationsteam versucht jedoch, den Einsatz des Motors zu begrenzen, um seine Lebensdauer zu verlängern. Nicht alle Bilder wurden in einer perfekt ausgerichteten Konfiguration aufgenommen. Dies führt dazu, dass sich die Farbüberlappung verringert und die Farbabdeckung reduziert wird.

THE IMAGING APPROACH

CaSSIS is a push-frame imaging system. It takes small images (called framelets) very quickly as the spacecraft flies over the surface. These framelets are transmitted to Earth, calibrated and then assembled into long swaths that cover typically 9.5 km × 40 km of the Martian surface per image. The time between each framelet is in the range of 350 to 400 milliseconds which is derived from the properties of the telescope and the detector.

CaSSIS is also designed to take stereo images. It does this by off-pointing the telescope by 10° from the nadir direction. A motor is used to rotate the telescope so that it is aligned with the ground-track but pointing forward. Once the image is acquired, the telescope is rotated by 180° about the nadir-pointing axis so that the telescope points backwards. It has 45 seconds to perform this rotation before the next image must be acquired in order to have overlap of the two images. This leads to a stereo convergence angle of 22.4° (on average and taking into account the curvature of Mars) and provides excellent stereo imagery of the surface. The rotation motor can be used to align every image (including non-stereo observations). However, the operation team tries to limit the use of the motor to increase its lifetime and not all images have been acquired in a perfectly aligned configuration. This tends to narrow the colour overlap reducing colour coverage.

PRINCIPE D'IMAGERIE

CaSSIS est un système d'imagerie dit « push-frame ». Il acquière de petites images (appelées « framelets ») très rapidement alors que la sonde survole la surface. Ces framelets sont transmises à la Terre, calibrées puis assemblées en de longues bandes qui couvrent généralement 9,5 km × 40 km de la surface martienne par image. L'intervalle de temps entre chaque framelet est de l'ordre de 350 à 400 millisecondes. Il dérive des propriétés du télescope, du détecteur et de l'altitude de la sonde.

CaSSIS est également conçu spécialement pour prendre des images stéréographiques. Pour ce faire, le télescope est incliné de 10° par rapport à la verticale (« nadir »). Un moteur est utilisé pour faire pivoter le télescope afin qu'il soit aligné avec la trace au sol en pointant vers l'avant. Une fois la première image acquise ainsi, le télescope est tourné de 180° autour de l'axe vertical de sorte qu'il pointe vers l'arrière. Le moteur dispose de 45 secondes pour effectuer cette rotation avant que l'image suivante ne soit acquise afin d'obtenir une superposition parfaite des deux images. Cela conduit à un angle de convergence de 22,4° (en moyenne et en tenant compte de la courbure de Mars) et fournit une excellente imagerie stéréographique de la surface. Le mécanisme de rotation peut également être utilisé pour aligner chaque image (y compris les observations non stéréo). Cependant, l'équipe d'opération tente de limiter l'utilisation du moteur pour augmenter sa durée de vie et toutes les images n'ont pas été acquises dans une configuration parfaitement alignée. Cela tend à limiter le chevauchement des filtres et ainsi réduire la superposition des couleurs.

L'APPROCCIO DI IMAGING

CaSSIS è un sistema di imaging a «spazzata». Prende piccole immagini (chiamate fotogrammi) molto rapidamente mentre il veicolo spaziale sorvola la superficie. Questi fotogrammi sono trasmessi alla Terra, calibrati e poi assemblati in lunghe strisce che coprono tipicamente 9,5 km × 40 km della superficie marziana per immagine. Il tempo che trascorre tra ogni fotogramma è dell'ordine di 350–400 millisecondi, che deriva dalle proprietà del telescopio e del rivelatore.

CaSSIS è anche progettato per acquisire immagini stereoscopiche (abbreviate «stereo»). Lo fa spostando il telescopio di 10° dalla direzione del nadir. Un motore ruota il telescopio in modo che sia allineato con la direzione di volo della navicella, ma puntato in avanti. Una volta acquisita l'immagine, il telescopio viene ruotato di 180° intorno all'asse di puntamento al nadir in modo che il telescopio punti all'indietro. Ha 45 secondi per eseguire questa rotazione prima che l'immagine successiva debba essere acquisita per avere la sovrapposizione delle due immagini.

Questo porta ad un angolo di convergenza stereo di 22,4° (in media e tenendo conto della curvatura di Marte) e fornisce eccellenti immagini stereo della superficie. Il motore di rotazione può essere utilizzato per allineare ogni immagine (comprese le osservazioni non stereo). Tuttavia, il team operativo cerca di limitare l'uso del motore per aumentarne la durata e non tutte le immagini sono state acquisite in una configurazione perfettamente allineata. Questo tende a restringere la porzione di immagine con una corretta sovrapposizione dei colori.

ExoMars TGO
CaSSIS ELU
Radiator Protective Cover
214-5400 S/N 02
REMOVE BEFORE FLIGHT

D   Das CaSSIS-Bilderfassungssystem auf der Werkbank an der Universität Bern. Das Teleskop ist die schwarze Struktur auf der rechten Seite, die an einer Trä-gerstruktur (goldfarben) mit einem Rotationsmechanismus befestigt ist. Die Elektronikbox befindet sich auf der linken Seite.

E   The CaSSIS imaging system sitting on the bench at the University of Bern. The telescope is the black structure to the right and is cantilevered off a support structure (gold-coloured) that contains a rotation mechanism. The electronics box is to the left.

F   Le système d'imagerie CaSSIS sur la table du laboratoire à l'Université de Bern. Le télescope est la structure noire sur la droite de l'image, fixée à sa structure de support (dorée) qui contient le mécanisme de rotation. L'électronique de contrôle se trouve à l'intérieur de la boîte dorée sur la gauche de l'image.

I   Il sistema di imaging CaSSIS posizionato su un banco di laboratorio all'Università di Berna. Il telescopio è la struttura nera a destra e sporge da una struttura di supporto (color oro) che contiene un meccanismo di rotazione. La scatola elettronica è a sinistra.

P   Система формирования изображений CaSSIS, установленная на стенде в Бернском университете. Телескоп – это черная конструкция справа, консольно закреплённая на несущей конструкции (золотого цвета), которая содержит механизм вращения. Блок электроники находится слева.

ПОДХОД К СЪЕМКЕ

CaSSIS – это push-frame система формирования изображений, использующая продольное сканирование многоэлементным площадным датчиком. Она создаёт очень быстро небольшие изображения (так называемые фреймлеты – «небольшие кадры»), по мере того как космический аппарат пролетает над снимаемой поверхностью. Эти кадры передаются на Землю, калибруются и затем собираются в длинные полосы, которые покрывают обычно 9,5 х 40 км марсианской поверхности. Промежуток времени между каждым кадром составляет от 350 до 400 миллисекунд, что определяется свойствами телескопа и детектора.

CaSSIS предназначен также и для получения стереоизображений. Это достигается смещением телескопа на 10° от направления надира. С помощью мотора телескоп поворачивается так, чтобы быть согласованным с поверхностным треком (описываемым TGO), но указывать вперед. После получения изображения телескоп поворачивается на 180° вокруг оси надира, чтобы быть направленным назад. На этот поворот отводится 45 секунд, после чего необходимо получить следующее изображение, для частичного наложения двух изображений друг на друга. Это создаёт угол стерео конвергенции в 22,4° (в среднем и с учетом кривизны Марса) и обеспечивает превосходную стереосъёмку поверхности. Ротационный механизм можно использовать для согласования с движением аппарата каждого изображения (не только при стереосъёмке). Однако операционная группа старается ограничивать использование мотора, дабы продлить срок его службы, и не все изображения получаются в идеальной конфигурации. Это приводит к сужению цветового наложения изображения и уменьшает цветовое покрытие поверхности.

DAS TELESKOP, DER DETEKTOR UND DIE FILTER

Das Teleskop ist eine F/6.52-Konstruktion aus carbonfaserverstärktem Kunststoff (CFK) und hat einen Hauptspiegel mit 13,5 cm Durchmesser und einer Brennweite von 880 mm. Es handelt sich um ein Vier-Spiegel-Off-Axis-System, das eine hervorragende Streulichtunterdrückung bietet. Die Struktur wurde auf ein geringes Gewicht ausgelegt, um die Gesamtmasse auf nur 5,5 kg zu reduzieren, von denen 1 kg allein auf den Hauptspiegel entfällt. Die CFK-Struktur verliert im Weltraum Feuchtigkeit und schrumpft leicht. Die Auswirkung auf die Position des Fokus wurde vorausberechnet und kompensiert.

Der Detektor ist ein hybrider CMOS-Baustein mit 2048 × 2048 quadratischen Pixeln in 10 μm Abstand. In Kombination mit dem Teleskop führt dies zu einem Abbildungsmassstab von ≈ 4,5 Metern pro Pixel aus der nominalen Umlaufbahn (Winkelskala = 11,4 μrad/px). Der Detektor ist in einem Detektorblock montiert, in dem auch die Elektronik für die Detektorauslesung untergebracht ist. Solche Detektoren funktionieren normalerweise am besten, wenn sie kalt sind, und deshalb wurde der Detektorblock mit einem Wärmeband an einen Kühler angeschlossen.

Eine Platte mit verschiedenen Farbbeschichtungen wurde direkt auf den Detektor gelegt, um die Farbfähigkeit zu gewährleisten. Es wurden drei Filter ausgewählt, die bei 499,9 nm (BLU), 836,2 nm (RED) und 936,7 nm (NIR) zentriert sind. Zudem wurde ein panchromatischer Filter bei 675,0 nm (PAN) eingesetzt. Es ist zu beachten, dass diese Filter nicht dem Standard-RGB-System einer handelsüblichen Kamera entsprechen, obwohl ein synthetisches RGB-Bild aus Bildern erstellt werden kann, die mit diesen Filtern aufgenommen wurden. Bilder, die durch die verschiedenen Filter aufgenommen wurden, können auf vielfältige Weise kombiniert werden, um die Vielfalt der Farben der Oberfläche hervorzuheben. Die Bandpässe der Filter wurden speziell ausgewählt, um die Unterschiede in den Farben der auf der Oberfläche zu erwartenden Mineralien hervorzuheben. Die hier gezeigten Bilder machen sich dies zunutze, um die lithologischen Unterschiede auf der Oberfläche deutlich sichtbar zu machen. Wir werden auch einige synthetische RGB-Bilder zeigen.

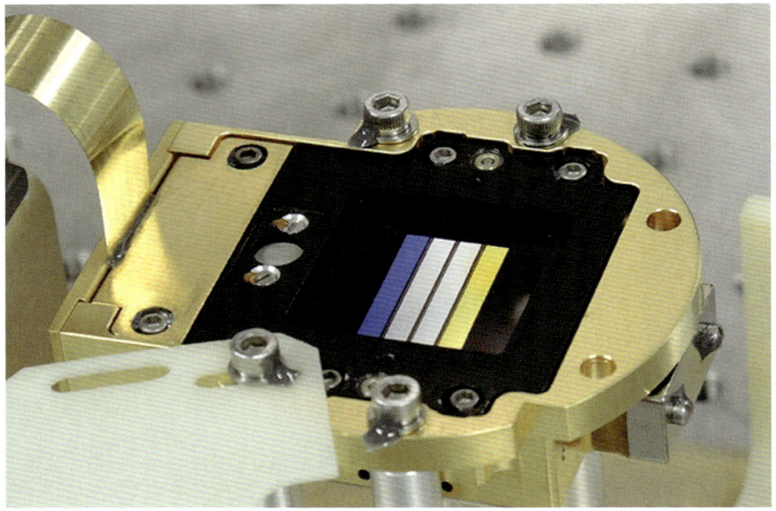

1

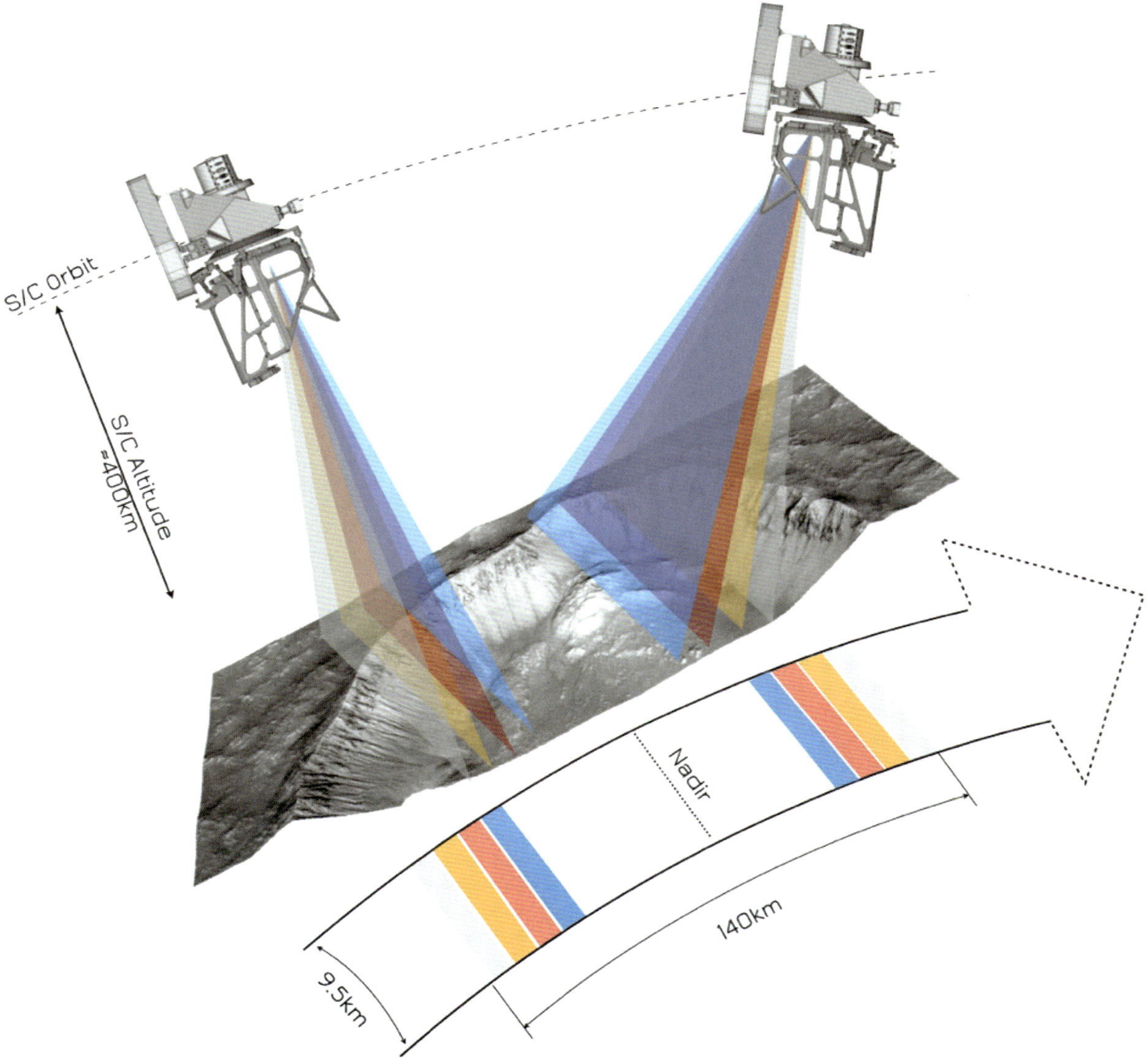

2

D 1 Der Detektorblock mit den Filtern, die direkt auf der aktiven Fläche des Detektors angebracht sind.

2 Das Bilderfassungsprinzip von CaSSIS. CaSSIS ist ein Push-Frame-Imager, bei dem kleine Bilder in vier Farben sehr schnell aufgenommen und dann am Boden zusammengesetzt werden. Ein Motor kann das Teleskop drehen, um zwei Bilder von der gleichen Position auf der Oberfläche zu erzeugen. Wenn das Timing richtig berechnet wird, überschneiden sich die beiden Bilder, und durch Nachbearbeitung kann ein Stereobild erstellt werden.

E 1 The detector block showing the filters which are mounted directly on top of the active area of the detector.

2 The imaging principle of CaSSIS. CaSSIS is a push-frame imager where small images around acquired in four colours very rapidly and then stitched together on ground. A motor can rotate the telescope to generate two images of the same position on the surface. If the timing is correctly calculated, then the two images will overlap and a stereo image can be made by post-processing.

F 1 Photo du bloc détecteur qui montre les filtres montés directement sur la partie active du capteur.

2 Principe d'imagerie de CaSSIS. CaSSIS est un imageur dit "push-frame" qui acquiert très rapidement de petites images à travers 4 filtres couleurs qui seront plus tard mosaïquées après leur réception sur Terre. Un moteur peut faire pivoter le télescope pour produire deux images d'un même point de la surface. Si l'intervalle de temps entre les deux images est correct, ces deux images seront parfaitement superposées et pourront servir à la reconstitution stéréoscopique des reliefs.

I 1 Il blocco del rilevatore con i filtri montati direttamente sulla superficie attiva del rilevatore.

2 Il principio di imaging di CaSSIS. CaSSIS è un sistema di imaging a spazzata, in cui vengono prese delle piccole immagini nei quattro filtri molto velocemente e poi collegate insieme sulla Terra. Un motore è in grado di ruotare il telescopio, creando due immagini della stessa posizione sulla superficie, ma creando un senso di tri-dimensionalità. Se la sincronizzazione è calcolata correttamente, allora le due immagini si possono sovrapporre e si può creare un'immagine stereoscopica.

P 1 Детекторный блок, на котором видны фильтры, установленные непосредственно на активной области детектора.

2 Принцип формирования изображений CaSSIS. CaSSIS – это система формирования изображений, использующая продольное сканирование многоэлементным площадным датчиком, где небольшие изображения получаются в четырех цветах очень быстро и затем сшиваются на земле. Мотор может вращать телескоп для получения двух изображений одной и той же области поверхности. Если правильно рассчитать время, то два изображения наложатся друг на друга, и при последующей обработке можно получить стереоизображение.

THE TELESCOPE, THE DETECTOR AND FILTERS

The telescope is an F/6.52 design made from carbon-fibre reinforced plastic (CFRP) and has a 13.5 cm diameter primary mirror with an 880 mm focal length. It is a four mirror, off-axis, system that provides excellent stray-light rejection. The structure was light-weighted to reduce the total mass to just 5.5 kg of which 1 kg is devoted to the primary mirror alone. The CFRP structure loses moisture in space and shrinks slightly. The effect on the position of the focus was pre-calculated and compensated for.

The detector is a hybrid CMOS device with 2048 × 2048 square pixels of 10 µm pitch. In combination with the telescope, this leads to an image scale of ≈ 4.5 metre per pixel from the nominal orbit (angular scale = 11.4 µrad/px). The detector is mounted in a detector block that also houses the electronics for the detector read-out. Such detectors normally operate best when cold and so the detector block was connected to a radiator with a thermal strap.

A plate with different coloured coatings was placed directly on top of the detector to provide the colour capability. Three filters were chosen centred at 499.9 nm (BLU), 836.2 nm (RED), and 936.7 nm (NIR) and a panchromatic filter at 675.0 nm (PAN) was also included. Note that these filters do not correspond to the standard RGB system that one might have in a commercial camera although a synthetic RGB image can be constructed from images acquired through these filters. Images through the different filters can be combined in numerous ways to emphasize the diversity of the colour of the surface. The filter bandpasses were chosen specifically to emphasize difference seen in the colours of minerals expected on the surface. The images herein take advantage of this to provide a strong visual display of lithological differences on the surface. We will also show some synthetic RGB images.

LE TÉLESCOPE, LE DÉTECTEUR
ET LES FILTRES

Le télescope est fabriqué en fibres de carbone renforcées de plastique (PRFC). Il possède un miroir primaire de 13,5 cm de diamètre, une distance focale de 880 mm et une ouverture f/6.52. Il s'agit d'un système à quatre miroirs, hors axe, qui offre un excellent rejet de la lumière parasite. La structure a été allégée pour réduire la masse totale à seulement 5,5 kg dont 1 kg pour le seul miroir primaire. La structure en fibres s'assèche rapidement dans le vide spatial et se contracte légèrement. L'effet sur la position de la mise au point a été pré-calculé et compensé.

Le capteur est de type CMOS hybride avec 2048 × 2048 pixels d'une taille physique de 10x10 μm. Combinés à la focale et à l'ouverture du télescope, cela conduit à une résolution d'image d'environ 4,5 mètres par pixel à partir de l'orbite nominale (échelle angulaire = 11,4 μrad/px). Le capteur est monté dans un bloc détecteur qui abrite également l'électronique de lecture. De tels capteurs fonctionnent normalement mieux lorsqu'ils sont froids et le bloc détecteur est donc connecté à un radiateur par une bride thermique.

Une fenêtre avec des revêtements de différentes couleurs a été placée directement sur le capteur pour lui fournir la capacité de distinguer les couleurs. Trois filtres couleur ont été choisis, centrés à 499,9 nm (BLU), 836,2 nm (RED) et 936,7 (NIR) ainsi qu'un filtre panchromatique centré à 675,0 nm (PAN). Notez que ces filtres ne correspondent pas au système RVB standard que l'on pourrait avoir dans un appareil photo commercial bien qu'une image RVB synthétique puisse être construite à partir d'images acquises grâce à ces filtres. Les images à travers les différents filtres peuvent être combinées de nombreuses manières pour souligner la diversité des couleurs de la surface. Ici, les filtres passe-bande ont été choisis spécifiquement pour souligner les différences de couleurs observées entre différents types de minéraux attendus à la surface. Les images couleurs de CaSSIS permettent donc de mettre particulièrement en évidence les différences lithologiques à la surface. Nous présentons également quelques images RVB synthétiques plus proches des couleurs naturelles que l'on observerait à l'œil.

IL TELESCOPIO, IL RILEVATORE E I FILTRI

Il telescopio è un F/6.52 realizzato in plastica rinforzata con fibra di carbonio (CFRP) e ha uno specchio primario di 13,5 cm di diametro con una lunghezza focale di 880 mm. Si tratta di un sistema a quattro specchi fuori asse, che fornisce un'eccellente riduzione della luce diffusa. La struttura è stata alleggerita per ridurre la massa totale a soli 5,5 kg di cui 1 kg è dedicato al solo specchio primario. La struttura CFRP perde umidità nello spazio e si restringe leggermente. L'effetto sulla posizione del fuoco del telescopio è stato pre-calcolato e compensato.

Il rivelatore è un CMOS ibrido con 2048 × 2048 pixel quadrati di 10 μm. In combinazione con il telescopio, questo corrisponde ad una scala di immagine di ≈ 4,5 metri per pixel dall'orbita nominale (scala angolare = 11,4 μrad/px). Il rivelatore è montato in un blocco che ospita anche l'elettronica per la lettura del rivelatore. Tali rivelatori normalmente funzionano meglio quando sono freddi e quindi il blocco rivelatore è stato collegato a un radiatore con una cinghia termica.

Una piastra con diversi rivestimenti colorati (filtri) è stata posta direttamente sopra il rivelatore per fornire la capacità di misurare colori diversi. Sono stati scelti tre filtri centrati a 499.9 nm (BLU), 836.2 nm (ROSSO), e 936.7 nm (NIR) ed è stato incluso anche un filtro pancromatico a 675.0 nm (PAN). Si noti che questi filtri non corrispondono al sistema RGB (red-green-blue) standard che si potrebbe avere in una telecamera commerciale, anche se è comunque possibile costruire un'immagine RGB sintetica da immagini acquisite attraverso questi filtri. Le immagini attraverso i diversi filtri possono essere combinate in numerosi modi per enfatizzare la diversità del colore della superficie. Le lunghezze di banda dei filtri sono stati scelte specificamente per enfatizzare la differenza di colore dei minerali previsti sulla superficie. Le immagini riportate in questo libro sfruttano questo principio per fornire una forte visualizzazione delle differenze litologiche sulla superficie. Mostreremo anche alcune immagini RGB sintetiche.

ТЕЛЕСКОП, ДЕТЕКТОР И ФИЛЬТРЫ

Телескоп представляет собой модель F/6.52, изготовленную из углепластика (CFRP), и имеет первичное зеркало диаметром 13,5 см с фокусным расстоянием 880 мм. Это четырехзеркальная внеосевая система, обеспечивающая превосходную защиту от засветки. Облегчённость каркаса позволила снизить общую массу до 5,5 кг, из которых 1 кг приходится только на первичное зеркало. Углепластик теряет влагу в космосе и немного сжимается. Влияние этого сжатия на положение фокуса было предварительно рассчитано и скомпенсировано.

Детектор представляет собой гибридное КМОП-устройство с матрицей из 2048 x 2048 квадратных пикселей с шагом 10 мкм. В сочетании с телескопом оно формирует изображения с масштабом в ≈ 4,5 метра на пиксель с номинальной орбиты (угловой масштаб = 11,4 мкрад/пкс). Детектор установлен в детекторном блоке, в котором также находится электроника для считывания данных. Такого рода детекторы обычно лучше всего работают в холодном состоянии, поэтому его блок был подключён к радиатору с помощью термоленты.

Пластина с различными цветными покрытиями была помещена непосредственно на детектор для обеспечения возможности цветной съёмки. Были выбраны три фильтра с центрами на 499,9 нм (BLU), 836,2 нм (RED) и 936,7 (NIR), также был включен и панхроматический фильтр на 675,0 нм (PAN). Обратите внимание, что эти фильтры не соответствуют стандартной системе RGB, которую можно встретить в коммерческой камере, хотя синтетическое RGB изображение может быть построено из изображений, сделанных через эти фильтры. Изображения, полученные через различные фильтры, можно комбинировать разными способами, чтобы подчеркнуть разнообразие цвета поверхности. Полосы пропускания фильтров были подобраны специально для того, чтобы лучше подчеркнуть различия в цветах минералов, ожидаемых на поверхности. В представленных здесь изображениях это используется для обеспечения яркого визуального отображения литологических различий на поверхности. Мы также приводим несколько синтетических RGB-изображений.

DER DATENSATZ UND DAS BILDARCHIV

In den ersten drei Jahren des Betriebs von CaSSIS wurden über 20 000 Bilder vom Mars aufgenommen. Allerdings sind nicht alle Bilder brauchbar. Im Jahr 2019 zum Beispiel gab es einen Staubsturm, der den Planeten vollständig einhüllte und die Aufnahme der Oberfläche verhinderte. In dieser Zeit wurden viele Bilder aufgenommen, um festzustellen, ob der Staubsturm nachlässt oder nicht. Für die Untersuchung der Farben auf der Oberfläche sind diese Bilder jedoch nicht brauchbar.

Einige Bilder waren zeitlich ungenau, was dazu führte, dass sie nicht vom Zielpunkt aus aufgenommen wurden. Geringfügige Probleme mit der Zeitmessung sowohl innerhalb von CaSSIS als auch auf der Raumsonde wurden nach sorgfältiger Analyse gefunden und bis Anfang 2021 weitgehend behoben. Der Rotationsmechanismus fiel 2019 aus und ist seitdem nur noch eingeschränkt nutzbar, obwohl das Betriebsteam einen Ansatz entwickelt hat, bei dem der Motor unter Umgehung des beschädigten Bereichs genutzt wird. Dies führt in den meisten Fällen zu einem ausgezeichneten Betrieb, aber einige Bilder können falsch ausgerichtet sein, was die Breite des Farbstreifens verringert.

Insgesamt schätzt man jedoch, dass 60–70 % der CaSSIS-Bilder wissenschaftlich nutzbar sind und etwa 3 % der Planetenoberfläche abdecken. Die Rohdaten und die um die instrumentellen Fehler korrigierten/kalibrierten Daten sind über das Planetary Science Archive (PSA) der Europäischen Weltraumorganisation (ESA) verfügbar und können von der Öffentlichkeit genutzt werden. Das CaSSIS-Team liefert die Daten sechs Monate nach ihrer Erfassung an das PSA. Diese Zeit wird normalerweise genutzt, um Kalibrierungs- und geometrische Rekonstruktionspipelines laufen zu lassen und die Bilder für den allgemeinen Gebrauch zu validieren.

THE DATASET AND THE IMAGE ARCHIVE

Over 20,000 images were acquired by CaSSIS during the first three years of operation at Mars. Not all images are useful, however. In 2019 for example, there was a dust storm that totally encircled the planet preventing imaging of the surface. Many images were acquired during this period to determine whether the dust storm was subsiding or not. For studying colour on the surface, these images are not useable.

Some images were not accurately timed resulting in them not being from the targeted point. Minor timing problems both within CaSSIS and on the spacecraft were found after careful analysis and were mostly corrected by early 2021. The rotation mechanism experienced a failure in 2019 and its use has been restricted since that time although the operations team have developed an approach that uses the motor while avoiding a damaged area. This results in excellent operation most of the time but some images can be misaligned which reduces the colour swath width.

In total, however, it is estimated that 60–70 % of CaSSIS images are scientifically useful covering around 3 % of the planet's surface. The raw data and data that have been corrected/calibrated for the instrumental defects are available through the European Space Agency's Planetary Science Archive (PSA) and are open to the public. The CaSSIS team delivers the data to the PSA six months after they have been acquired. This time is normally used to run calibration and geometric reconstruction pipelines and validate the images for general use.

JEU DE DONNÉES ET ARCHIVAGE DES IMAGES

Plus de 20 000 images ont été acquises par CaSSIS au cours de ses trois premières années d'exploitation scientifique. Toutes ces images ne sont cependant pas utiles. En 2019, par exemple, une tempête de poussière globale a totalement encerclé la planète, empêchant d'en observer la surface. De nombreuses images ont néanmoins été prises au cours de cette période afin de déterminer si la tempête de poussière diminuait d'intensité. Ces images ne sont pas utilisables pour étudier les couleurs de la surface.

Certaines images ont parfois été acquises un peu trop tôt ou trop tard, et ne montrent donc pas la zone initialement visée. Des problèmes mineurs de synchronisation à la fois au sein de CaSSIS et sur la sonde ont été découverts après une analyse minutieuse et ont pour la plupart été corrigés au début de 2021. Le mécanisme de rotation a connu une défaillance en 2019 et son utilisation a été restreinte depuis lors. Mais l'équipe d'opérations a développé une nouvelle stratégie qui permet l'utilisation du moteur tout en évitant une zone endommagée du mécanisme. Il en résulte un excellent fonctionnement la plupart du temps, mais quelques images peuvent occasionnellement être mal alignées, ce qui réduit la largeur de la zone imagée en couleurs.

On estime que 60 à 70 % de l'ensemble des images CaSSIS sont scientifiquement utiles. Ces images couvrent ensemble environ 3 % de la surface de la planète. Les données brutes ainsi que les données calibrées et corrigées des défauts instrumentaux sont disponibles publiquement via l'archive de sciences planétaires (PSA) de l'Agence Spatiale Européenne. L'équipe CaSSIS délivre les données au PSA six mois après leur acquisition. Ce temps est normalement utilisé pour exécuter les scripts d'étalonnage et de reconstruction géométrique et valider les images en vue de leur utilisation publique.

IL SET DI DATI E L'ARCHIVIO DELLE IMMAGINI

Oltre 20.000 immagini sono state acquisite da CaSSIS durante i primi tre anni di attività su Marte. Tuttavia, non tutte le immagini sono utili. Nel 2019, ad esempio, c'è stata una tempesta di polvere che ha circondato completamente il pianeta impedendo l'imaging della superficie. Sono state acquisite molte immagini durante questo periodo per determinare se la tempesta di polvere si stesse placando o meno, ma queste immagini non sono utilizzabili per studiare i colori della superficie.

Alcune immagini non sono state sincronizzate con precisione e quindi non ritraggono il punto sulla superficie previsto. Sono stati riscontrati dopo un'attenta analisi alcuni problemi di sincronizzazione minori sia all'interno di CaSSIS che sul veicolo spaziale e sono stati per lo più corretti all'inizio del 2021. Il meccanismo di rotazione ha subito un guasto nel 2019 e da allora il suo utilizzo è stato limitato, sebbene il team operativo abbia sviluppato un approccio all'utilizzo del motore che evita la zona danneggiata. Ciò si traduce in un funzionamento eccellente per la maggior parte del tempo, ma alcune immagini possono essere disallineate, il che riduce l'ampiezza della zona in cui i colori si sovrappongono.

In totale, tuttavia, si stima che il 60–70 % delle immagini CaSSIS siano scientificamente utili, coprendo circa il 3 % della superficie del pianeta. I dati grezzi e i dati che sono stati corretti/calibrati per i difetti strumentali sono disponibili tramite il Planetary Science Archive (PSA) dell'Agenzia spaziale europea e sono aperti al pubblico. Il team di CaSSIS consegna i dati al PSA sei mesi dopo che sono stati acquisiti. Questo tempo viene normalmente utilizzato per eseguire processi di calibrazione e ricostruzione geometrica e convalidare le immagini per un uso generale.

НАБОР ДАННЫХ И АРХИВ ИЗОБРАЖЕНИЙ

За первые три года работы на марсианской орбите CaSSIS получил более 20 000 изображений. Однако, не все изображения одинаково полезны. Например, в 2019 году случилась пыльная буря, полностью охватившая планету и не позволявшая получать изображения поверхности. В течение данного периода было сделано много снимков, чтобы определить, утихла ли пыльная буря. Для изучения цветов поверхности эти изображения не годятся.

Также, время получения некоторых изображений не было рассчитано достаточно точно, что привело к их отклонению от целевой точки. Незначительные проблемы с синхронизацией как в CaSSIS, так и на орбитальном аппарате были выявлены после тщательного анализа и, в основном, исправлены к началу 2021 года. В 2019 году произошёл сбой в механизме вращения, и с тех пор его использование было ограничено, хотя оперативная группа разработала подход, который использует мотор, избегая при этом повреждённого участка.

Это приводит к отличной работе в большинстве случаев, но некоторые изображения могут быть смещены, что уменьшает ширину цветовой полосы.

В целом, однако, можно предположить, что 60–70 % изображений CaSSIS имеют научную ценность, покрывая, при этом, около 3 % поверхности планеты. Необработанные данные и данные, которые были скорректированы / откалиброваны на предмет дефектов аппаратуры, доступны в Архиве планетарных наук (PSA) Европейского космического агентства и открыты для общественности. Команда CaSSIS передаёт данные в PSA через шесть месяцев после того, как они были получены. Это время обычно используется для выполнения калибровки и геометрической реконструкции, и утверждения изображений для общего использования.

EINE AUSWAHL VON BILDERN

A SELECTION OF IMAGES

UNE SÉLECTION D'IMAGES

UNA SELEZIONE DI IMMAGINI

ПОДБОРКА ИЗОБРАЖЕНИЙ

Die folgenden Seiten zeigen etwa 180 der besten Bilder, die mit CaSSIS aufgenommen wurden. Der Mars wird oft als «roter Planet» bezeichnet, und in der Tat ist er auf den meisten CaSSIS-Bildern rot, wenn diese Bilder so bearbeitet werden, dass sie simulieren, wie die Oberfläche für das menschliche Auge aussehen würde. Wir können die CaSSIS-Bilder jedoch bearbeiten, um die Farbe zu verbessern (wir dehnen das Histogramm jedes Farbbandes) und so die Farbvielfalt hervorzuheben. Dies ist für die Wissenschaft von grossem Nutzen, da so verschiedene Arten von Mineralien auf der Oberfläche sichtbar gemacht werden können. Ausserdem liefert es bemerkenswert farbenfrohe Bilder, die für das Auge schön sind. Es ist jedoch wichtig, daran zu denken, dass die Bilder nicht die wahre Farbe des Mars zeigen, sondern Farbunterschiede auf der Oberfläche aufzeigen, die die Marsgeomorphologen im Hinblick auf die Mineralogie und Lithologie der Oberfläche zu interpretieren versuchen können.

Jede Bildunterschrift enthält eine kurze Beschreibung des Bildes (in fünf Sprachen) sowie den Breitengrad, den Längengrad und die Region auf dem Mars, in der es aufgenommen wurde. Sie weist auch auf die Interpretation hin, die wir mit den aufgenommenen Daten vornehmen können. Zudem zeigen wir die Richtung nach Norden und stellen eine 1-Kilometer-Skala zur Verfügung. Die Bilder wurden in der Regel gedreht, um sie an das Format des Buches anzupassen, was dazu führt, dass Norden in den meisten Fällen ungefähr links liegt. Die Richtung zur Sonne wird ebenfalls angezeigt.

Beachten Sie auch, dass das Auge manchmal trügt! Man kann den Eindruck gewinnen, dass eine Struktur eine topografische Erhebung ist, obwohl es sich in Wirklichkeit um eine Senke handelt. Wir haben versucht, eine Warnung zu geben, wenn ein Bild diesen falschen Eindruck stark vermittelt. Zwei Dinge sollten beachtet werden.

Erstens: Krater sind immer Vertiefungen, und wenn Sie den Eindruck haben, dass ein Krater in Wirklichkeit eine Kuppel ist, dann werden Sie in der Regel getäuscht! Zweitens: Wenn Sie das Gefühl haben, dass Sie die Topografie nicht richtig sehen, drehen Sie das Buch um und betrachten Sie das Bild von oben statt von unten. Das hilft Ihrem Auge oft, die richtige Perspektive zu finden.

Die Bilder sind nach Themen geordnet, aber es besteht keine Notwendigkeit, die Bilder in einer bestimmten Reihenfolge zu betrachten. Stöbern und geniessen!

The following pages show around 180 of some of the best images obtained by CaSSIS. Mars is often referred to as "the Red Planet" and, indeed, it is red in most CaSSIS images when those images are processed to simulate how the surfaces would look to the human eye. However, we can process CaSSIS images to enhance the colour (we stretch the histogram of each colour band) and thereby make the colour diversity stand out. This is extremely useful scientifically because it can reveal different types of minerals exposed on the surface. It also provides remarkably colourful images that are beautiful to the eye. It is important to remember, however, that the images do not show the true colour of Mars but reveal colour differences on the surface that planetary geomorphologists can try to interpret in terms of the surface mineralogy and lithology.

Each caption provides a short description of the image (in 5 languages) and the latitude, longitude, and region on Mars where it was obtained. It will also draw attention to the interpretation that we can place on the acquired data. We also show the direction pointing north and provide a 1 kilometre scalebar. The images have usually been rotated to fit the format of the book and results in north being roughly to the left in most cases. The direction to the Sun is also shown.

Also be aware that the eye can sometimes deceive! One can get the impression that a structure is a topographic high when in fact it is a depression. We have tried to give a warning when there is an image that gives this incorrect impression strongly. Two things should be noted. Firstly, craters are always depressions and if you get an impression that a crater is actually a dome, then you are usually being fooled! Secondly, if you get a feeling that you are not seeing the topography correctly, turn the book around and look at the picture from the top instead of the bottom. This often helps your eye to get the perspective correct.

The images are presented in themes but there is no need to view the pictures in a specific order. Browse and enjoy!

Les pages suivantes montrent environ 180 des meilleures images obtenues par CaSSIS. Mars est souvent appelée « la planète rouge » et, en effet, elle apparaît rouge dans la plupart des images CaSSIS lorsque ces images sont traitées pour simuler l'apparence des surfaces vues à l'œil nu. Cependant, nous pouvons traiter les images CaSSIS pour améliorer les contrastes (nous étirons l'histogramme de chaque canal) et faire ainsi ressortir la diversité des couleurs. Ceci est extrêmement utile scientifiquement car les couleurs peuvent révéler différents types de minéraux exposés à la surface. Le procédé fournit également des images remarquablement colorées qui sont agréables à regarder. Il est important de se rappeler, cependant, que les images ne montrent pas les vraies couleurs de Mars mais révèlent la variabilité spectrale de la surface que les géologues planétaires peuvent interpréter en termes de minéralogies et de lithologies de la croûte.

Chaque légende fournit une brève description de l'image (en 5 langues) et la latitude, la longitude et le nom de la région sur Mars où elle a été obtenue. Les légendes proposent également des éléments d'interprétation des structures et couleurs observées. Sur chaque image, nous montrons une barre d'échelle de 1 kilomètre et indiquons la direction du nord géographique ainsi que la position du soleil illuminant le paysage. Les images ont en effet généralement été pivotées pour s'adapter au mieux au format du livre et le nord se trouve sur la gauche dans la plupart des cas.

Notez également que l'œil peut parfois tromper ! On peut avoir parfois l'impression qu'une structure est en relief positif alors qu'il s'agit en fait d'une dépression. Nous avons essayé d'indiquer les cas où les images apparaissent trompeuses. Deux conseils utiles. Premièrement, les cratères sont toujours des dépressions et si vous avez l'impression qu'un cratère est en fait un dôme, alors vous faites sans doute erreur ! Deuxièmement, si vous avez l'impression que vous n'interprétez pas correctement la topographie, observez l'image en retournant le livre. Cela aide souvent votre œil à trouver la perspective correcte.

Les images sont présentées par thèmes mais il n'est pas nécessaire de les regarder dans un ordre particulier. Parcourez les pages à votre guise et admirez la beauté singulière de Mars !

Le pagine seguenti mostrano circa 180 di alcune delle migliori immagini ottenute da CaSSIS. Marte viene spesso definito «il pianeta rosso» e, in effetti, è rosso nella maggior parte delle immagini di CaSSIS quando tali immagini vengono elaborate per simulare come apparirebbero le superfici all'occhio umano. Tuttavia, possiamo elaborare le immagini CaSSIS per migliorare il contrasto di colore (estendendo l'istogramma di ciascuna banda di colore) e quindi far risaltare la diversità dei colori della superficie. Questo è estremamente utile scientificamente perché può rivelare diversi tipi di minerali esposti sulla superficie, ma fornisce anche immagini straordinariamente colorate che sono belle da guardare e osservare. È importante ricordare, tuttavia, che le immagini non mostrano il vero colore di Marte ma rivelano differenze di colore sulla superficie che i geomorfologi planetari possono cercare di interpretare in termini di mineralogia e litologia di superficie.

Ogni didascalia fornisce una breve descrizione dell'immagine (in cinque lingue), la latitudine, la longitudine, la regione su Marte in cui è stata ottenuta e un' interpretazione scientifica dei dati acquisiti. Vengono mostrate anche la direzione del nord e una barra di scala di 1 chilometro. Le immagini sono state generalmente ruotate per adattarsi al formato del libro e il nord si trova nella maggior parte dei casi a sinistra. Viene mostrata anche la direzione e l'elevazione del Sole.

Si faccia attenzione al fatto che l'occhio a volte può ingannare! Si può avere l'impressione che una struttura sia più alta del resto dell'immagine quando in realtà è una depressione. Abbiamo cercato di dare un avvertimento quando c'è un'immagine che dà fortemente questa impressione errata. Si devono notare due cose. In primo luogo, i crateri sono sempre depressioni e se si ha l'impressione che un cratere sia in realtà una cupola, di solito è il vostro occhio che vi sta ingannando! In secondo luogo, se hai la sensazione di non vedere correttamente la topografia, gira il libro e guarda l'immagine dall'alto anziché dal basso. Questo spesso aiuta l'occhio a ottenere la prospettiva corretta.

Le immagini sono presentate per temi, ma non è necessario visualizzare le immagini in un ordine specifico. Sfogliate liberamente e buon divertitimento!

На следующих страницах представлено около 180 лучших изображений, полученных CaSSIS. Марс часто называют «красной планетой», и действительно, он красный на большинстве снимков CaSSIS, если эти изображения обрабатываются для имитации того, как поверхность выглядела бы для человеческого глаза. Однако мы можем обработать изображения CaSSIS так, чтобы усилить цветность (мы растягиваем гистограммы каждой цветовой полосы) и, тем самым, сделать цветовое разнообразие более заметным. Это чрезвычайно полезно с научной точки зрения, поскольку позволяет выявлять различные типы минералов, находящихся на поверхности. Это также позволяет получить удивительно красочные изображения, радующие глаз. Важно помнить, однако, что изображения не показывают истинный цвет Марса, но выявляют цветовые различия на поверхности, которые планетарные геоморфологи могут попытаться интерпретировать с точки зрения минералогии и литологии поверхности.

Каждая подпись содержит краткое описание изображения (на 5 языках), а также широту, долготу и регион на Марсе, где оно было получено. Она также обращает внимание на интерпретацию, которую мы можем дать полученным данным. Кроме того, мы указываем направление на север, приводим масштабную линейку в 1 километр и указываем положение солнца. Изображения, как правило, были повёрнуты, чтобы соответствовать формату книги, в результате чего север, в большинстве случаев, оказывается примерно слева.

Также помните, что глаз иногда может ошибаться! У смотрящего может сложиться впечатление, что структура является топографической возвышенностью, в то время как на самом деле это впадина (и наоборот). Мы старались предупреждать в случаях, когда изображение явно производит неверное впечатление. Следует отметить два момента. Во-первых, кратеры – это всегда впадины, и если у вас создается впечатление, что кратер – это купол, то вы, как правило, обманываетесь! Во-вторых, если у вас возникло ощущение, что вы неправильно видите рельеф, переверните книгу и посмотрите еще раз на снимок. Это часто помогает глазу правильно оценить перспективу.

Изображения сгруппированы по темам, но нет необходимости знакомиться с ними в определенном порядке. Желаем Вам приятного просмотра!

1 DELTAS, FLÜSSE, ERDRUTSCHE UND FÄCHER

DELTAS, FLOWS, LANDSLIDES
AND FANS

DELTAS, COULÉES, GLISSEMENTS
DE TERRAIN ET ÉVENTAILS

DELTA, FLUSSI, FRANE E
CONI DI DEIEZIONE

ДЕЛЬТЫ, ПОТОКИ, ОПОЛЗНИ
И ВЕЕРНЫЕ ОТЛОЖЕНИЯ

ERDRUTSCH IN TERRA CIMMERIA

Hier ist der innere Rand eines grossen Kraters zusammengebrochen, möglicherweise als Folge eines späteren kleineren Einschlags direkt in den Kraterrand. Der Standort liegt westlich von Durius Vallis, welches ein ziemlich dichtes Entwässerungsnetz aufweist, was auf frühere fluviale Aktivitäten in diesem Gebiet hindeutet.

LANDSLIDE IN TERRA CIMMERIA

Here, the internal rim of a large crater has collapsed, possibly as a result of a later smaller impact directly into the crater rim. The site is west of Durius Vallis which has a fairly dense drainage network indicating past fluvial activity in the area.

GLISSEMENT DE TERRAIN DANS LA RÉGION DE TERRA CIMMERIA

Ici, la bordure interne d'un grand cratère s'est effondrée, probablement à la suite d'un impact ultérieur plus petit directement dans cette bordure. Le site se trouve à l'ouest de Durius Vallis qui possède un réseau de drainage assez dense indiquant une activité fluviale passée dans la région.

FRANA IN TERRA CIMMERIA

In questa immagine si po' osservare il bordo interno di un grande cratere che è crollato, probabilmente come risultato di un successivo impatto più piccolo direttamente sul bordo del cratere. Il sito è a ovest di Durius Vallis, che presenta una rete di drenaggio abbastanza densa ed indica una passata attività fluviale nella zona.

ОПОЛЗЕНЬ НА КИММЕРИЙСКОЙ ЗЕМЛЕ

На снимке видны разрушения внутреннего вала большого кратера. Вероятно, разрушение произошло в результате более позднего удара небольшого метеорита. Это место расположено на западе от Долины Дуэро, где находится довольно густая дренажная сеть, указывающая на речную активность в этом районе в прошлом.

1.1 LANDSLIDE IN TERRA CIMMERIA

ACQUISITION TIME: 2021-01-01T23:57:47.053
LONGITUDE: 169.4 E   LATITUDE: 21.3 S
REGION: CIMMERIA

EXHUMIERTES DELTA IM EBERSWALDER KRATER

Es handelt sich um ein gut untersuchtes exhumiertes Delta im Krater Eberswalde, das sich durch den Fluss von flüssigem Wasser in ein Kraterseesystem gebildet hat. Es steht in der engeren Auswahl als Ziel für Landemissionen, da es Sedimentgestein enthält und die Möglichkeit besteht, dass diese Gesteine Hinweise auf Leben enthalten. Beachten Sie die helle Schichtung an den Enden der Ströme, die wahrscheinlich auf hydratisierte Mineralien zurückzuführen ist.

EXHUMED DELTA IN EBERSWALDE CRATER

This is a well-studied exhumed delta in Eberswalde crater, which formed by the flow of liquid water into a crater lake system. It has been shortlisted as a target for landed missions because of the presence of sedimentary rocks and the possibility that those rocks might retain evidence of life. Note the light-toned layering at the ends of the flows, probably due to hydrated minerals.

DELTA EXHUMÉ DANS LE CRATÈRE EBERSWALDE

Un delta exhumé très étudié dans le cratère Eberswalde, formé par l'écoulement d'eau liquide dans un lac au fond du cratère. Il a été présélectionné comme site d'atterrissage possible pour de futures missions en raison de la présence de roches sédimentaires et de la possibilité que ces roches puissent conserver des traces de vie passée. Notez les couches de couleur claire aux extrémités des méandres, probablement riches en minéraux hydratés.

DELTA AFFIORANTE NEL CRATERE EBERSWALDE

Questo è un delta molto studiato nel cratere Eberswalde, che si è formato dal flusso di acqua liquida in un sistema di laghi craterici. È stato selezionato come obiettivo per future missioni sulla superficie a causa della presenza di rocce sedimentarie e la possibilità che queste rocce possano conservare prove di vita passata. Si noti la stratificazione chiara alle estremità dei canali, probabilmente dovuta a minerali idratati.

ДЕЛЬТА ВЫСОХШЕЙ РЕКИ В КРАТЕРЕ ЭБЕРСВАЛЬДЕ

Это хорошо изученная дельта высохшей реки в кратере Эберсвальде, которая образовалась в результате поступления воды в систему кратерного озера. Дельта была включена в список целей для посадочных миссий из-за наличия осадочных пород и вероятности, что в этих породах могли сохраниться свидетельства жизни. Обратите внимание на светлые слои у концов потоков. Вероятно, там находятся гидратированные минералы.

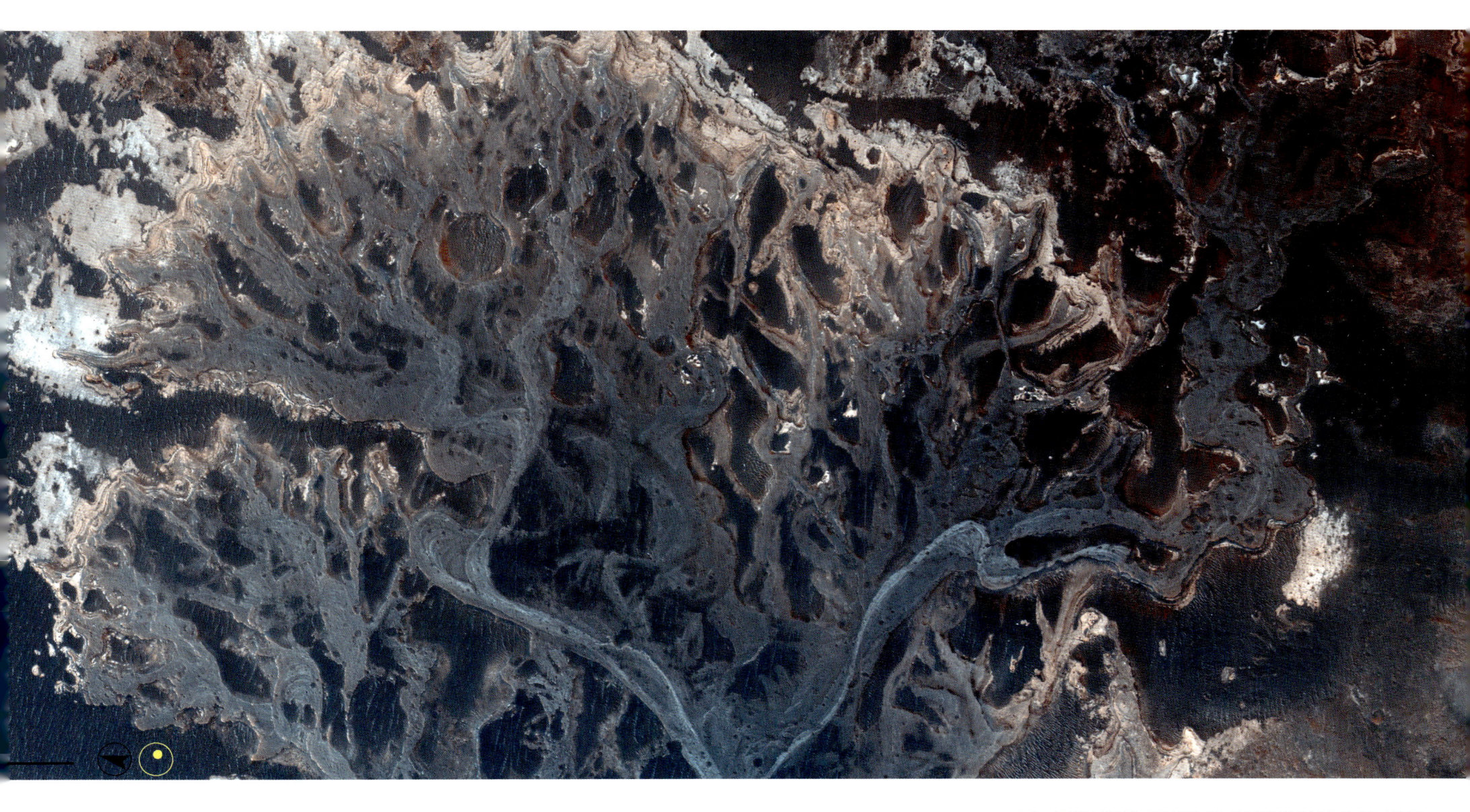

1.2   EXHUMED DELTA IN EBERSWALDE CRATER

ACQUISITION TIME: 2018-11-16T21:40:42.766
LONGITUDE: 326.3 E   LATITUDE: 23.6 S
REGION: MARGARITIFER

FLUVIALE UND VULKANISCHE ABLAGERUNGEN IM JEZERO-KRATER

Jezero ist ein Krater mit einem Durchmesser von 45 km, der als Landeplatz für den Mars-2020-Rover der NASA, Perserverance, ausgewählt wurde. Der Krater enthält diese Schwemmdelta-Ablagerung (Mitte unten), die reich an Ton ist. Es wird vermutet, dass Jezero einst überflutet wurde, als sich auf dem Mars noch Talnetzwerke bildeten. Beachten Sie den gewundenen Kanal im Osten (oben).

FLUVIAL AND VOLCANIC DEPOSITS IN JEZERO CRATER

Jezero is a 45 km diameter crater that was selected as the landing site for NASA's Mars 2020 rover, Perserverance. The crater contains this fan-delta deposit (centre bottom) which is rich in clays. It is thought that Jezero was once flooded at a time when valley networks were forming on Mars. Note the sinuous channel to the east (top).

DÉPÔTS FLUVIAUX ET VOLCANIQUES DANS LE CRATÈRE JEZERO

Jezero est un cratère de 45 km de diamètre qui a été choisi comme site d'atterrissage pour le rover Mars 2020 de la NASA, Perserverance. Le cratère contient des dépôts sédimentaires deltaïques (en bas au centre) riches en argiles. On pense que Jezero a été autrefois inondé, à une époque où les réseaux de vallées fluviales se formaient sur Mars. Notez le chenal sinueux à l'est (en haut).

DEPOSITI FLUVIALI E VULCANICI NEL CRATERE JEZERO

Jezero è un cratere di 45 km di diametro che è stato scelto come sito di atterraggio per il rover della NASA Mars 2020, Perserverance. Il cratere contiene questo deposito di un delta a ventaglio (in basso al centro) che è ricco di argille. Si pensa che Jezero sia stato inondato in un'epoca in cui le reti di valli si stavano formando su Marte. Si noti il canale sinuoso a est (in alto).

ФЛЮВИАЛЬНЫЕ И ВУЛКАНИЧЕСКИЕ ОТЛОЖЕНИЯ В КРАТЕРЕ ЕЗЕРО

Езеро – это кратер диаметром 45 км, который был выбран в качестве места посадки марсохода НАСА «Персерванс». В кратере (внизу в центре) находятся отложения в форме веерной дельты, богатые глиной. Считается, что Езеро ранее, когда на Марсе формировались сети долин, был затоплен. Обратите внимание на извилистый канал на востоке (вверху).

1.3 FLUVIAL AND VOLCANIC DEPOSITS IN JEZERO CRATER

ACQUISITION TIME: 2018-10-22T02:43:16.568
LONGITUDE: 77.3 E   LATITUDE: 18.8 N
REGION: SYRTIS MAJOR

ERDRUTSCHABLAGERUNG IN COPRATES CHASMA

Das Coprates Chasma ist einer cer grossen Canyons im Vallis-Marineris-System. An der Nordwand des Canyons scheint es zu einem Erdrutsch gekommen zu sein, der zu dieser fächerartigen Ablagerung auf dem Canyonboden führte. In der Nähe gibt es Hinweise auf Tone und Sulfate. Die Farbunterschiede in der Erdrutschablagerung könnten auf diese Mineralien hinweisen.

LANDSLIDE DEPOSIT IN COPRATES CHASMA

Coprates Chasma is one of the major canyons in the Vallis Marineris system. The north wall of the canyon appears to have suffered a landslide resulting in this fan-like deposit on the canyon floor. There is evidence of clays and sulphates close by. The colour variations in the landslide deposit may be indications of these minerals.

DÉPÔT DE GLISSEMENT DE TERRAIN DANS COPRATES CHASMA

Coprates Chasma est l'un des principaux canyons du système de Vallis Marineris. La paroi nord du canyon semble avoir subi un glissement de terrain qui a formé ce cône détritique sur le fond du canyon. Il y a des traces d'argiles et de sulfates à proximité. Les couleurs des roches au sein du glissement de terrain sont des indicateurs de la présence de ces minéraux.

DEPOSITO DI UNA FRANA NEL COPRATES CHASMA

Il Coprates Chasma è uno dei principali canyon del sistema Vallis Marineris. La parete nord del canyon sembra aver subito una frana che ha portato a questo deposito a ventaglio sul pavimento del canyon. Ci sono prove di argille e solfati nelle vicinanze e le variazioni di colore nel deposito della frana possono essere indicazioni di questi minerali.

ОПОЛЗНЕВЫЕ ОТЛОЖЕНИЯ В КАНЬОНЕ КОПРАТА

Каньон Копрата – один из самых больших каньонов в системе долин Маринера. На его северной стене, по-видимому, произошел сход оползня, сформировавшего веерообразные отложения на дне каньона. Вариации цвета в оползневых отложениях могут указывать на глины и сульфаты, признаки которых были обнаружены поблизости.

1.4   LANDSLIDE DEPOSIT IN COPRATES CHASMA

ACQUISITION TIME: 2018-10-29T15:36:11.888
LONGITUDE: 301.0 E   LATITUDE: 13.2 S
REGION: VALLES MARINERIS

FÄCHERARTIGES DELTA AUF DEM BODEN DES HOLDEN-KRATERS

Der Holden-Krater hat einen Durchmesser von etwa 140 km. Der Rand im Südwesten scheint von Wasser aus dem Uzboi Vallis angeschnitten worden zu sein, das durchbrach und ein Schwemmdelta innerhalb von Holden bildete. Das Bild zeigt den nördlichen Rand dieser Ablagerung. Man beachte die Schichtung an den hellen Rändern und die invertierten, kanalartigen Strukturen unten rechts.

FAN-LIKE DELTA ON THE FLOOR OF HOLDEN CRATER

Holden Crater is around 140 km in diameter. The rim to the south-west appears to have been cut by water from Uzboi Vallis that broke through and produced a fan-like delta within Holden. The image shows the northern edge of this deposit. Note the layering in the light-coloured edges and the inverted channel-like structures to the lower right.

CÔNE DELTAÏQUE AU FOND DU CRATÈRE HOLDEN

Le cratère Holden a un diamètre d'environ 140 km. Sa bordure sud-ouest semble avoir été dégradée par l'écoulement d'eau dans le chenal Uzboi Vallis qui a produit un cône détritique dans Holden. L'image montre la bordure nord des dépôts deltaïques. Notez la stratification dans les couches de couleur claire et les structures en forme de chenal inversé en bas à droite.

DEPOSITO DI DEIEZIONE SUL PAVIMENTO DEL CRATERE HOLDEN

Il cratere Holden ha un diametro di circa 140 km. Il bordo del cratere a sud-ovest sembra essere stato sfondato dall'acqua di Uzboi Vallis che ha prodotto un delta lobato all'interno del cratere stesso. L'immagine mostra il bordo settentrionale di questo deposito. Si noti la stratificazione nei bordi chiari e le strutture simili a canali invertiti in basso a destra.

ВЕЕРНАЯ ДЕЛЬТА НА ДНЕ КРАТЕРА ХОЛДЕН

Диаметр кратера Холден составляет около 140 км. На снимке показан его северный край. Скорее всего, вода из долины Узбой прорвалась в кратер и образовала веерообразную дельту внутри него, срезав обод на юго-западе. Обратите внимание на слоистость в светлых краях отложений и каналоподобные структуры с перевёрнутым рельефом (топографическая инверсия) в правом нижнем углу.

1.5   FAN-LIKE DELTA ON THE FLOOR OF
      HOLDEN CRATER

ACQUISITION TIME: 2021-03-13T19:50:14.748
LONGITUDE: 325.4 E   LATITUDE: 26.3 S
REGION: MARGARITIFER

STROMLINIENFÖRMIGE INSELN IN ARES VALLIS

Dies ist ein Bild von zwei strömungsgesäumten Inseln im Abflusskanal von Ares Vallis. Das durch den Kanal fliessende Wasser stiess auf die Krater, verlangsamte sich und versuchte, um sie herum zu fliessen. Die langsamere Strömung war weniger erodierend und hinterliess die «Schweife» hinter den Kratern.

STREAM-LINED ISLANDS IN ARES VALLIS

This is an image of two stream-lined islands in the Ares Vallis outflow channel. The water flowing through the channel came up against the craters, slowed, and tried to flow around them. The slower flow was less eroding leaving the "tails" behind the craters.

ÎLES ÉRODÉES PAR UN COURS D'EAU DANS LA VALLÉE D'ARES

Deux anciennes îles érodées par l'eau ayant coulé au fond du chenal de débâcle Ares Vallis. L'écoulement d'eau a dû contourner les cratères et a ainsi été ralenti, protégeant les régions en aval de l'érosion et produisant ces morphologies caractéristiques.

ISOLE ALLINEATE NELLA ARES VALLIS

Questa è un'immagine di due isolette allineate nel canale di deflusso dell'Ares Vallis. L'acqua che scorre attraverso il canale si è scontrata con i crateri, ha rallentato e ha cercato di scorrere intorno ad essi. Il flusso d'acqua più lento ha eroso meno la superficie, lasciando delle «code» dietro i crateri.

ОСТРОВА ДОЛИНЫ АРЕС

Это изображение двух островов в канале оттока – Долина Ареса. Вода, текущая по каналу, сталкивалась с кратерами, замедлялась и пыталась течь вокруг них. Этот более медленный поток обладал меньшей эрозионной силой, что привело к образованию «хвостов» за кратерами.

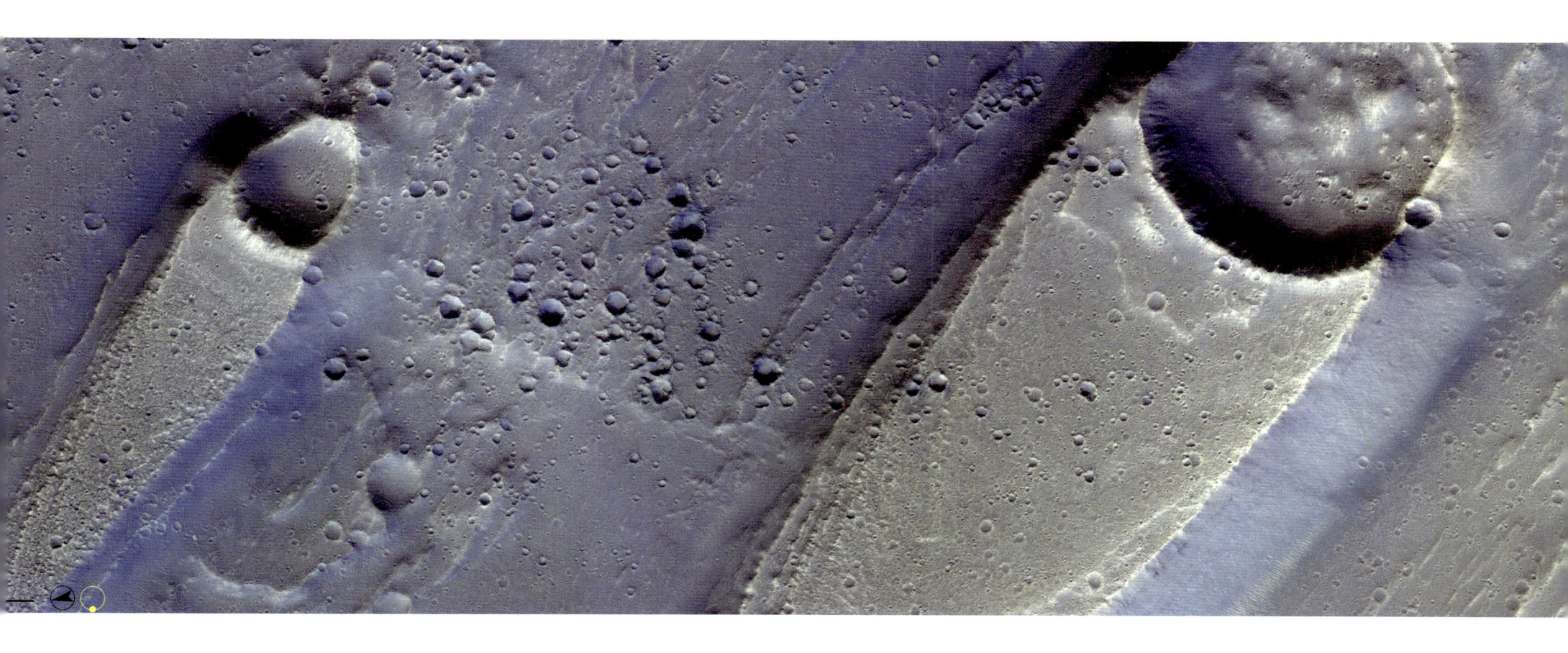

1.6 STREAM-LINED ISLANDS IN ARES VALLIS

ACQUISITION TIME: 2021-05-06T05:59:57.183
LONGITUDE: 329.9 E   LATITUDE: 15.5 N
REGION: CHRYSE

ERDRUTSCH IN COPRATES CHASMA

Dies ist ein Bild eines Schuttfächers auf dem Boden des Coprates Chasma. Er könnte das Ergebnis eines Einsturzes der rechten Schluchtwand und eines damit einhergehenden Erdrutsches sein. Dabei wurde bemerkenswert farbiges Material freigelegt. Die türkise Farbe wird normalerweise mit mafischen Mineralien in CaSSIS-Produkten in Verbindung gebracht. Beachten Sie, dass die Schluchtwand ebenfalls bunt gefärbt ist, was auf unterschiedliche Lithologien hinweist.

LANDSLIDE IN COPRATES CHASMA

This is an image of a debris fan on the floor of Coprates Chasma. It may be the result of collapse of the canyon wall to the right with accompanying landslide. It has exposed remarkably colourful material. The turquoise colour is normally associated with mafic minerals in CaSSIS products. Notice that the canyon wall is also variegated indicating different lithologies.

GLISSEMENT DE TERRAIN DANS LA RÉGION DE COPRATES CHASMA

Image d'un cône détritique sur le sol du canyon Coprates Chasma. Il a pu être formé suite à l'effondrement de la paroi du canyon à droite, accompagné d'un glissement de terrain. Il a exposé des matériaux remarquablement colorés. Les teintes turquoises sont généralement associées aux minéraux mafiques dans les images CaSSIS. Remarquez que la paroi du canyon montre également des couleurs diverses, indiquant une grande variété de lithologies.

FRANA NEL COPRATES CHASMA

Questa è un'immagine di un conoide di deiezione sul pavimento del Coprates Chasma. Può essere il risultato del crollo della parete del canyon a destra con conseguente frana. Questa, ha esposto materiale notevolmente colorato. Nelle immagini prodotte da CaSSIS, il colore turchese è normalmente associato a minerali mafici. Si noti che anche la parete del canyon è variegata, indicando diverse litologie.

ОПОЛЗЕНЬ В КАНЬОНЕ КОПРАТ

Это изображение веера обломков (коллювиального конуса) на дне каньона Копрата. Вероятно, произошел обвал правой стены каньона с сопутствующим оползнем. Благодаря этому обнажился удивительно красочный материал. Бирюзовый цвет в фильтрах CaSSIS обычно показывает мафические минералы (железо-магнезиальные минералы, присутствующие в магматических горных породах). Обратите внимание, что стена каньона также пестрая, это указывает на различное литологическое строение.

1.7 LANDSLIDE IN COPRATES CHASMA

ACQUISITION TIME: 2021-01-25T12:44:01.280
LONGITUDE: 297.4 E   LATITUDE: 13.4 S
REGION: MARINERIS

FLIESSSTRUKTUREN IN ATHABASCA VALLIS

Das Bild zeigt eine strömungsgeprägte Insel an der Kreuzung eines Kanals mit dem Hauptkanal des Athabasca Vallis in Elysium Planitia. Das Athabasca-Vallis-System entstand spät in der Geschichte des Mars (in der späten amazonischen Periode), wenn auch der genaue Entstehungsmechanismus noch nicht ganz klar ist. Es scheint wahrscheinlich, dass eine Flüssigkeit (entweder Wasser, Lava oder eine Kombination aus beidem) in den Spalten der Cerberus Fossae ihren Ursprung hatte und nach Süden floss, wobei sie – wie hier zu sehen – stromlinienförmige Inseln bildete, während die Flüssigkeit auf Krater traf, die Hindernisse für den Fluss bildeten.

FLOW STRUCTURES IN ATHABASCA VALLIS

The image shows a stream-lined island at a junction of a channel with the main channel of Athabasca Vallis in Elysium Planitia. The Athabasca Valles system was formed late in Martian history (the late Amazonian period) although the exact formation mechanism is still somewhat unclear. It seems probable that a fluid (either water, lava, or a combination of the two) originated in the Cerberus Fossae fissures and flowed south creating stream-lined islands, as observed here, as the fluid ran up against craters that formed obstacles to the flow.

STRUCTURES D'ÉCOULEMENT DANS LA VALLÉE D'ATHABASCA

L'image montre une île érodée par d'anciens écoulements massifs d'un fluide à la jonction d'un chenal secondaire avec le chenal principal d'Athabasca Vallis dans Elysium Planitia. Le système de vallées d'Athabasca s'est formé tard dans l'histoire martienne (fin de la période amazonienne) mais son mécanisme exact de formation n'est toujours pas bien compris. Il est probable qu'un fluide (soit de l'eau, soit de la lave, soit une combinaison des deux) ait pris sa source dans les fissures de Cerberus Fossae et se soit écoulé vers le sud, érodant les chenaux et créant des îles, comme on l'observe ici, lorsque le fluide s'est heurté à des cratères qui formaient des obstacles à l'écoulement.

STRUTTURE DI FLUSSO NELLA ATHABASCA VALLIS

L'immagine mostra un'isola modellata da un qualche tipo di flusso alla giunzione tra un canale laterale e il canale principale di Athabasca Vallis in Elysium Planitia. Il sistema delle Athabasca Valles si è formato tardi nella storia marziana (il tardo periodo Amazzonico) anche se l'esatto meccanismo di formazione è ancora poco chiaro. Come si può osservare qui, sembra probabile che un fluido (acqua, lava o una combinazione dei due) abbia avuto origine nelle fessure di Cerberus Fossae e sia fluito verso sud creando delle isole allineate, quando il fluido si è imbattuto in crateri che rappresentavano degli ostacoli alla corrente.

ПОТОКОВЫЕ СТРУКТУРЫ ДОЛИН АТАБАСКИ

На снимке виден «обтекаемый» остров на пересечении канала с главным руслом долин Атабаски равнины Элизий. Система долины сформировалась в поздней марсианской истории (поздний амазонийский период), хотя точный механизм формирования до сих пор остается неясным. Кажется вероятным, что поток воды, лавы или их смеси возник в расщелинах борозд Цербера и направился на юг. Когда же кратеры оказывались на пути потока, образовывались «обтекаемые» острова, как показано на этом снимке.

1.8 FLOW STRUCTURES IN ATHABASCA VALLIS

ACQUISITION TIME: 2018-10-03T02:13:49.760
LONGITUDE: 155.9 E   LATITUDE: 8.7 N
REGION: ELYSIUM

LINIENFÖRMIGES TALBODENMATERIAL

Dieses Bild zeigt lineares Talfüllungsmaterial in Arabia Terra unmittelbar westlich der Region Ismeniae Fossae und in der Nähe der Dichotomie-Grenze. Die Lineamente deuten darauf hin, dass das Material um Hindernisse herum in Täler geflossen ist. Es wird vermutet, dass das Material eisreich ist, aber von einer Schuttschicht oder von Ablagerungsmaterial bedeckt ist, das z. B. durch die Sublimation von Eis entstanden ist. Diese Gebiete sind vorrangige Ziele für künftige Raumfahrtmissionen, die nach Wassereis nahe der Oberfläche suchen.

LINEATED VALLEY FLOOR MATERIAL

This image shows lineated valley fill material in Arabia Terra immediate to the west of the Ismeniae Fossae region and close to the dichotomy boundary. The lineaments suggest flow of material down valleys around obstacles. It is thought that the material is ice-rich but covered with a debris layer or lag material produced following sublimation of ice, for example. These areas are prime targets for future spacecraft missions looking for water ice close to the surface.

MATÉRIAUX STRIÉS DE FOND DE VALLÉE

Cette image montre une unité à la texture de surface striée remplissant une vallée d'Arabia Terra, à l'ouest de la région d'Ismeniae Fossae et près de la dichotomie globale martienne. Les linéaments suggèrent un écoulement du matériau dans les vallées autour d'obstacles. On pense que le matériau est riche en glace mais recouvert d'une couche rocheuse ou éventuellement d'un résidu sec de sublimation de la glace. Ces zones sont des cibles de choix pour les futures missions spatiales qui rechercheront de la glace d'eau près de la surface.

MATERIALE DI FONDOVALLE LINEATO

Questa immagine mostra depositi lineari vallivi in Arabia Terra immediatamente ad ovest della regione di Ismeniae Fossae e vicino al confine della dicotomia. I lineamenti suggeriscono il flusso di materiale all'interno delle valli intorno agli ostacoli. Si pensa che il materiale sia ricco di ghiaccio ma coperto da uno strato di detriti o da materiale depositato prodotto, per esempio, in seguito alla sublimazione del ghiaccio. Queste aree sono obiettivi primari per future missioni spaziali alla ricerca di ghiaccio d'acqua vicino alla superficie.

ЛИНЕЙНЫЕ ОТЛОЖЕНИЯ НА ДНЕ ДОЛИНЫ

На этом изображении показан линейно залегающий материал заполнения долин земли Аравия к западу от области Борозд Исмена, вблизи границы дихотомии. Линеаменты свидетельствуют о течении материала, обходящем препятствия, вниз по долинам. Предполагается, что материал был насыщен льдом, но покрыт слоем обломков или остаточным веществом, образовавшимся, например, вследствие сублимации льда. Такие области являются первоочередными целями будущих миссий космических аппаратов, направленных на поиск водяного льда вблизи поверхности.

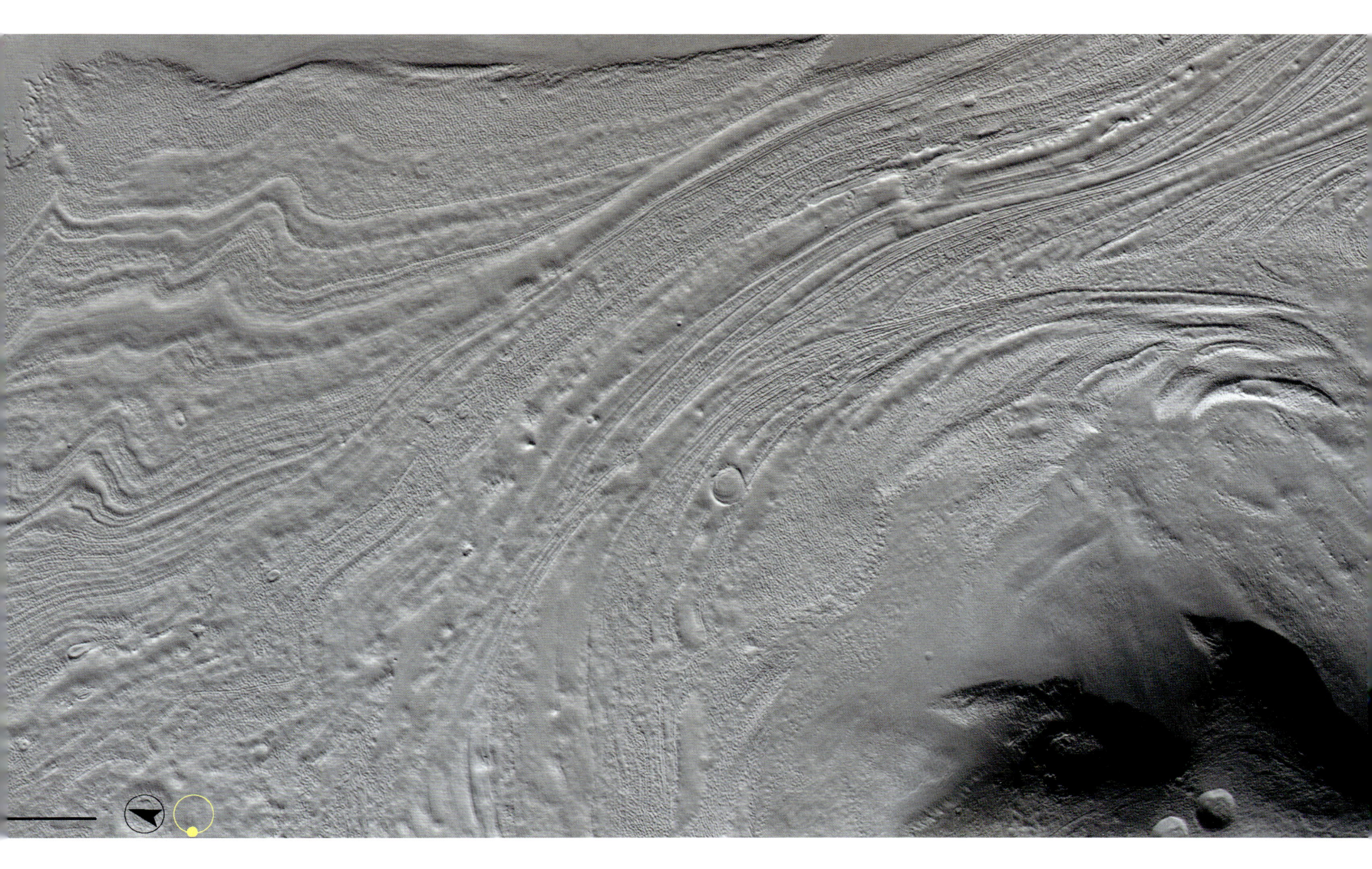

1.9 LINEATED VALLEY FLOOR MATERIAL

ACQUISITION TIME: 2021-03-06T10:35:42.360
LONGITUDE: 35.2 E LATITUDE: 40.3 N
REGION: ARABIA

ERDRUTSCHE UND ERODIERTE SCHICHTEN

Dies ist ein Bild von hell getönten Schichten und einer wahrscheinlichen Erdrutschablagerung in der Region Libya Montes in Tyrrhena Terra. Es befindet sich in einer Senke, die geschichtet wirkt und zu erodieren scheint. Das synthetische RGB-Bild weist verschiedene Farben auf, die auf unterschiedliche Mineralogien und möglicherweise freiliegendes Grundgestein hinweisen.

LANDSLIDE AND ERODED LAYERS

This is an image of light-toned layers and a probable landslide deposit in the Libya Montes region of Tyrrhena Terra. It is situated within a depression that appears to be layered and eroding. There are different colours in the synthetic RGB image indicating different mineralogies and possibly exposed bedrock.

GLISSEMENT DE TERRAIN ET STRATES ÉRODÉES

Image montrant des strates de couleur claire et un dépôt résultant probablement d'un glissement de terrain dans la région de Libya Montes dans Tyrrhena Terra. Le terrain est situé dans une dépression qui semble être stratifiée et activement érodée. Les différentes couleurs indiquent une variété de minéralogies dans le substrat rocheux exposé.

FRANA E STRATIFICAZIONI EROSE

Questa è un'immagine di stratificazioni di colore chiaro e di un probabile deposito di una frana nella regione Libia Montes di Tyrrhena Terra. Si trova all' interno di una depressione che sembra essere stratificata ed sottoposta a costante erosione. Ci sono diversi colori nell'immagine composta nei filtri RGB che indicano diverse mineralogie e evidenziano il pavimento roccioso.

ОПОЛЗЕНЬ И ЭРОДИРОВАННЫЕ СЛОИ

На этом снимке видны слои светлого материала и возможных оползневых отложений, находящиеся в пределах многоуровневой эродированной впадины, в районе Гор Ливии Тирренской Земли. На синтезированном RGB-изображении присутствуют различные цвета, указывающие на разнообразие минералов и, возможно, обнаженную коренную породу.

1.10 LANDSLIDE AND ERODED LAYERS

ACQUISITION TIME: 2019-01-01T20:22:10.800
LONGITUDE: 83.6 E LATITUDE: 0.1 N
REGION: TYRRHENA

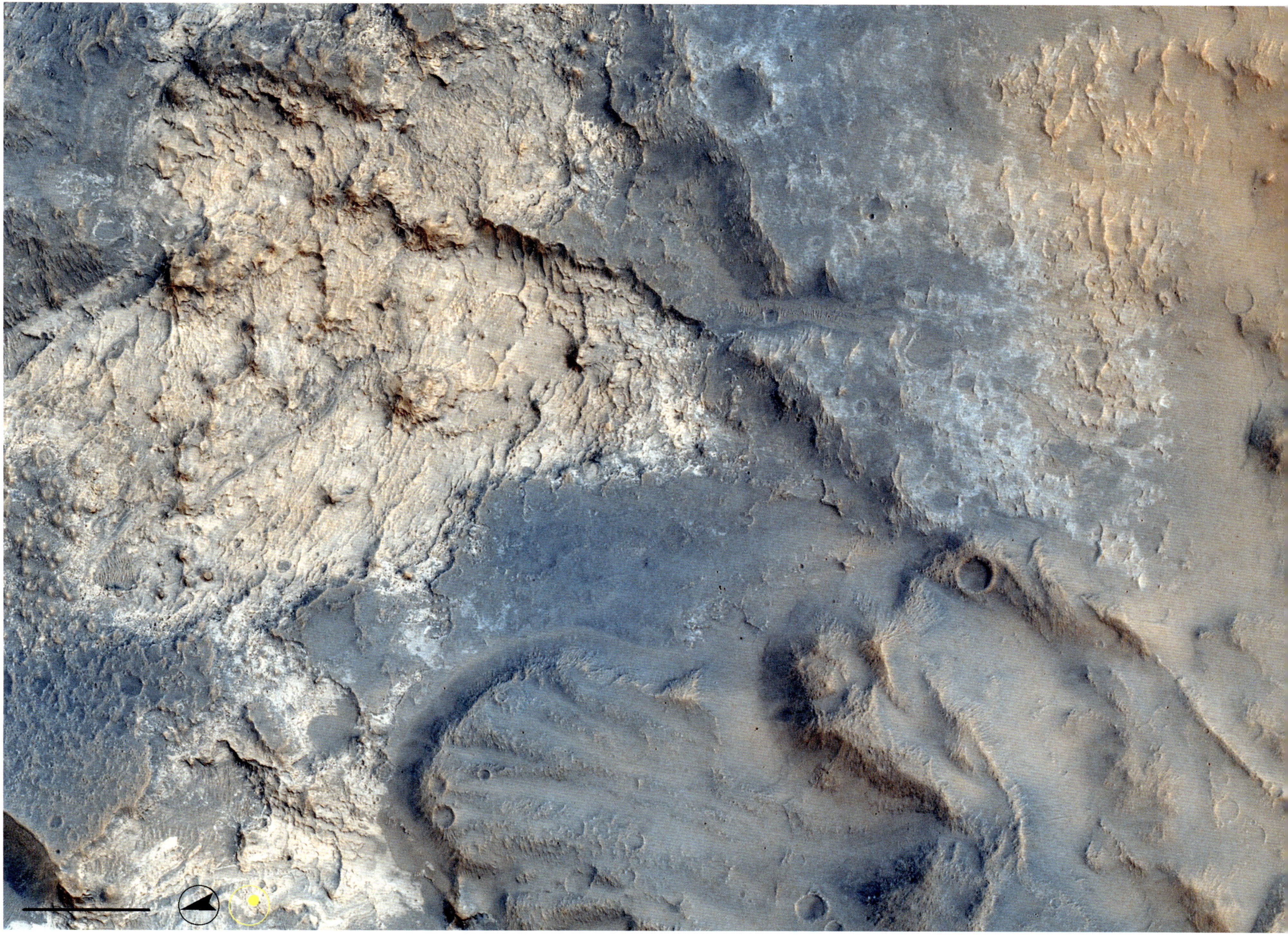

2 TÄLER UND RINNEN

VALLEYS AND CHANNELS

VALLÉES ET CHENAUX

VALLI E CANALI

ДОЛИНЫ И КАНАЛЫ

RINNE MIT ANGRENZENDEN MINERAL-AUFSCHLÜSSEN IM ANTONIADI-KRATER

Ein Kanal und freiliegendes Grundgestein in der südöstlichen Ecke des Kraters Antoniadi in Syrtis Major Planum. Antoniadi ist ein ≈ 390 km grosser Krater und wurde aufgrund von Hinweisen auf veränderte Mineralien als möglicher zukünftiger Landeplatz vorgeschlagen. Ausserdem gibt es Hinweise darauf, dass Antoniadi einst Flüsse und Seen enthielt. Hier sehen wir einen Kanal, aber es ist nicht klar, ob der Entstehungsmechanismus in diesem Fall flüssiges Wasser beinhaltete. Auf der Nordseite des Kanals sind Bereiche zu erkennen, in denen die oberste Schicht verloren gegangen zu sein scheint und eine hellere, grünere Oberfläche zum Vorschein kommt, die wahrscheinlich mit dem Grundgestein zusammenhängt.

CHANNEL WITH ADJACENT MINERAL EXPOSURES IN ANTONIADI CRATER

A channel and exposed bedrock in the south-east corner of Antoniadi Crater in Syrtis Major Planum. Antoniadi is a ≈ 390 km crater and has been proposed as a possible future landing site because of the evidence of altered minerals. Moreover, there is evidence that Antoniadi once contained rivers and lakes. Here we see a channel but it is not clear that the formation mechanism involved liquid water in this specific case. The north side of the channel shows areas where a top layer appears to have been lost revealing a lighter-toned greener surface which is probably related to the bedrock.

CHENAL ET MINÉRAUX DANS LE CRATÈRE ANTONIADI

Un chenal et le substrat rocheux exposé au sud-est du cratère Antoniadi dans la plaine de Syrtis Major. Antoniadi est un cratère d'environ 390 km et a été proposé comme un possible futur site d'atterrissage en raison de la présence de minéraux altérés, preuves qu'Antoniadi contenait autrefois des rivières et des lacs. Il n'est cependant pas certain que de l'eau liquide soit impliquée dans la formation du chenal visible sur cette image. Le versant nord du chenal expose des zones où une couche supérieure semble avoir été érodée, révélant la surface plus verte et plus claire de la roche-mère.

CANALE CON VICINE CHIAZZE SUPERFICIALI DI MINERALI NEL CRATERE ANTONIADI

Un canale e il basamento roccioso superficiale nell'angolo sud-est del cratere Antoniadi in Syrtis Major Planum. Antoniadi è un cratere di ≈ 390 km ed è stato proposto come possibile sito di atterraggio futuro a causa della presenza di minerali alterati. Inoltre, ci sono prove che Antoniadi una volta conteneva fiumi e laghi. Qui vediamo un canale, ma non è chiaro se il meccanismo di formazione abbia coinvolto acqua liquida in questo specifico caso. Il lato nord del canale mostra aree in cui uno strato superficiale di materiale sembra essere stato perso, rivelando una superficie di colore verde più chiaro che è probabilmente legata al basamento roccioso.

КАНАЛ И ОБНАЖЕННАЯ КОРЕННАЯ ПОРОДА КРАТЕРА АНТОНИАДИ

Канал и обнаженная коренная порода в юго-восточном углу кратера Антониади на Плато Большой Сирт. Антониади — это кратер размером 390 км, который был предложен для посадки будущих космических миссий благодаря наличию измененных минералов. Более того, есть свидетельства, что в этом кратере когда-то были реки и озера. Неизвестно, участвовала ли вода в формировании этого канала. На северной стороне этого канала видны участки, где обнажилась поверхность, представленная в светло-зеленых тонах. Вероятно, это коренная порода.

2.1 CHANNEL WITH ADJACENT MINERAL EXPOSURES IN ANTONIADI CRATER

ACQUISITION TIME: 2018-05-31T05:02:54.00
LONGITUDE: 62.7 E   LATITUDE: 19.4 N
REGION: SYRTIS MAJOR

RINNE MIT FREIGELEGTEM HELLEN MATERIAL IN HELLESPONTUS

Dieses Bild zeigt Kanäle in der Region Hellespontus Montes südöstlich des 175 km grossen Kraters Terby. Hadriacus Cavi, eine Reihe von 65 km langen und 15 km breiten, ostwestlich ausgerichteten Vertiefungen, liegt direkt im Norden. Die Region scheint geologisch komplex zu sein. Hier sehen wir eine Vielzahl verschiedener Farben auf der Oberfläche, die auf eine unterschiedliche Mineralogie hinweisen. In einem der Kanäle (unten links) wurde helles Material freigelegt, und es gibt Anzeichen für eine Schichtung.

CHANNEL WITH EXPOSED LIGHT-TONED MATERIAL IN HELLESPONTUS

This image is of channels in the Hellespontus Montes region to the south-east of the 175 km size Terby Crater. Hadriacus Cavi, a set of 65 km long 15 km wide east-west oriented depressions, lies just to the north. The region seems to be geologically complex. Here we see a variety of different colours on the surface indicating different mineralogy. Within one of the channels (bottom left), light-toned materials have been exposed and there is some evidence of layering.

CHENAL AVEC MATÉRIAUX CLAIRS EXPOSÉS DANS LA RÉGION D'HELLESPONTUS

Cette image représente des chenaux dans la région de Hellespontus Montes, au sud-est du cratère Terby, d'un diamètre de 175 km. Hadriacus Cavi, un ensemble de dépressions orientées est-ouest de 65 km de long et 15 km de large, se trouve juste au nord. La région semble être géologiquement complexe. Ici, nous voyons une grande variété de couleurs à la surface indiquant une minéralogie variée. Dans l'un des chenaux (en bas à gauche), des matériaux clairs ont été exposés avec des signes de stratification de la roche.

CANALE CON MATERIALE CHIARO IN SUPERFICIE A HELLESPONTUS

Questa è una immagine di alcuni canali nella regione di Hellespontus Montes a sud-est del cratere Terby di 175 km. Hadriacus Cavi, un insieme di depressioni orientate est-ovest lunghe 65 km e larghe 15 km, si trova appena a nord. La regione sembra essere geologicamente complessa. Qui vediamo una varietà di colori diversi sulla superficie che indicano una diversa mineralogia. All'interno di uno dei canali (in basso a sinistra), sono stati esposti materiali di colore chiaro e c'è qualche evidenza di stratificazione.

КАНАЛ С ОБНАЖЕНИЯМИ СВЕТЛЫХ ТОНОВ ГОРНОЙ ЦЕПИ ГЕЛЛЕСПОНТ

Это изображение каналов в окрестностях горной цепи Геллеспонт на юго-востоке от 175-километрового кратера Терби. Чуть севернее находится район Котловин Адриана – ряд впадин длиной 65 км и шириной 15 км, ориентированный с востока на запад. Этот район представляется геологически сложным. Разнообразие цветов поверхности указывает на богатую минералогию. В одном из каналов (слева внизу) обнажились материалы светлых тонов, и есть некоторые признаки стратификации (расслоения).

2.2   CHANNEL WITH EXPOSED LIGHT-TONED MATERIAL IN HELLESPONTUS

ACQUISITION TIME: 2020-12-13T22:08:09.070
LONGITUDE: 77.5 E   LATITUDE: 29.3 S
REGION: HELLESPONTUS

AUSGEDEHNTES RINNENSYSTEM BEI REULL VALLIS

Der Rand des ≈66 km breiten Greg-Kraters in Promethei Terra ist ziemlich degradiert und weist viele Rinnen auf, die manchmal als Gullys bezeichnet werden. Diese Strukturen deuten jedoch eher auf eine fluviale Strömung hin als auf die Trockenmechanismen, die wir an anderen Kraterwänden sehen. Der Krater befindet sich unmittelbar östlich von Reull Vallis, das anscheinend von Wasser geformt wurde und eine ungewöhnliche Konzentration von gletscherähnlichen Eisformen aufweist, die darauf hindeuten, dass Eis von den Innenwänden abfliesst.

EXTENSIVE CHANNEL SYSTEM NEAR REULL VALLIS

The rim of the ≈66 km wide Greg Crater in Promethei Terra is quite degraded and shows a lot of channels that are sometimes referred to as gullies. However, these structures suggest fluvial flow rather than the dry mechanisms we see on other crater walls. The crater is immediately east of Reull Vallis which appears to have been carved by water and presents an unusual concentration of glacier-like forms suggesting ice flowing off the inner walls.

VASTE SYSTÈME DE CHENAUX PRÈS DE REULL VALLIS

Au sein de Promethei Terra, le bord du cratère Greg, d'un diamètre d'environ 66 km, est assez dégradé et présente de nombreux petits chenaux également appelés ravines. Ces structures particulières suggèrent une érosion aqueuse plutôt que les mécanismes secs à l'oeuvre sur les parois d'autres cratères. Le cratère se trouve immédiatement à l'est de Reull Vallis, qui semble avoir été sculptée par l'eau et présente une concentration inhabituelle de formations glaciaires s'écoulant des parois internes.

ESTESO SISTEMA DI CANALI VICINO A REULL VALLIS

Il bordo del cratere Greg, largo ≈66 km, in Promethei Terra, è abbastanza degradato e mostra molti canali che sono talvolta indicati come calanchi. Tuttavia, queste strutture suggeriscono un flusso fluviale piuttosto che i meccanismi che non coinvolgono liquidi e che osserviamo sulle pareti di altri crateri. Il cratere è immediatamente ad est di Reull Vallis, il quale sembra essere stato scavato dall'acqua e presenta un'insolita concentrazione di conformazioni simili a ghiacciai che scorrono dalle pareti interne.

ОБШИРНАЯ СИСТЕМА КАНАЛОВ В ОКРЕСТНОСТЯХ ДОЛИНЫ РЕУЛЛ

Обод кратера Грег шириной около 66 км на земле Прометея сильно разрушен. По нему проходит множество каналов, которые иногда называют оврагами. Однако эти структуры указывают скорее на флювиальный поток, нежели на сухие процессы, которые мы наблюдаем на стенах других кратеров. Кратер находится к востоку от долины Реулл, которая, вероятно, была выточена водой и представляет собой необычную концентрацию ледяных форм, подобных ледникам, сходящих с внутренних стен.

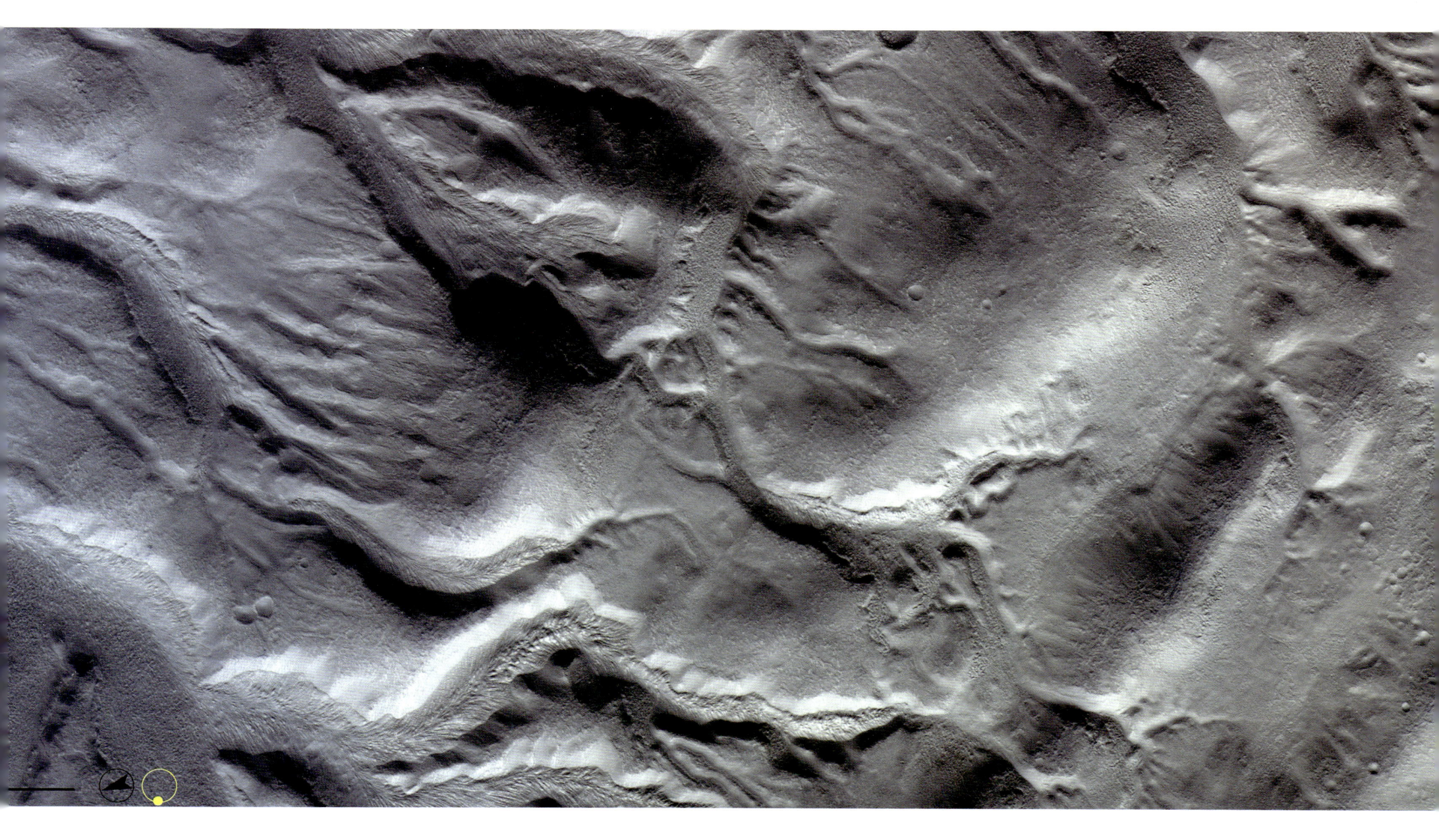

2.3   EXTENSIVE CHANNEL SYSTEM NEAR REULL VALLIS

ACQUISITION TIME: 2018-05-19T19:43:00.000
LONGITUDE: 112.3 E   LATITUDE: 39.3 S
REGION: PROMETHEI

RINNEN IM NETZ DER GRANICUS VALLES UND TINJAR VALLES

Die Granicus Valles sind ein Netz von Tälern in Elysium Planitia westlich von Elysium Mons. Man nimmt an, dass es sich dabei um Abflusskanäle handelt, ähnlich wie bei den grossen Kanälen wie Ares Vallis, Simud und Tiu Valles, die in Chryse Planitia münden. Die Kanäle könnten das Ergebnis von Lahars sein – Schlammströmen aus vulkanischem Material, das aus den Elysium-Vulkanschloten stammt. Die Rinnen könnten aber auch durch vulkanisches Aufschmelzen des Permafrosts und anschliessendes Abfliessen des Schlamms entstanden sein.

CHANNELS IN THE GRANICUS VALLES AND TINJAR VALLES NETWORK

The Granicus Valles are a network of valleys in Elysium Planitia to the west of Elysium Mons. They are thought to be outflow channels similar to the large channels such as Ares Vallis, Simud and Tiu Valles that flow into Chryse Planitia. The channels may have been the result of lahars – mudflow of volcanic material arising from the Elysium volcanic vents. However, the channels might also have been produced by volcanic melting of permafrost followed by a flow of the slurry.

CHENAUX DANS LE RÉSEAU DES VALLÉES DE GRANICUS ET DE TINJAR

Les vallées de Granicus constituent un réseau de vallées dans Elysium Planitia à l'ouest d'Elysium Mons. On pense qu'il s'agit de chenaux d'écoulement similaires aux grands chenaux tels qu'Ares Vallis, Simud et Tiu Valles qui se jettent dans Chryse Planitia. Ces chenaux pourraient être le résultat de lahars, des coulées de boue et autres matériaux volcaniques provenant des cheminées volcaniques d'Elysium. Cependant, les chenaux pourraient également avoir été produits par la fonte du pergélisol suivie d'un écoulement de la boue.

CANALI IN GRANICUS VALLES E TINJAR VALLES

Le Granicus Valles sono una rete di valli in Elysium Planitia a ovest di Elysium Mons. Si pensa che siano canali di deflusso simili ai più grandi Ares Vallis, Simud e Tiu Valles che scorrono in Chryse Planitia. I canali potrebbero essere il risultato di «lahar» (colate di fango composte di materiale vulcanico) provenienti dalle bocche vulcaniche dell'Elysium. Tuttavia, i canali potrebbero anche essere stati prodotti dallo scioglimento vulcanico del permafrost seguito da colate di fango.

КАНАЛЫ СЕТИ ДОЛИН ГРАНИК И ТИНДЖАР

Сеть долин Граник лежит на равнине Элизий к западу от горы Элизий. Считается, что они являются каналами оттока, подобными крупным каналам – сетям долин Арес, Симунд и Тиу, текшим в направлении равнины Хриса. На данный момент существует две версии образованиях этих каналов: в результате лахаров – грязевых потоков вулканического материала, вытекавших из жерл вулкана Элизий, или в результате вулканического таяния вечной мерзлоты создавшего грязевые потоки.

2.4 CHANNELS IN THE GRANICUS VALLES AND TINJAR VALLES NETWORK

ACQUISITION TIME: 2018-11-12T07:23:30.467
LONGITUDE: 135.7 E LATITUDE: 28.0 N
REGION: ELYSIUM

EIN FRISCH AUSSEHENDES TAL, DAS IN DAS ERIDANIA-BECKEN EINMÜNDET

Zwischen Terra Cimmeria und Terra Sirenum befindet sich ein relativ tief gelegenes Gebiet, das als Eridania-Becken bezeichnet wird. Es gibt signifikante mineralogische Hinweise darauf, dass dies ein riesiger Binnensee mit einer maximalen Tiefe von ≈ 2400 m gewesen sein könnte. Das hier zu sehende Tal befindet sich an der Südwestseite von Eridania, nordwestlich des Bjerknes-Kraters, und könnte zur Füllung des Sees in der späten Noachischen Periode der Marsgeschichte beigetragen haben.

FRESH-LOOKING VALLEY DISCHARGING INTO ERIDANIA BASIN

There is a relatively low-lying area between Terra Cimmeria and Terra Sirenum referred to as Eridania Basin. There is significant mineralogical evidence that this may have been a vast inland lake, with maximum depths of ≈ 2400 m. The valley seen here is on the south-west side of Eridania, north-west of Bjerknes Crater, and may have contributed to filling the lake in the late Noachian period of Martian history.

VALLÉE D'APPARENCE FRAÎCHE SE DÉVERSANT DANS LE BASSIN D'ERIDANIA

Il existe une zone faiblement élevée entre Terra Cimmeria et Terra Sirenum, appelée bassin d'Eridania. Des indices minéralogiques convaincants prouvent qu'il s'agissait peut-être d'un vaste lac intérieur, avec des profondeurs maximales d'environ 2400 m. La vallée que l'on voit ici se trouve sur le bord sud-ouest d'Eridania, au nord-ouest du cratère Bjerknes, et pourrait avoir contribué à remplir le lac à la fin de la période noachienne de l'histoire martienne.

UNA RECENTE VALLE CHE SCARICA NEL BACINO DI ERIDANIA

C'è un'area relativamente bassa tra la Terra Cimmeria e la Terra Sirenum chiamata Bacino di Eridania. Esiste una significativa evidenza mineralogica che questo bacino potrebbe essere stato un vasto lago interno, con una profondità massima di ≈ 2400 m. La valle in figura è sul lato sud-ovest di Eridania, a nord-ovest del cratere Bjerknes, e potrebbe aver contribuito a riempire il lago nel tardo periodo Noachiano della storia marziana.

МОЛОДАЯ ДОЛИНА, ВПАДАЮЩАЯ В БАССЕЙН ЭРИДАНИИ

Между Киммерийской землёй и землёй Сирен есть относительно низменная область, называемая бассейном Эридании. Существуют значительные минералогические свидетельства того, что в ней мог располагаться обширный внутриконтинентальный водоём с максимальной глубиной до 2400 м. Долина, которую мы видим здесь, находится на юго-западной стороне Эридании, к северо-западу от кратера Бьеркнес. Возможно, она участвовала в наполнении этого озера в поздний нойский период марсианской истории.

2.5   FRESH-LOOKING VALLEY DISCHARGING INTO ERIDANIA BASIN

ACQUISITION TIME: 2020-12-13T16:18:06.058
LONGITUDE: 167.6 E   LATITUDE: 39.9 S
REGION: CIMMERIA

KANAL BEI SAMARA VALLES

Das Bild zeigt ein Tal, das Teil des Samara/Himera-Talsystems ist, das einst in das Margaritifer-Becken floss. Diese Täler sind besonders interessant, weil sie möglicherweise erst relativ spät in der Geschichte des Mars aktiv waren. Es gibt Hinweise, die nahelegen, dass Wasser während der Hesperianischen Periode in vulkanische Lavaebenen eingeschnitten hat.

CHANNEL NEAR SAMARA VALLES

The image shows a valley that is part of the Samara/Himera valley system that once flowed into Margaritifer Basin. These valleys are particularly interesting because they may have been active relatively late in Martian history. This is suggested by evidence that water has incised volcanic lava plains formed during the Hesperian period on Mars.

CHENAL PRÈS DE SAMARA VALLIS

L'image montre une vallée qui fait partie du système de vallées Samara/Himera qui s'écoulait autrefois dans le bassin de Margaritifer. Ces vallées sont particulièrement intéressantes car elles ont pu être actives relativement tard dans l'histoire martienne. Ceci est suggéré par les indices que l'eau a incisé les plaines volcaniques formées pendant la période Hespérienne sur Mars.

CANALE VICINO A SAMARA VALLES

L'immagine mostra una valle che fa parte del sistema di valli di Samara/Himera che un tempo scorreva nel bacino di Margaritifer. Queste valli sono particolarmente interessanti perché potrebbero essere state attive relativamente tardi nella storia marziana, cosa suggerita dal fatto che l'acqua ha inciso le pianure di lava vulcanica formatesi durante il periodo Hesperiano su Marte.

КАНАЛ СИСТЕМЫ ДОЛИН САМАРА

На снимке показана долина, являющаяся частью системы долин Самара/Химера, которые когда-то впадали в бассейн Жемчужной Земли. Эти долины особенно интересны, поскольку они могли быть активны в поздней марсианской истории. На это указывает тот факт, что вода прорезала вулканические лавовые равнины, сформировавшиеся во время марсианского Гейсперийского периода.

2.6   CHANNEL NEAR SAMARA VALLES

ACQUISITION TIME: 2018-10-01T12:43:13.694
LONGITUDE: 338.9 E   LATITUDE: 22.9 S
REGION: NOACHIS

HER DESHER VALLIS UND DER ANGRENZENDE KRATER

Dieses Bild von Her Desher Vallis in Noachis Terra zeigt das Tal selbst auf der linken Seite, aber auch einen Krater auf der rechten Seite, der ebenso bunt ist. Dies deutet darauf hin, dass die in Her Desher Vallis gefundenen Minerale (z. B. Schichtsilikate) in diesem Gebiet weit verbreitet und allgegenwärtig sind. Dies deutet auf eine Zeit hin, in der flüssiges Wasser auf der Marsoberfläche floss und somit das Basaltgestein veränderte.

HER DESHER VALLIS AND ADJACENT CRATER

This image of Her Desher Vallis in Noachis Terra shows the valley itself to the left but also a crater to the right that is equally colourful. This indicates that the minerals (e.g. phyllosilicates) found in Her Desher Vallis are pervasive and ubiquitous in the area, pointing to a period when liquid water was flowing on the surface of Mars, hence modifying its basaltic bedrocks.

HER DESHER VALLIS ET CRATÈRE ADJACENT

Cette image de Her Desher Vallis dans Noachis Terra montre la vallée elle-même à gauche mais aussi un cratère à droite qui est tout aussi coloré. Cela indique que les minéraux (par exemple les phyllosilicates) trouvés dans Her Desher Vallis sont répandus et omniprésents dans la région, ce qui indique une période où de l'eau liquide coulait à la surface de Mars, altérant ainsi ses roches basaltiques.

HER DESHER VALLIS E UN CRATERE NELLE VICINANZE

Questa immagine di Her Desher Vallis in Noachis Terra mostra la valle stessa a sinistra e anche un cratere a destra che è altrettanto colorato. Questo indica che i minerali (ad esempio i fillosilicati) trovati nella Her Desher Vallis sono pervasivi e onnipresenti nella zona, segnalando un periodo in cui l'acqua liquida scorreva sulla superficie di Marte, modificando così le sue rocce basaltiche.

ДОЛИНА ДЕШЕР И ПРИЛЕГАЮЩИЙ КРАТЕР

На этом снимке долины Дешер земли Ноя слева видна сама долина, а справа — не менее красочный кратер. Это свидетельствует о том, что минералы (например, филлосиликаты), найденные в долине Дешер, широко распространены и повсеместно встречаются в этом районе. Это, в свою очередь, указывает на период, когда жидкая вода текла по поверхности Марса, изменяя его базальтовые породы.

2.7 HER DESHER VALLIS AND ADJACENT CRATER

ACQUISITION TIME: 2019-03-03T18:48:03.533
LONGITUDE: 312.3 E   LATITUDE: 25.2 S
REGION: NOACHIS

TÄLER IN TEMPE TERRA

Dieses Talsystem stammt aus einer Region westlich der Idaeus Fossae in Tempe Terra. Ein Teil des 300 km langen Moa-Valles-Systems liegt im Süden (rechts). Es wurde vorgeschlagen, dass das Moa-Valles-System durch relativ späte (frühamazonische) hydrologische Aktivitäten entstanden ist. Diese Art von dendritischen Entwässerungsmustern ist am häufigsten in flacherem Gelände anzutreffen und dort, wo die Struktur der Oberfläche hinsichtlich der Festigkeit einheitlich ist. Beachten Sie jedoch die Querschnitte, die in der Mitte des Bildes auftreten.

VALLEYS IN TEMPE TERRA

This valley system is from a region to the west of Idaeus Fossae in Tempe Terra. Part of the 300 km long Moa Valles system is to the south (right). It has been argued that the Moa Valles system was produced by relatively late (early Amazonian) hydrological activity. These types of dendritic drainage patterns are most common in flatter terrains and where the structure of the surface is uniform in terms of strength. Note, however, the cross-cutting that occurs in the centre of the frame.

VALLÉES DE TEMPE TERRA

Ce système de vallées est situé dans une zone à l'ouest d'Idaeus Fossae dans la région de Tempe Terra. Une partie du système de vallées de Moa, long de 300 km, se trouve au sud (à droite). Il est possible que le système de vallées de Moa ait été produit par une activité hydrologique relativement tardive (début de l'Amazonien). Ces types de systèmes de drainage dendritiques sont plus courants sur les terrains plus plats et lorsque les propriétés mécaniques de la surface sont uniformes. Notez, cependant, la fracture transversale qui traverse la région au centre de l'image.

LE VALLI IN TEMPE TERRA

Questo sistema di valli proviene da una regione a ovest di Idaeus Fossae in Tempe Terra. Parte del sistema di Moa Valles, lungo 300 km, si trova a sud (a destra). È stato ipotizzato che il sistema Moa Valles sia stato prodotto da un'attività idrologica relativamente tardiva (inizio del periodo Amazzoniano). Queste drenaggi con forme dendritiche sono più comuni nei terreni più piatti e dove la struttura della superficie è uniforme in termini di resistenza. Si noti, tuttavia, il taglio trasversale che si verifica al centro del fotogramma.

ДОЛИНЫ ЗЕМЛИ ТЕМПЕ

Это система долин к западу от Борозд Идея на земле Темпе. Часть системы долин Моа длиной 300 км находится на юге (справа). Считается, что долины Моа возникли в результате относительно поздней (ранний амазонийский период) водной активности. Такие следы ветвистых каналов наиболее распространены на более плоских участках и там, где наблюдается однородная устойчивость поверхности. Обратите внимание на пересечение каналов в центре кадра.

2.8 VALLEYS IN TEMPE TERRA

ACQUISITION TIME: 2021-02-27T03:09:39.969
LONGITUDE: 306.7 E   LATITUDE: 38.3 N
REGION: TEMPE

DAO VALLIS

Das Dao Vallis ist ein Tal an der Grenze zwischen Hellas Planitia, Promethei Terra und Tyrrhena Terra, das von Osten her in das Hellas-Becken entwässert haben dürfte. Es entspringt an den südlichen Hängen des Hadriacus Mons und erstreckt sich zusammen mit seinem Nebenfluss, dem Niger Vallis, über eine Länge von etwa 1200 km. Es wurde diskutiert, ob dieses Talsystem durch Lava oder durch Gletscherprozesse entstanden sein könnte, aber es scheint, dass flüssiges Wasser die beste Erklärung für die beobachteten Merkmale liefert. Auf diesem Bild sind Anzeichen für den Einsturz des Ufers (oben in der Mitte) und eine tektonische Struktur, die das Tal durchschneidet (unten rechts und in der Mitte), zu sehen.

DAO VALLIS

Dao Vallis is a valley on the border between Hellas Planitia, Promethei Terra and Tyrrhena Terra and would have drained into Hellas Basin from the east. It runs from the southern slopes of Hadriacus Mons and, together with its tributary, Niger Vallis, extend for about 1200 km. There has been discussion of whether this valley system could have been produced by lava or by glacial processes but it seems that liquid water provides the best explanation for the observed features. In this particular image, we can see evidence of collapse of the bank (top centre) and a tectonic structure cutting across the valley (bottom right and centre).

DAO VALLIS

Dao Vallis est une vallée située à la frontière entre Hellas Planitia, Promethei Terra et Tyrrhena Terra et qui aurait débouché dans le bassin d'Hellas depuis l'est. Elle part des pentes méridionales de Hadriacus Mons et, avec son affluent, Niger Vallis, s'étend sur environ 1 200 km. On s'est demandé si ce système de vallées pouvait avoir été produit par de la lave ou par des processus glaciaires, mais il semble que l'eau liquide soit la meilleure explication des caractéristiques observées. Sur cette image particulière, nous pouvons voir des preuves de l'effondrement du banc (en haut au centre) et une structure tectonique coupant la vallée (en bas à droite et au centre).

DAO VALLIS

Dao Vallis è una valle al confine tra Hellas Planitia, Promethei Terra e Tyrrhena Terra e avrebbe drenato nel bacino di Hellas da est. Ha origine sulle pendici meridionali dell'Hadriacus Mons e, insieme al suo affluente, Niger Vallis, si estende per circa 1.200 km. Si è dibattuto a lungo se questo sistema di valli possa essere stato prodotto dalla lava o da processi glaciali, ma sembra che l'acqua liquida fornisca la migliore spiegazione per le caratteristiche osservate. In questa particolare immagine, possiamo vedere l'evidenza del collasso di un lido (in alto al centro) e una struttura tettonica che taglia la valle (in basso a destra e al centro).

ДОЛИНА ДАО

Долина Дао – находится на границе между равниной Эллада и землями Прометея и Тирренской. Она, возможно, впадала в бассейн Эллада с востока. Долина начинается с южных склонов Адриатической горы и вместе со своим притоком – долиной Нигер, простирается примерно на 1 200 км. Одно время шла дискуссия о том, могла ли эта система долин быть образована лавой или ледниковыми процессами, но, похоже, что жидкая вода лучше всего объясняет наблюдаемые черты. На этом снимке мы видим свидетельства обрушения берега (вверху в центре) и тектоническую структуру, проходящую через долину (внизу справа и в центре).

2.9   DAO VALLIS

ACQUISITION TIME: 2020-12-29T03:40:27.851
LONGITUDE: 93.7 E   LATITUDE: 33.9 S
REGION: HELLAS

WARREGO VALLES

Warrego Valles ist ein älteres Talsystem von 190 km Länge im nördlichsten Teil von Aonia Terra in der Region Thaumasia Fossae. Es wurde so interpretiert, dass es möglicherweise einen Oberflächenabfluss im Noachischen Zeitalter zeigt, und liefert damit einen Hinweis darauf, dass der Mars einst wärmer und feuchter war und wahrscheinlich Regen- oder Schneeniederschläge verzeichnete. Der hier cargestellte Talboden ist recht rau und strukturiert.

WARREGO VALLES

Warrego Valles is an older valley system 190 km long in the northernmost part of Aonia Terra in the Thaumasia Fossae region. It has been interpreted as indicating possible surface run-off in the Noachian time frame, hence providing evidence that Mars may have once been warmer, wetter, and likely with rain or snow precipitation. The valley floor seen here is quite rough and structured.

WARREGO VALLES

Warrego Valles est un ancien système de vallées de 190 km de long dans la partie la plus septentrionale d'Aonia Terra dans la région de Thaumasia Fossae. Il a été interprété comme indiquant un possible écoulement d'eau en surface au cours de la période noachienne, fournissant ainsi la preuve que Mars a pu être autrefois plus chaude, plus humide, et probablement avec des précipitations de pluie ou de neige. Le fond de la vallée montré ici est assez rugueux et texturé.

WARREGO VALLES

Warrego Valles è un antico sistema di valli lungo 190 km nella parte più settentrionale di Aonia Terra nella regione di Thaumasia Fossae. È stato interpretato come indicativo di un possibile deflusso superficiale nel periodo Noachiano, fornendo così la prova che Marte potrebbe essere stato un tempo più caldo, più umido e probabilmente con precipitazioni di pioggia o neve. Il fondovalle qui presentato è piuttosto irregolare e strutturato.

ДОЛИНЫ УОРРЕГО

Долины Уоррего – древняя, 190-километровая система, находящаяся в самой северной части земли Аонид, в районе борозд Тавмазии. Согласно принятым интерпретациям, она свидетельствует о возможном существовании поверхностного стока в нойский период, тем самым, подтверждая гипотезу, что Марс когда-то был теплее и более влажным с вероятными осадками в виде дождя или снега. Видимое здесь дно долины довольно неровное и структурированное.

2.10 WARREGO VALLES

ACQUISITION TIME: 2018-05-20T09:25:57.000
LONGITUDE: 268.2 E LATITUDE: 42.7 S
REGION: AONIA

GRABEN IM LADON-BECKEN

Dieses Bild zeigt einen Graben im Ladon-Becken, einem Teil von Margaritifer Terra. Das gesamte Gebiet ist tonhaltig, und dieses Bild zeigt Anzeichen einer wässrigen Umwandlung (Mitte rechts). Es ist umstritten, ob das Deckmaterial auf diesem Bild (Mitte und Mitte links) vulkanischen oder sedimentären Ursprungs ist. Die Grabenwände zeigen unterschiedlich gefärbte Schichtungen.

GRABEN IN LADON BASIN

This image is of a graben in the Ladon Basin part of Margaritifer Terra. The whole area is clay-rich and this image shows evidence of aqueous alteration (centre right). It is contentious whether the capping material in this image (centre and centre left) is volcanic or sedimentary in origin. The graben walls show different coloured layering.

GRABEN DANS LE BASSIN DE LADON

Cette image montre un graben du bassin de Ladon dans la région de Margaritifer Terra. Toute la zone est riche en argile et cette image montre des preuves d'altération aqueuse de la roche (centre droit). La question de l'origine sédimentaire ou volcanique de la roche en surface (au centre et au centre gauche) est cependant toujours controversée. Les parois du graben présentent des strates de couleur variée.

GRABEN NEL BACINO LADON

Questa immagine ritrae un graben nel bacino di Ladon, parte di Margaritifer Terra. L'intera area è ricca di argilla e questa immagine mostra prove di alterazione acquosa (centro destra). È controverso se il materiale che ha coperto la superficie in questa immagine (centro e centro sinistra) sia di origine vulcanica o sedimentaria. Le pareti del graben mostrano stratificazioni di colori diversi.

ГРАБЕН В БАССЕЙНЕ ЛАДОН

Это изображение грабена в бассейне Ладон, являющемся частью Жемчужной земли. Территория здесь богата глиной. На снимке справа в центре видны признаки изменения поверхности под воздействием воды. Остаётся неясным, является ли покровный материал (центр и центр слева) вулканическим или осадочным по происхождению. Стены грабена демонстрируют различные цветовые слои.

2.11 GRABEN IN LADON BASIN

ACQUISITION TIME: 2018-05-06T12:52:21.000
LONGITUDE: 332.1 E   LATITUDE: 18.1 S
REGION: MARAGRITIFER

MÜNDUNG DES UZBOI VALLIS

Dieses Bild stammt von der Stelle, an welcher Uzboi Vallis in den Holden-Krater in Margaritifer Terra eintritt. Dieses verbesserte, synthetische RGB-Bild zeigt einen Kanal, der den Boden des Kraters durchschneidet. Es zeigt ebenfalls hell getönte, geschichtete Ablagerungen, welche freigelegt wurden. Es gibt Hinweise darauf, dass sich solche Schichten auf dem Boden von Uzboi gebildet haben, als der Abfluss durch den Einschlag, der den Holden-Krater bildete, blockiert wurde. Danach überflutete der See im Uzboi Vallis den Rand des Holden-Kraters, und das Wasser begann, sowohl den Rand als auch die Ablagerungen zu erodieren, sodass sie in ihrer heutigen Form sichtbar wurden.

MOUTH OF UZBOI VALLIS

This image is from the point where Uzboi Vallis enters Holden Crater in Margaritifer Terra. This enhanced synthetic RGB image shows a channel cutting through the floor of the crater and exposures of light-toned layered deposits. Evidence suggests that such layers were formed on the Uzboi floor when drainage was blocked by the impact forming Holden crater. Afterwards the lake in Uzboi Vallis flowed over the rim of Holden and water started eroding both the rim and the deposits, exposing them as they are visible today.

EMBOUCHURE DE LA VALLÉE D'UZBOI

Cette image provient du point d'entrée d'Uzboi Vallis dans le cratère Holden (région de Margaritifer Terra). Cette image montre un chenal découpant le fond du cratère et des dépôts stratifiés de couleur claire exposés en surface. Il existe des preuves que de telles couches se soient formées sur le sol d'Uzboi lorsque le drainage a été bloqué par l'impact formant le cratère Holden. Le lac de la vallée d'Uzboi a ensuite submergé le bord d'Holden et l'eau a commencé à éroder à la fois le bord et les dépôts, les exposant tels qu'ils sont visibles aujourd'hui.

IMBOCCATURA DI UZBOI VALLIS

Particolare della regione in cui Uzboi Vallis entra nel cratere Holden in Margaritifer Terra. Questa immagine sintetica RGB mostra un canale che taglia il pavimento del cratere e la presenza di depositi stratificati dai toni chiari. Le evidenze suggeriscono che tali strati si sono formati sul pavimento di Uzboi quando il drenaggio è stato bloccato dall'impatto che ha formato il cratere Holden. In seguito il lago di Uzboi Vallis ha superato il bordo di Holden e l'acqua ha iniziato a erodere sia il bordo che i depositi, esponendoli superficialmente nel modo in cui sono visibili oggi.

УСТЬЕ ДОЛИНЫ УЗБОЙ

Это место, в котором долина Узбой входит в кратер Холден Жемчужной земли. На изображении в цветах RGB виден канал, прорезающий дно кратера, и обнажения многослойных отложений светлых тонов. Есть основания полагать, что такие слои образовались на дне долины, когда сток был перекрыт ударом метеорита, сформировавшим кратер Холден. После того, как уровень озера в долине Узбой превысил высоту обода кратера, вода начала размывать обод и отложения, обнажив их так, как они видны сегодня.

2.12   MOUTH OF UZBOI VALLIS

ACQUISITION TIME: 2018-05-31T10:40:38.000
LONGITUDE: 325.1 E   LATITUDE: 27.1 S
REGION: MARAGRITIFER

GEWUNDENES TAL IN ARABIA TERRA

Dies ist ein Bild einer gewundenen Talstruktur im Norden von Arabia Terra in der Nähe der Dichotomie-Grenze. Hier gibt es mehrere leicht ungewöhnliche Merkmale. Der Hauptkanal mündet in eine unregelmässige Struktur von Senken (Mitte rechts). Beachten Sie die abrupte Änderung der Tiefe des Kanals unten rechts und die orthogonal zum Kanal verlaufenden Bergrücken in der Nähe dieses Punktes. Dieses besondere Tal ist namenlos und wurde offenbar noch nicht eingehend erforscht.

SINUOUS VALLEY IN ARABIA TERRA

This is an image of a sinuous valley structure in northern Arabia Terra close to the dichotomy boundary. There are several slightly unusual features here. The main channel crosses into an irregular depression structure (centre right). Note that there is an abrupt change in depth of the channel to the lower right and note also the ridges orthogonal to the channel near this point. This particular valley is unnamed and does not appear to have been extensively researched.

VALLÉE SINUEUSE DANS LA REGION D'ARABIA TERRA

Image de la structure d'une vallée sinueuse d'Arabia Terra, près de la limite de la dichotomie globale martienne. Plusieurs caractéristiques légèrement inhabituelles sont à noter ici. Le chenal principal traverse une dépression irrégulière (centre droit). Notez qu'il y a un changement abrupt dans la profondeur du chenal en bas à droite et notez également les crêtes orthogonales au chenal près de ce point. Cette vallée particulière n'est pas nommée et ne semble pas avoir fait l'objet de recherches approfondies.

UNA SINUOSA VALLE IN ARABIA TERRA

Questa è un'immagine della struttura sinuosa di una valle nel nord dell'Arabia Terra vicino al confine della dicotomia. Ci sono diverse caratteristiche piuttosto insolite qui. Il canale principale attraversa una depressione con struttura irregolare (al centro a destra). Si noti che c'è un brusco cambiamento nella profondità del canale in basso a destra e ci sono delle creste ortogonali al canale vicino a questo punto. Questa particolare valle è senza nome e non sembra essere stata studiata a fondo.

ИЗВИЛИСТАЯ ДОЛИНА ЗЕМЛИ АРАВИЯ

Это изображение извилистой долиной структуры на севере земли Аравия вблизи границы дихотомии. Здесь есть несколько особенностей. Основной канал пересекает впадину неправильной формы (в центре справа). Обратите внимание на хребты в правом нижнем углу, ортогональные каналу, и на резкое изменение глубины этого канала. Данная долина не имеет названия, и все еще не была тщательно исследована.

2.13 SINUOUS VALLEY IN ARABIA TERRA

ACQUISITION TIME: 2021-03-14T17:05:46.794
LONGITUDE: 352.2 E   LATITUDE: 36.3 N
REGION: ARABIA

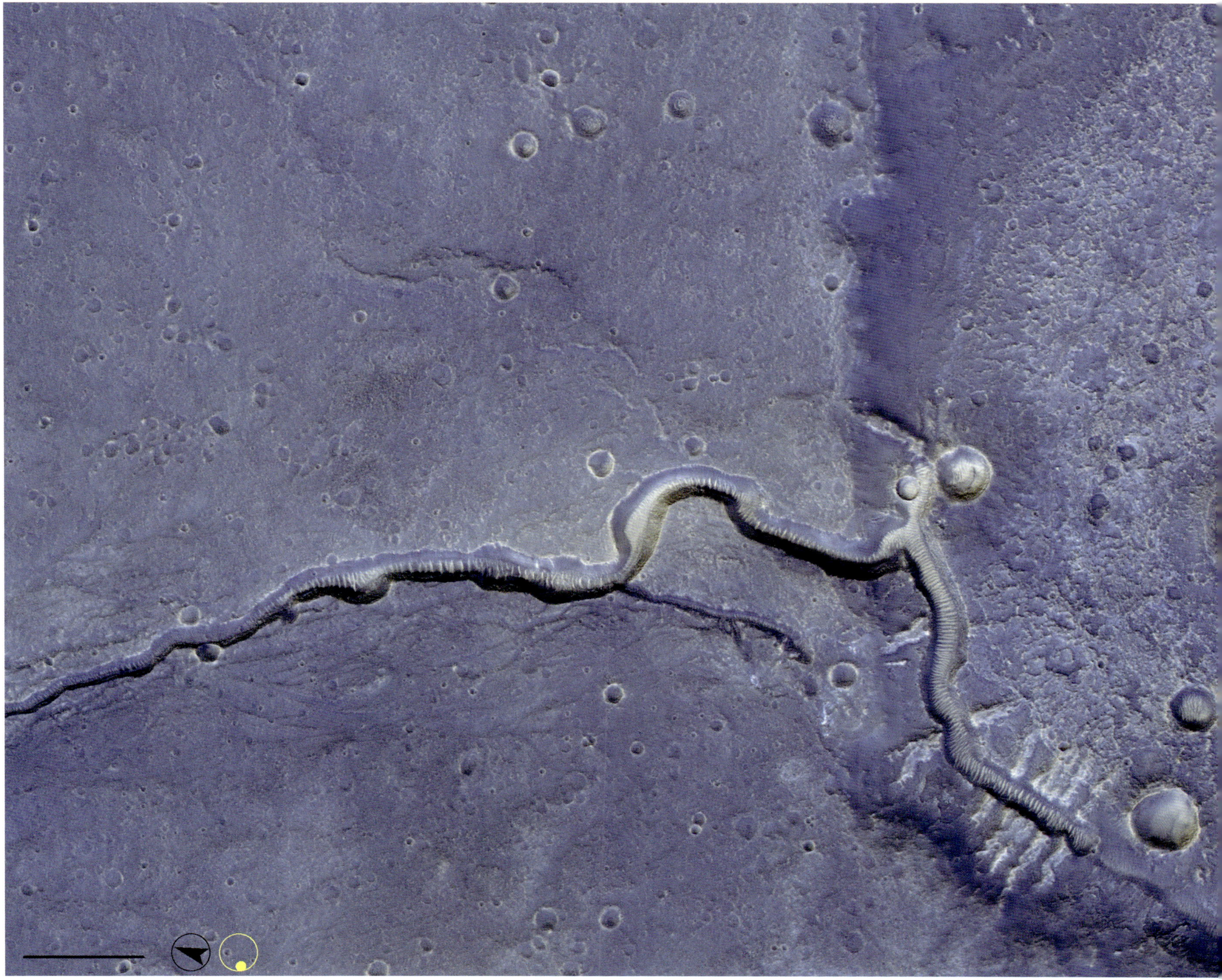

DIE STUMMELIGEN NEBENFLÜSSE DES NIRGAL VALLIS

Nirgal Vallis ist formell Teil von Noachis Terra. Es liegt nördlich des Argyre-Beckens und hätte in das Uzboi Vallis gemündet, bevor das im Tal fliessende Wasser den Rand des Holden-Kraters durchbrochen hätte. Das Tal ist etwa 610 km lang und für seine stumpfen Nebenflüsse bekannt, die auf diesem verbesserten synthetischen RGB-Bild von CaSSIS zu sehen sind. Es wird vermutet, dass die stummelige Beschaffenheit der Nebenflüsse eher auf die Freisetzung von Grundwasser durch einen Prozess wie Klifferosion als auf Niederschläge zurückzuführen ist.

THE STUBBY TRIBUTARIES OF NIRGAL VALLIS

Nirgal Vallis is formally part of Noachis Terra. It is north of the Argyre Basin and would have discharged into Uzboi Vallis before the water associated with the valley would have breached the rim of Holden Crater. The valley is around 610 km long and is known for the stubby tributaries that are seen in this enhanced synthetic RGB image from CaSSIS. It has been suggested that the stubby nature of the tributaries is a result of the release of ground-water by a process such as sapping rather than as a result of precipitation.

LES AFFLUENTS TRONQUÉS DE NIRGAL VALLIS

Nirgal Vallis fait partie de la région de Noachis Terra. Elle se trouve au nord du bassin d'Argyre et se serait déversée dans Uzboi Vallis avant que l'eau associée à la vallée n'atteigne le bord du cratère Holden. La vallée fait environ 610 km de long et est connue pour ses affluents qui sont visibles sur cette image CaSSIS. Il a été suggéré que l'apparence tronquée des affluents est le résultat de la libération des eaux souterraines par un processus de sapement plutôt que par des précipitations atmosphériques.

I TOZZI AFFLUENTI DI NIRGAL VALLIS

Nirgal Vallis è formalmente parte di Noachis Terra. Si trova a nord del bacino Argyre e sarebbe sfociata nella Uzboi Vallis prima che l'acqua della valle rompesse il bordo del cratere Holden. La valle è lunga circa 610 km ed è nota per i tozzi affluenti che si vedono in questa immagine composa RGB di CaSSIS. È stato suggerito che la natura tozza degli affluenti sia il risultato del rilascio di acqua freatica da un processo di infiltrazione piuttosto che il risultato di precipitazioni.

КОРОТКИЕ ПРИТОКИ ДОЛИНЫ НЕРГАЛ

Долина Нергал формально является частью земли Ноя и находится к северу от бассейна Аргир. Вероятно, она впадала в долину Узбой, пока ее мощные потоки не прорвались через край кратера Холден, сформировав в нем озеро. Длина долины составляет около 610 км, она известна своими короткими притоками, которые видны на этом увеличенном синтезированном RGB изображении с CaSSIS. Предполагается, что укороченный вид притоков является результатом выхода подземных вод, а не выпадения осадков.

2.14   THE STUBBY TRIBUTARIES OF NIRGAL VALLIS

ACQUISITION TIME: 2018-09-04T02:43:14.000
LONGITUDE: 317.3 E   LATITUDE: 27.7 S
REGION: NOACHIS

RUBICON VALLES

Die Rubicon Valles liegen an der Westflanke des Alba Mons im Norden des Tharsis-Anstiegs. In einigen Gebieten sind die Lavaströme von Alba Patera vermutlich nur etwa 0,5 Milliarden Jahre alt, obwohl sie im Allgemeinen auf ein höheres Alter (≈ 3,4 Milliarden Jahre) geschätzt werden. Einzelne Ströme waren von geringer Viskosität, was zu leichten Neigungen führte, welche sich über ein sehr grosses Gebiet mit einer Länge von mehr als 300 km erstrecken. Die Rubicon-Täler scheinen in die Lavaströme eingeschnitten zu sein, und man nimmt an, dass diese Täler durch das Abfliessen von flüssigem Wasser entstanden sind.

RUBICON VALLES

Rubicon Valles are on the western flank of Alba Mons in the north of the Tharsis rise. In some areas the lava flows from Alba Patera are thought to be only around 0.5 billion years old although they are generally estimated to be older (≈ 3.4 billion years old). Individual flows were of low viscosity resulting in gentle slopes over a very large area with long (> 300 km). Rubicon Valles appear to be cut into the lava flows and it is thought that run-off of liquid water produced these valley networks.

RUBICON VALLES

Les vallées du Rubicon se trouvent sur le flanc ouest d'Alba Mons, au nord du dôme de Tharsis. Dans certaines zones, on pense que les coulées de lave d'Alba Patera n'ont qu'environ 0,5 milliard d'années, bien que l'on estime généralement qu'une majorité d'entre elles sont bien plus anciennes (≈ 3,4 milliards d'années). Les coulées de lave avaient une faible viscosité, ce qui a conduit à former des pentes douces sur une zone très étendue et longue (> 300 km). Les vallées du Rubicon semblent être taillées dans les coulées de lave et on pense que le ruissellement d'eau liquide a produit ces réseaux de vallées.

RUBICON VALLES

Le Rubicon Valles si trovano sul fianco occidentale dell'Alba Mons nel nord dell'altura di Tharsis. In alcune zone, si pensa che le colate di lava di Alba Patera abbiano solo circa 0,5 miliardi di anni, anche se generalmente si stima che siano più antiche (≈ 3,4 miliardi di anni). Le singole colate avevano una bassa viscosità con la conseguente formazione di pendii dolci su un'area molto ampia e lunga (> 300 km). Le Rubicon Valles sembrano essere scavate nelle colate di lava e si pensa che il deflusso di acqua liquida abbia prodotto queste reti di valli.

ДОЛИНЫ РУБИКОН

Долины Рубикон находятся на западной стороне вулкана Альба на севере плато Фарсида. Считается, что в некоторых местах возраст лавовых потоков патеры (неровный кратер) Альба составляет всего около 0,5 млрд. лет, хотя в целом они оцениваются как более древние (≈ 3,4 млрд. лет). Отдельные потоки лавы имели низкую вязкость, что привело к образованию пологих склонов на большой территории с протяженностью порядка 300 км. Так как долины Рубикон врезаны в лавовые потоки, считается, что эти сети долин создал сток жидкой воды.

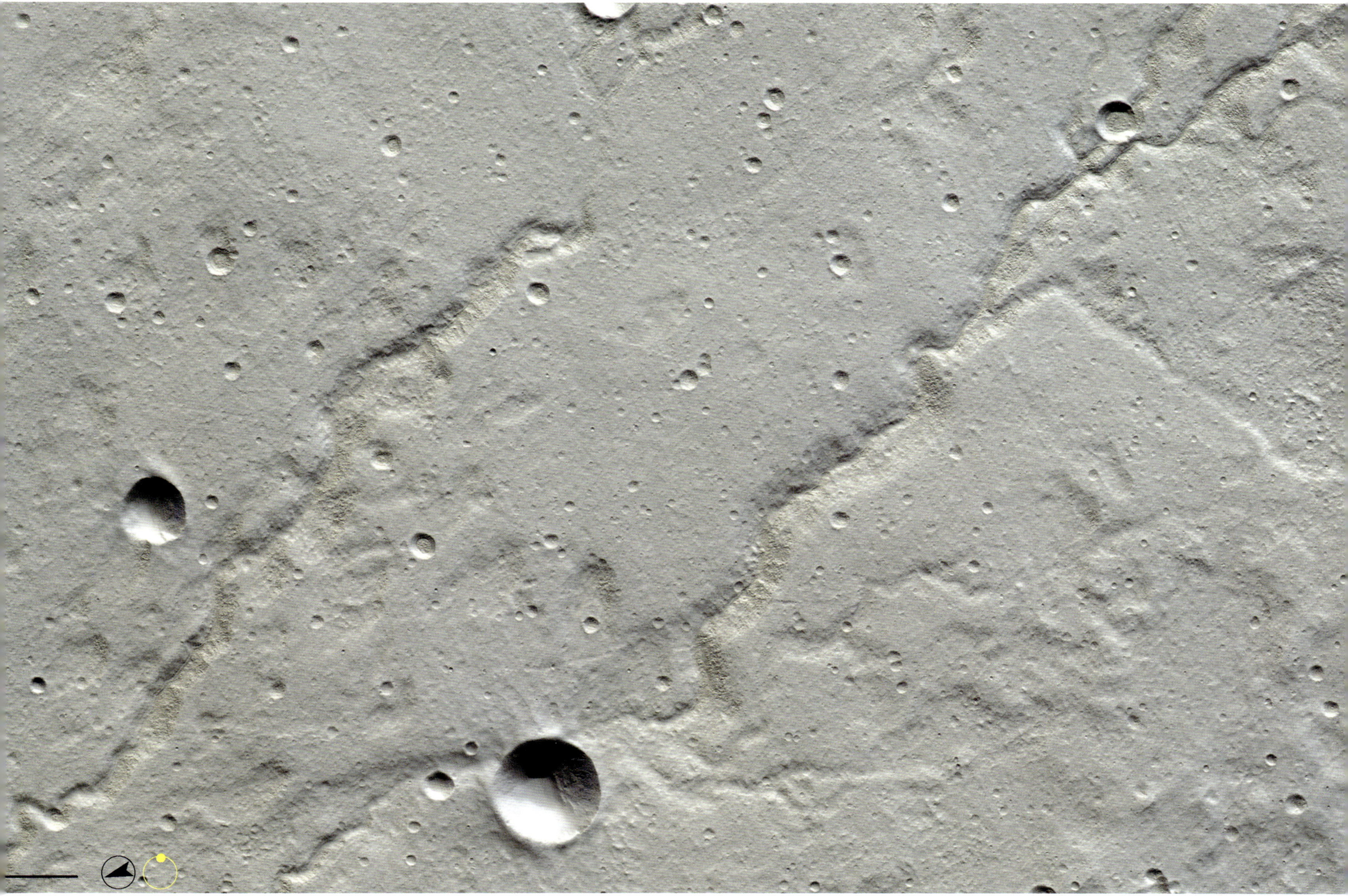

3 KRATERINNERES UND ERHEBUNGEN

CRATER INTERIORS AND UPLIFTS

INTÉRIEURS DE CRATÈRES ET SOULÈVEMENTS

AL'INTERNO DEI CRATERI E SOLLEVAMENTI CENTRALI

ВНУТРЕННИЕ ОБЛАСТИ И ЦЕНТРАЛЬНЫЕ ПОДНЯТИЯ КРАТЕРОВ

DIE ZENTRALE ERHEBUNG DES RITCHEY-KRATERS

Das Innere von Kratern enthält oft wichtige und farbenprächtige Gesteinsaufschlüsse, sowohl von den zentralen Hebungen des tiefen Gesteins als auch von Materialien wie Sedimenten und Lava, die am Boden abgelagert wurden. Dieses Beispiel stammt aus dem Ritchey-Krater mit 79 km Durchmesser. Tone, die durch die Wechselwirkung von Mineralien mit flüssigem Wasser entstehen, wurden im Krater und in der zentralen Erhebung gefunden, die beim ersten Einschlag entstanden ist. Das Bild zeigt eine extreme Mineralienvielfalt in der zentralen Erhebung, wobei die blaugrünen Farben eisenreiche Eruptivgesteine zeigen. Ein kleiner Krater, der später in der Erhebung entstanden ist, zeigt hell getöntes Material. Ein grosser Teil des Grundgesteins ist durch den energiereichen Einschlag und die anschliessende Rückprallbewegung zerklüftet. In relativ niedrig liegenden Bereichen sind auch einige unregelmässig geformte Dünen zu sehen.

THE CENTRAL UPLIFT OF RITCHEY CRATER

Crater interiors often include important and colourful bedrock exposures, both from the central uplifts of deep bedrock and materials like sediments and lava deposited on the floor. This example is from the 79 km diameter, Ritchey Crater. Clays, which are produced by minerals interacting with liquid water, have been found within the crater and in the central uplift that was produced during the initial impact. The image shows extreme mineral diversity within the central uplift, in which the blue-green colors represent igneous rocks rich in iron. A small crater that subsequently hit the uplift reveals light-toned material. Much of the bedrock is fractured by the high-energy impact event and subsequent rebound. Some irregularly-shaped dunes can also be seen in relatively low areas.

LE SOULÈVEMENT CENTRAL DU CRATÈRE RITCHEY

Les intérieurs de cratères exposent souvent de larges affleurements très colorés du substrat rocheux, provenant à la fois des soulèvements centraux du substrat profond et des matériaux tels que les sédiments et la lave déposés sur le sol. Cet exemple provient du cratère Ritchey, d'un diamètre de 79 km. Des argiles, qui sont produites par l'interaction de la roche avec de l'eau liquide, ont été trouvées dans le cratère et dans le soulèvement central produit lors de l'impact initial. L'image montre une grande diversité minéralogique dans le soulèvement central, dans lequel les teintes bleu-vert représentent des roches ignées riches en fer. Un petit cratère qui a ensuite touché le soulèvement révèle des roches de couleur claire. Une grande partie du substrat rocheux est fracturée par l'impact à haute énergie et le rebond ultérieur. On peut également voir quelques dunes de forme irrégulière dans des zones relativement basses.

IL SOLLEVAMENTO CENTRALE DEL CRATERE RITCHEY

Spesso si possono vedere all' interno dei crateri colorate esposizioni di basamenti rocciosi, sia dai sollevamenti centrali, sia dai pavimenti rocciosi sub-superficiali e materiali quali sedimenti e lava ivi depositati. Questo scatto proviene dal cratere Ritchey, di 79 km di diametro. È stato trovato materiale argilloso, che è prodotto da minerali che interagiscono con l'acqua liquida, all'interno del cratere e nel sollevamento centrale che è stato prodotto durante l'impatto iniziale.

L'immagine mostra un'estrema diversità di minerali all'interno del sollevamento centrale, in cui i colori blu-verde rappresentano rocce ignee ricche di ferro. Un piccolo cratere, che ha colpito successivamente il sollevamento, rivela materiale di colore chiaro. Gran parte del pavimento roccioso è fratturato dall'evento di impatto ad alta energia e dal successivo rimbalzo. Si possono osservare anche alcune dune di forma irregolare in aree relativamente depresse.

ЦЕНТРАЛЬНОЕ ПОДНЯТИЕ КРАТЕРА РИЧИ

Внутренняя область кратеров часто включает важные и красочные обнажения горных пород: в центре – поднятие коренных пород, на дне – отложения осадочных пород и лавы. Это пример кратера Ричи диаметром 79 км. Глины, которые образуются в результате взаимодействия минералов с жидкой водой, были обнаружены как в самом кратере, так и в его центральном поднятии, появившемся во время первоначального удара. Снимок CaSSIS показывает чрезвычайное разнообразие минералов в этой центральной структуре, причём магматические породы, богатые железом, окрашены в сине-зелёные цвета. Небольшой кратер, впоследствии пробивший поднятие, проявляет материал светлых тонов. Основная часть горных пород покрыта сеткой трещин, связанных с сильным ударом метеорита и последующим рикошетом. На относительно низких участках также видны дюны неправильной формы.

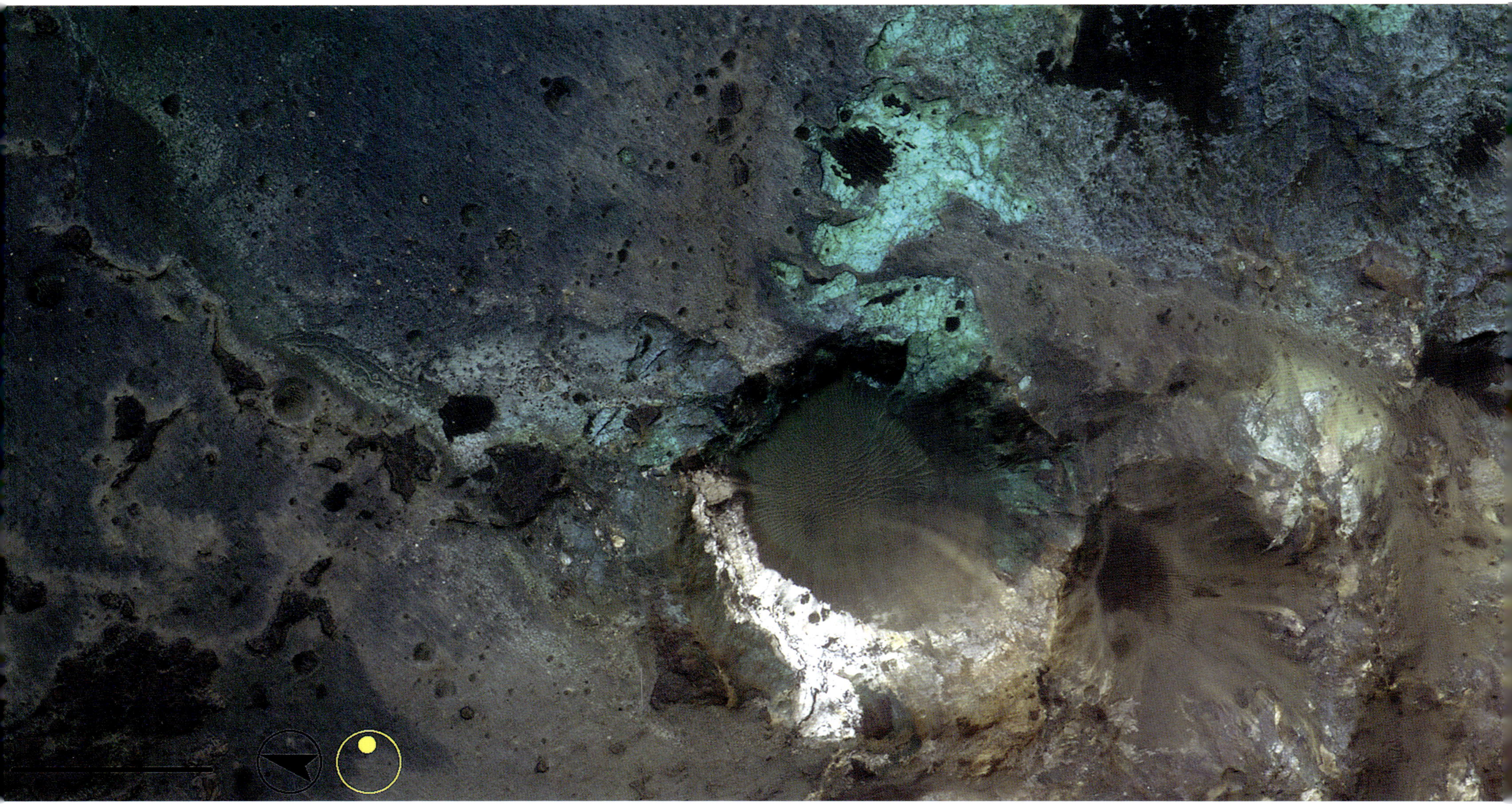

3.1 THE CENTRAL UPLIFT OF RITCHEY CRATER

ACQUISITION TIME: 2020-12-20T09:22:24.405
LONGITUDE: 309.0 E   LATITUDE: 27.9 S
REGION: NOACHIS

ZENTRALER GIPFEL DES ALGA-KRATERS IN NOACHIS TERRA

Der Alga-Krater, eine Struktur mit einem Durchmesser von 19 km in Noachis Terra, weist einen bemerkenswerten, 200 m hohen zentralen Gipfel auf (oben rechts zu sehen), der auf eine signifikante Mineralienvielfalt hinweist. Ein Teil des Kraterrands im Norden (links) wurde ebenfalls aufgenommen. Man sieht weitere örtliche Aufschlüsse des «grünen» Minerals auf der Spitze des Kraterrands, an seiner inneren Basis und um einen kleinen Krater (< 500 m Durchmesser) ausserhalb des Rands. Olivinhaltiges Gestein, das eine grünliche Farbe aufweist, wurde im nordwestlichen Teil der Erhebung gefunden.

CENTRAL PEAK OF ALGA CRATER IN NOACHIS TERRA

Alga Crater, a 19 km diameter structure in Noachis Terra, shows a remarkable 200 m high central peak (seen to the right above) indicating significant mineral diversity. Part of the crater rim to the north (left) has also been included where one can see other localised exposures of the «green» mineral on the top of the crater rim, at its internal base, and around a small (< 500 m diameter) crater outside the rim. Olivine-bearing bedrock has been found in the north-west part of the uplift, and produces a greenish colour

PIC CENTRAL DU CRATÈRE ALGA DANS NOACHIS TERRA

Le cratère Alga, une structure d'impact de 19 km de diamètre dans la région de Noachis Terra, présente un pic central remarquable de 200 m de haut (à droite de cette image) indiquant une diversité minéralogique importante. Une partie du bord du cratère au nord (à gauche) est également incluse où l'on peut voir d'autres affleurements localisés d'une formation «verte» sur le sommet du bord du cratère, à sa base interne, et autour d'un petit cratère (< 500 m de diamètre) à l'extérieur de la paroi. Un substrat rocheux contenant de l'olivine, un minéral de couleur verdâtre, a été trouvé dans la partie nord-ouest du soulèvement.

IL PICCO CENTRALE DEL CRATERE ALGA IN NOACHIS TERRA

Il cratere Alga, una struttura di 19 km di diametro in Noachis Terra, mostra un notevole picco centrale alto 200 m (visto a destra nell'immagine qui sopra) il cui colore indica una significativa diversità minerale. L'immagine include anche una parte del bordo del cratere a nord (a sinistra) dove si possono vedere altre esposizioni localizzate del minerale «verde» sulla cima del bordo del cratere, alla sua base interna, e intorno a un piccolo (< 500 m di diametro) cratere fuori dal bordo. É stato scoperto un pavimento roccioso che contiene olivina nella parte nord-ovest del sollevamento, producendo così un colore verdastro.

ЦЕНТРАЛЬНЫЙ ПИК КРАТЕРА АЛГА НА ЗЕМЛЕ НОЯ

На земле Ноя в кратере Алга диаметром 19 км есть достойный внимания центральный пик (поднятие) высотой 200 м (справа), демонстрирующий значительное разнообразие минералов. На северной части кратерного вала (в левой половине снимка) также можно увидеть другие обнажения «зеленого минерала»: на гребне, у внутреннего основания вала и за его пределами, вокруг небольшого кратера (диаметром < 500 м). Коренная порода, богатая оливином, была обнаружена в северо-западной части поднятия (слева внизу). Она имеет зеленоватый цвет на снимке.

3.2 CENTRAL PEAK OF ALGA CRATER IN NOACHIS TERRA

ACQUISITION TIME: 2021-01-28T11:24:51.433
LONGITUDE: 333.3 E   LATITUDE: 24.6 S
REGION: NOACHIS

ZERKLÜFTETES GRUNDGESTEIN IM ZENTRALGIPFEL EINES KRATERS

Der zentrale Gipfel eines 51 km grossen Kraters, der südöstlich des Huygens-Kraters an der westlichsten Seite von Tyrrhena Terra liegt. Das grünlich gefärbte Material ist stark zerklüftet. Es gibt parallele Dünenstrukturen, die die Erhebung auf den meisten Seiten umgeben. Im Norden (links) gibt es auch einige Aufschlüsse aus hellerem Material.

FRACTURED BEDROCK IN THE CENTRAL PEAK OF A CRATER

The central peak of a 51 km crater that lies south-east of Huygens crater on the westmost side of Tyrrhena Terra. The greenish coloured material is heavily fractured. There are parallel dune structures surrounding the uplift on most sides. There are also some exposures of lighter-toned material to the north (left).

SUBSTRAT ROCHEUX FRACTURÉ DANS LE PIC CENTRAL D'UN CRATÈRE

Image du pic central d'un cratère de 51 km de diamètre, au sud-est du cratère Huygens, à l'ouest de la région de Tyrrhena Terra. Le matériau de couleur verdâtre est fortement fracturé. Des structures dunaires parallèles entourent le soulèvement sur la plupart des côtés. Il y a aussi quelques expositions de matériaux plus clairs au nord (à gauche).

PAVIMENTO ROCCIOSO FRATTURATO NEL PICCO CENTRALE DI UN CRATERE

Il picco centrale di un cratere di 51 km che si trova a sud-est del cratere Huygens sul lato più occidentale di Tyrrhena Terra. Il materiale di colore verdastro è pesantemente fratturato e ci sono strutture dunali parallele che circondano il sollevamento sulla maggior parte dei lati. Del materiale più chiaro si trova esposto in superficie a nord (a sinistra)

РАЗДРОБЛЕННАЯ КОРЕННАЯ ПОРОДА В ЦЕНТРАЛЬНОМ ПИКЕ КРАТЕРА

На самом западном краю Тирренской земли к юго-востоку от кратера Гюйгенс находится 51-километровый кратер, центральный пик которого виден на этом снимке. Материал зеленоватого цвета сильно раздроблен, а параллельные цепочки дюн окружают поднятие кратера почти со всех сторон. Обнажение породы светлого тона виднеется на севере (слева).

3.3   FRACTURED BEDROCK IN THE CENTRAL PEAK OF A CRATER

ACQUISITION TIME: 2018-11-03T09:33:06.089
LONGITUDE: 62.6 E   LATITUDE: 18.4 S
REGION: TYRRHENA TERRA

ZENTRALE ERHEBUNG IN AONIA TERRA

Die zentrale Hebung (unten links) in einem ≈ 20 km grossen Krater südöstlich des Kraters Tábor in der Region Aonia Terra, die an Argyre Planitia angrenzt. Die Erhebung zeigt kleinräumige Farbveränderungen, die auf Veränderungen in der Mineralogie hinweisen. Der Südrand des Kraters ist auf der rechten Seite zu sehen. Er ist mit Aufschlüssen verschiedener Minerale gesprenkelt, was darauf hindeutet, dass es in diesem Gebiet unregelmässige Mineralschichten nahe der Oberfläche gibt.

CENTRAL UPLIFT IN AONIA TERRA

The central uplift (lower left) in an ≈ 20 km crater to the southeast of Tábor Crater in the Aonia Terra region adjacent to Argyre Planitia. The uplift shows small scale changes in colour indicating changes in mineralogy. The south rim of the crater can be seen to the right. It is speckled with exposures of different minerals suggesting that there are irregular layers of minerals close to the surface throughout the area

LE SOULÈVEMENT CENTRAL DANS LA RÉGION D'AONIA TERRA

Le soulèvement central (en bas à gauche) d'un cratère d'environ 20 km de diamètre au sud-est du cratère Tábor dans la région d'Aonia Terra adjacente à Argyre Planitia. Le soulèvement présente des variabilités de couleur à petite échelle indiquant des différences de minéralogie. Le bord sud du cratère est visible à droite. Il apparaît parsemé d'affleurements exposant différentes minéralogies, ce qui suggère qu'il existe des couches irrégulières de composition variée près de la surface dans toute la région.

IL SOLLEVAMENTO CENTRALE DI AONIA TERRA

Il sollevamento centrale (in basso a sinistra) in un cratere di ≈ 20 km a sud-est del cratere Tábor nella regione di Aonia Terra adiacente ad Argyre Planitia. Il sollevamento mostra cambiamenti di colore su piccola scala che indicano cambiamenti nella mineralogia. Il bordo sud del cratere si trova sulla destra. È punteggiato da diversi minerali esposti in superficie che suggeriscono la presenza di stratificazioni irregolari di minerali vicino alla superficie in tutta l'area.

ЦЕНТРАЛЬНОЕ ПОДНЯТИЕ (КРАТЕРА) В ЗЕМЛЕ АОНИД

Слева внизу расположено центральное поднятие 20-километрового кратера, что к юго-востоку от кратера Табор земли Аонид недалеко от равнины Аргир. Мелкие цветные пятна здесь – свидетельства разнообразия минерального состава пород. Справа виден южный край кратера, испещренный выходами различных пород. Это указывает на неравномерную структуру их слоёв вблизи поверхности по всей территории.

3.4 CENTRAL UPLIFT IN AONIA TERRA

ACQUISITION TIME: 2021-01-12T06:14:58.592
LONGITUDE: 302.3 E   LATITUDE: 36.6 S
REGION: AONIA

BUNTE ZENTRALE ERHEBUNG UND GRUBE EINES KRATERS IN MARGARITIFER TERRA

Dieses Bild zeigt das Zentrum eines Kraters mit einem Durchmesser von 44 km in Margaritifer Terra südlich von Pyrrhae Chaos im Ladon-Becken. Es handelt sich um eine Erhebung mit einer zentralen Grube. Die Wände, die die Grube umgeben, weisen viele unterschiedliche Farben auf, was auf eine Reihe von Mineralien schliessen lässt. Es scheint sich um chaotisch angeordnete Blöcke von unterschiedlicher Farbe zu handeln. Diese Art von gebrochener Struktur wird als Megabrekzie bezeichnet.

COLOURFUL CENTRAL UPLIFT AND PIT OF A CRATER IN MARGARITIFER TERRA

This image shows the centre of a 44 km diameter crater in Margaritifer Terra just south of Pyrrhae Chaos in the Ladon Basin area. There is an uplift with a central pit. The walls surrounding the pit have very diverse colours suggesting a range of mineralogies. There seem to be chaotically-arranged blocks of varying colours. This type of broken structure is referred to as megabreccia.

SOULÈVEMENT CENTRAL COLORÉ ET FOSSE D'UN CRATÈRE DANS MARGARITIFER TERRA

Cette image montre le centre d'un cratère de 44 km de diamètre dans la région de Margaritifer Terra juste au sud de Pyrrhae Chaos dans la région du bassin Ladon. On observe un soulèvement avec une fosse centrale. Les parois entourant la fosse ont des couleurs très diverses suggérant une large gamme de minéralogies. Des blocs rocheux aux couleurs variées sont disposés de façon chaotique. On nomme « mégabreccia » ce type de structure disloquée

SOLLEVAMENTO CENTRALE COLORATO E FOSSA DI UN CRATERE IN MARGARITIFER TERRA

Questa immagine mostra il centro di un cratere di 44 km di diametro in Margaritifer Terra appena a sud di Pyrrhae Chaos nella zona del Bacino Ladon. Si nota un sollevamento con una fossa centrale. Le pareti che circondano la fossa hanno colori molto diversi che suggeriscono una vasta gamma di mineralogie. Sembra che ci siano blocchi disposti caoticamente di vari colori. Questo tipo di struttura fratturata viene definito megabreccia.

РАЗНОЦВЕТНОЕ ЦЕНТРАЛЬНОЕ ПОДНЯТИЕ И ЯМА КРАТЕРА ЖЕМЧУЖНОЙ ЗЕМЛИ

На этом снимке показан центр 44-километрового кратера, расположенного на Жемчужной земле к югу от хаоса Пирры в районе бассейна Ладон. В нём есть кольцевое поднятие с центральной ямой. Стены, окружающие яму, окрашены очень пёстро, что говорит о разнообразии минерального состава. Похоже, что здесь есть хаотично расположенные обломки пород разного цвета. Такой тип разрушенной структуры называется мегабрекчия.

3.5 COLOURFUL CENTRAL UPLIFT AND PIT OF A CRATER IN MARGARITIFER TERRA

ACQUISITION TIME: 2021-01-23T09:38:30.170
LONGITUDE: 331.0 E LATITUDE: 13.4 S
REGION: MARGARITIFER

ZENTRALE ERHEBUNG DES KRATERS OSTROV

Der Ostrov-Krater hat einen Durchmesser von 73 km und liegt im westlichen Teil von Noachis Terra. Er zeigt in den CaSSIS-Produkten bemerkenswerte Farben, die oft gut mit mafischen Mineralien in den Infrarotspektrometerdaten korrelieren. Es wird angenommen, dass Olivin eine Komponente dieser Hebung ist. Es ist jedoch zu beachten, dass es eine beträchtliche Vielfalt an Farben gibt. Nördlich (links) der Hebung befindet sich eine schmale lineare Mulde und geschichtetes Material.

CENTRAL UPLIFT OF OSTROV CRATER

Ostrov Crater is a 73 km diameter crater in the western part of Noachis Terra and shows remarkable colours in CaSSIS products that often correlate well with mafic minerals in infrared spectrometer data. Olivine is thought to be a component of this uplift. Note however that there is considerable diversity of colour. There is also a narrow linear trough and layered material to the north (left) of the uplift.

SOULÈVEMENT CENTRAL DU CRATÈRE OSTROV

Le cratère Ostrov est un cratère de 73 km de diamètre situé dans la partie occidentale de Noachis Terra. Il présente des couleurs remarquables dans les images CaSSIS qui sont souvent corrélées avec les minéraux mafiques révélés par spectrométrie infrarouge. On pense que l'olivine est un minéral important de la roche composant ce soulèvement. On note cependant une diversité de couleur considérable. On observe également un fossé droit et étroit et une stratification au nord (à gauche) du soulèvement.

SOLLEVAMENTO CENTRALE DEL CRATERE OSTROV

Il cratere Ostrov è un cratere di 73 km di diametro nella parte occidentale di Noachis Terra e mostra notevoli colori nelle immagini prodotte da CaSSIS che spesso si correlano bene con i dati dello spettrometro infrarosso e che indicano la presenza di particolari minerali. Si pensa che l'olivina sia uno dei componenti di questo sollevamento. Si noti comunque che c'è una notevole diversità di colore. Inoltre, si può osservare anche una stretta depressione lineare e materiale stratificato a nord (a sinistra) del sollevamento.

ЦЕНТРАЛЬНОЕ ПОДНЯТИЕ КРАТЕРА ОСТРОВ

Кратер Остров диаметром 73 км, лежащий в западной части земли Ноя, проявляет удивительные цвета в фильтрах CaSSIS. Такие оттенки в данных инфракрасного спектрометра часто соответствуют мафическим минералам (то есть содержащим много магния и железа). Считается, что в этом поднятии присутствует оливин. Однако, обратите внимание на значительные вариации цвета. На севере (слева) виден узкий прямолинейный желоб и расслоенные породы.

3.6 **CENTRAL UPLIFT OF OSTROV CRATER**

ACQUISITION TIME: 2021-01-21T08:26:14.063
LONGITUDE: 331.8 E   LATITUDE: 27.0 S
REGION: NOACHIS

ZERKLÜFTETER KRATERBODEN BEI CYDONIA MENSAE

Dieser Krater befindet sich an der Dichotomie-Grenze zwischen Arabia Terra und Cydonia Mensae. Es ist ein komplexer Krater. Ein Teil der zentralen Erhebung ist oben rechts zu sehen. Das zerklüftete Gelände an der Basis des inneren Randes ist ungewöhnlich. Beachten Sie auch die helle Schicht im Kraterrand selbst auf der linken Seite des Bildes. Die Zerklüftung ist möglicherweise das Ergebnis eines Zusammenbruchs von wasserreichem Füllmaterial.

FRACTURED CRATER FLOOR NEAR CYDONIA MENSAE

This crater is on the dichotomy boundary where Arabia Terra meets Cydonia Mensae. It is a complex crater. Part of the central uplift can be seen to the top right. The fractured terrain at the base of the internal rim is unusual. Note also the light-toned layer in the rim itself to the left of the image. The fracturing may be the result of collapse of water-rich in-filling material.

FOND DE CRATÈRE FRACTURÉ PRÈS DE CYDONIA MENSAE

Ce cratère se trouve sur la limite de la dichotomie globale martienne, à la frontière entre Arabia Terra et Cydonia Mensae. C'est un cratère complexe. On peut voir une partie du soulèvement central en haut à droite. Le terrain fracturé à la base du rebord interne est inhabituel. Notez également la couche claire au sein du rebord, à gauche de l'image. La fracturation peut être le résultat de l'effondrement de formations initialement riches en eau.

PAVIMENTO FRATTURATO DI UN CRATERE VICINO A CYDONIA MENSAE

Questo cratere si trova sul confine dicotomico dove l'Arabia Terra incontra Cydonia Mensae. È un cratere complesso e parte del sollevamento centrale può essere visto in alto a destra. Il terreno fratturato alla base del bordo interno è insolito. Si noti lo strato di colore chiaro nel margine stesso a sinistra dell'immagine. La frattura può essere il risultato del collasso di materiale di riempimento ricco d'acqua.

ИЗЛОМАННЫЙ РЕЛЬЕФ КРАТЕРА ВБЛИЗИ СТОЛОВЫХ ГОР КИДОНИИ

Этот сложный кратер находится в области дихотомии, где земля Аравия встречается со столовыми горами Кидонии. Часть центрального поднятия видна справа вверху. Основание внутреннего кольца имеет необычно изломанный рельеф. Такое множество трещин может быть результатом обрушения заполняющего материала, насыщенного водой. Обратите внимание также на светлый слой внешнего обода кратера в левой части снимка.

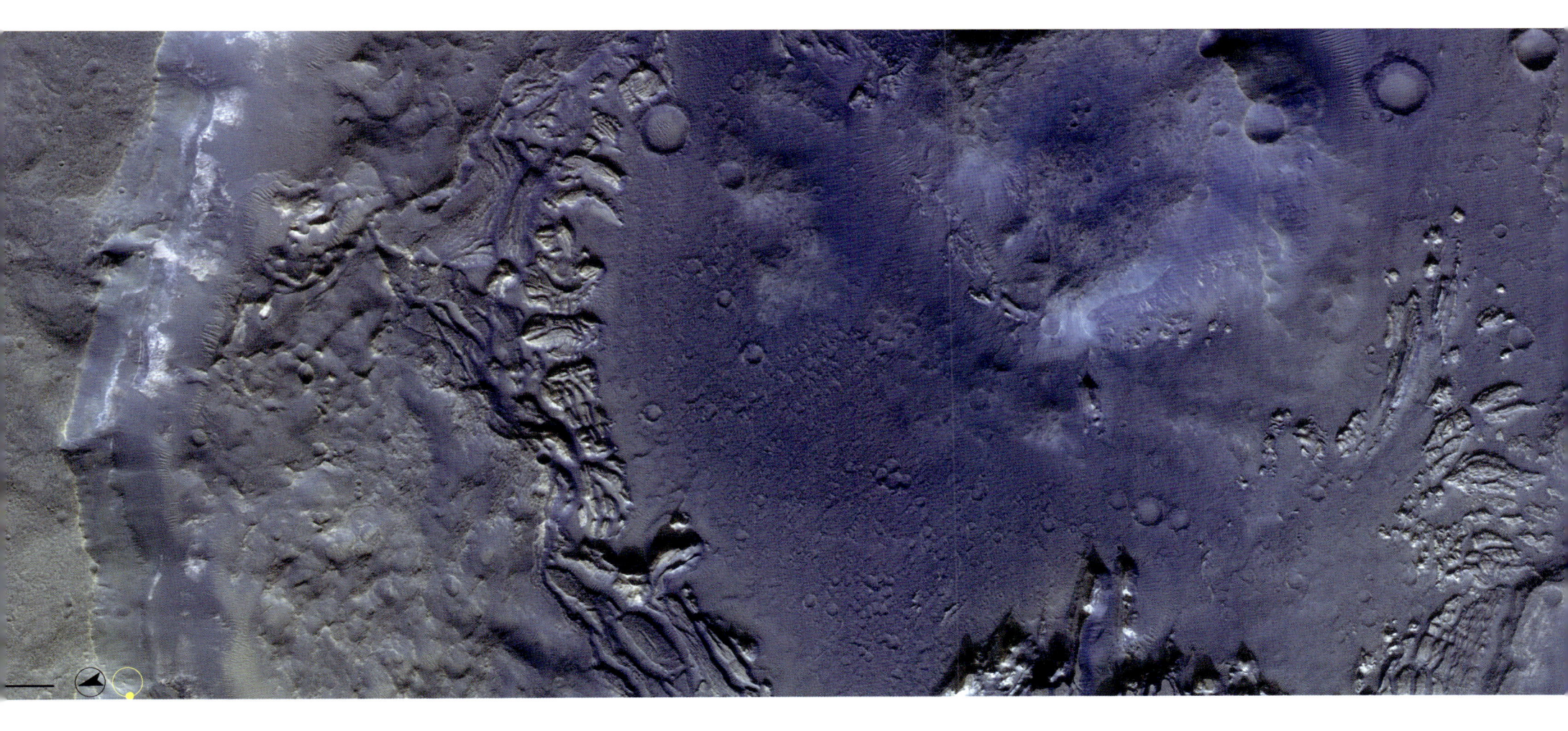

3.7   FRACTURED CRATER FLOOR NEAR CYDONIA MENSAE

ACQUISITION TIME: 2021-05-05T04:34:09.126
LONGITUDE: 351.8 E   LATITUDE: 35.7 N
REGION: ARABIA

MUSCHELGELÄNDE INNERHALB EINES KRATERS IN UTOPIA

In den Ebenen von Utopia Planitia ist ein muschelartiges Gelände zu sehen, das als Hinweis für die Sublimation von oberflächennahem Eis durch einen thermokarstähnlichen Mechanismus gesehen wird. Dieser Krater befindet sich ebenfalls in Utopia Planitia, und sein Inneres weist einige (aber nicht alle) der Merkmale des muschelartigen Geländes auf. Material scheint verloren gegangen zu sein, möglicherweise in mehreren Episoden, sodass an einigen Stellen kleine Buckel zurückblieben. Die strömungsähnliche Form des Auswurfs wurde als Hinweis für Bodeneis oder Wasser zum Zeitpunkt des Einschlags interpretiert.

SCALLOP TERRAIN WITHIN A CRATER IN UTOPIA

Scallop-like terrain is seen in the plains of Utopia Planitia and has been interpreted as evidence of the sublimation of near-surface ice in a thermokarst-like mechanism. This crater is also in Utopia Planitia and its interior shows some (but not all) of the charactaristics of the scallop-like terrain. Material seems to have been lost, possibly in several episodes, leaving behind small hummocks in places. The flow-like shape of the ejecta has been interpreted as evidence for ground ice or water at the time of impact.

TERRAIN FESTONNÉ DANS UN CRATÈRE D'UTOPIA

Un terrain présentant des formes de coquilles est observé dans les plaines d'Utopia Planitia et a été interprété comme la preuve de la sublimation de glace proche de la surface dans un contexte thermokarstique. L'intérieur de ce cratère présente certaines des caractéristiques générales des terrains à forme de coquille (mais pas toutes). Les formations superficielles semblent avoir été érodés, probablement en plusieurs épisodes, laissant derrière eux de petits hummocks par endroits. La forme fluide des éjectas a été interprétée comme une preuve de la présence de glace ou d'eau dans le sol au moment de l'impact.

SCALLOPS ALL'INTERNO DI UN CRATERE IN UTOPIA

Nelle pianure di Utopia Planitia si vede un terreno a smerli ed è stato interpretato come prova della sublimazione del ghiaccio vicino alla superficie, coivolto in un meccanismo simile al termocarismo. Anche questo cratere si trova in Utopia Planitia e il suo interno mostra alcune (ma non tutte) caratteristiche simili a scallops. Del materiale sembra essere stato perso, possibilmente in diversi episodi, lasciando dietro di sé piccole gobbe in alcuni punti. La forma dell'ejecta, simile a una colata di materiale, è stata interpretata come prova della presenza di ghiaccio o acqua al momento dell'impatto.

ГРЕБЕНЧАТЫЙ РЕЛЬЕФ ВНУТРИ КРАТЕРА РАВНИНЫ УТОПИЯ

«Гребенчатый» рельеф, наблюдающийся в некоторых местах равнины Утопия, интерпретируется как свидетельство сублимации (перехода из твёрдой фазы в газообразную) подповерхностного льда в процессе термокарста (проседания пород из-за вытаивания или сублимации льда). Данный кратер также лежит на равнине Утопия и показывает некоторые (но не все) признаки подобного рельефа. По-видимому, материал был утрачен в несколько этапов и оставил после себя небольшие торосы. «Текучая» форма выбросов ударного кратера может свидетельствовать о наличии грунтового льда или воды во время столкновения.

3.8 **SCALLOP TERRAIN WITHIN A CRATER IN UTOPIA**

ACQUISITION TIME: 2021-02-22T15:06:57.734
LONGITUDE: 101.2 E   LATITUDE: 44.6 N
REGION: UTOPIA

MINERALISCHE VIELFALT IM TAYTAY-KRATER

Der Taytay-Krater ist ein Krater von 18,2 km Durchmesser in Margaritifer Terra. Ares Vallis und Aram Chaos befinden sich in der Nähe (im Südwesten). In und um den Krater wurde Olivin entdeckt. Die hellgrünen Farbtöne, die hier zu sehen sind, scheinen mit der Olivinverteilung übereinzustimmen, die von anderen Instrumenten festgestellt wurde. Olivin ist ein Magnesium-Eisen-Silikat, das jedoch schnell verwittert und sich verändert, was darauf hindeutet, dass das Material hier nicht über einen längeren Zeitraum mit Wasser in Berührung gekommen ist.

MINERAL DIVERSITY IN TAYTAY CRATER

Taytay Crater is a 18.2 km diameter crater in Margaritifer Terra. Ares Vallis and Aram Chaos are near-by (to the south-west). Olivine has been detected in and around the crater. The light green hues seen here appear to correlate with the olivine distribution seen by other instruments. Olivine is a magnesium-iron silicate but it weathers and alters quickly suggesting that the material here has not been in contact with water for any significant length of time.

DIVERSITÉ MINÉRALE DANS LE CRATÈRE TAYTAY

Le cratère Taytay est un cratère de 18,2 km de diamètre dans la région de Margaritifer Terra. Ares Vallis et Aram Chaos sont proches (au sud-ouest). De l'olivine a été détectée dans et autour du cratère. Les teintes vert clair observées ici semblent correspondre à la distribution de l'olivine vue par d'autres instruments. L'olivine est un silicate de magnésium et de fer qui vieillit et s'altère rapidement en présence d'eau, ce qui suggère que cette roche n'a jamais été en contact prolongé avec de l'eau liquide.

DIVERSITÀ MINERALE NEL CRATERE TAYTAY

Il cratere Taytay è un cratere di 18,2 km di diametro in Margaritifer Terra. Ares Vallis e Aram Chaos sono qui vicino (a sud-ovest). É stata rilevata olivina all'interno e intorno al cratere. Le sfumature verde chiaro sembrano essere correlate alla distribuzione dell'olivina rilevate da altri strumenti. L'olivina è un silicato di magnesio e ferro, ma si altera rapidamente in presenza di acqua, il che indica che il materiale che si vede in questa immagine non è stato in contatto con l'acqua per un periodo di tempo significativo.

МИНЕРАЛЬНОЕ РАЗНООБРАЗИЕ В КРАТЕРЕ ТАЙТАЙ

Тайтай – это кратер диаметром 18,2 км в Жемчужной земле. Рядом (на юго-западе) находятся долина Арес и хаос Арам. В кратере и вокруг него обнаружен оливин. Судя по всему, светло-зеленые оттенки, наблюдаемые здесь, хорошо соотносятся с распределением оливина, наблюдаемым другими приборами. Оливин – это магниево-железный силикат, который быстро выветривается и изменяется. Это позволяет предположить, что материал здесь не был в контакте с водой в течение какого-то значительного времени.

3.9 MINERAL DIVERSITY IN TAYTAY CRATER

ACQUISITION TIME: 2021-05-07T05:31:59.235
LONGITUDE: 340.3 E   LATITUDE: 7.2 N
REGION: MARGARITIFER

DOPPELSTÖCKIGES KRATERINNERES IN MERIDIANI

Dieser kleine Krater (≈ 3 km Durchmesser) mit dem Namen Ada befindet sich in der Region Meridiani östlich der Landestelle des NASA-Rovers Opportunity. Er ist ungewöhnlich wegen der steilen Wände und der zweistufigen Struktur im Inneren, die vielleicht auf einen Stärkekontrast des Ziels zurückzuführen ist. Die untere (westliche) Seite des Kraters ist sehr hell (einige Teile der Oberfläche sättigten den Detektor). Die Sonne stand zum Zeitpunkt der Aufnahme (10.18 Uhr Ortszeit) im Osten und verursachte helle Reflexionen am Westrand.

BI-LEVEL CRATER INTERIOR IN MERIDIANI

This small (≈ 3 km diameter) crater named Ada is in the Meridiani region east of where NASA's Opportunity rover landed. It is unusual because of the steep walls and the bi-level structure seen in the interior, perhaps due to a strength contrast of the target. The lower (west) side of the crater is very bright (some parts of the surface saturated the detector). The Sun was in the east at the time of imaging (10:18 local time) producing bright reflections from the west rim.

INTÉRIEUR DE CRATÈRE À DEUX NIVEAUX DANS MERIDIANI

Ce petit cratère (≈ 3 km de diamètre) nommé Ada se trouve dans la région de Meridiani, à l'est de l'endroit où s'est posé le rover Opportunity de la NASA. Il est inhabituel en raison de ses parois abruptes et de la structure à deux niveaux observée à l'intérieur, peut-être due à une variabilité de dureté des roches impactées. Le côté inférieur (ouest) du cratère est très clair (certaines parties de la surface ont saturé le détecteur). Le soleil était à l'est au moment de l'image (10:18 heure locale) produisant des réflexions brillantes sur le bord ouest.

INTERNO DEL CRATERE A DUE LIVELLI IN MERIDIANI

Questo piccolo (≈ 3 km di diametro) cratere chiamato Ada si trova nella regione Meridiani a est di dove il rover Opportunity della NASA è atterrato. È un cratere insolito a causa delle pareti ripide e della struttura a due livelli che si trova all'interno, forse dovuta a un contrasto di resistenza del materiale bersaglio. Il lato inferiore (ovest) del cratere è molto chiaro (alcune parti della superficie hanno saturato il rilevatore). Il Sole era a est al momento dell'acquisizione delle immagini (10:18 ora locale) producendo riflessi luminosi sul bordo occidentale.

ДВУХУРОВНЕВЫЙ КРАТЕР НА ПЛАТО МЕРИДИАНА

Этот небольшой (примерно 3 км з диаметре) кратер под названием Ада находится на плато Меридиана к востоку от места посадки марсохода «Оппортьюнити» космического агентства NASA. Он необычен из-за крутых стен и двухуровневой структуры, наблюдаемой в его внутренней части. Возможно, изображение таково из-за высокой контрастности объекта. Нижняя (западная) сторона кратера очень яркая, некоторые участки поверхности даже пересветили детектор. Во время съемки (10:18 по местному времени) Солнце находилось на востоке, ярко освещая западный вал кратера.

3.10 BI-LEVEL CRATER INTERIOR IN MERIDIANI

ACQUISITION TIME: 2021-02-10T15:55:55.115
LONGITUDE: 356.7 E   LATITUDE: 3.6 S
REGION: MERIDIANI

KRATER UND KREISFÖRMIGE STRUKTUR IN TYRRHENA

Dieses Bild stammt aus der Region Tyrrhena Terra. In der Nähe dieses Ortes gibt es gut belichtete chloridreiche Ablagerungen (wie NaCl, Kochsalz). Leicht violette Farben in CaSSIS-Bildern sind mögliche Indikatoren für Chloride, und es gibt davon einige kleine Flecken auf der rechten Seite dieses Bildes. Neben dem Krater mit den bräunlichen Auswürfen befindet sich jedoch ein kreisförmiges Merkmal, das wahrscheinlich ein gefüllter Krater war, der jetzt vollständig erodiert ist. Beachten Sie das Tal unten links, das auf einen früheren Wasserfluss hindeutet.

CRATER AND CIRCULAR STRUCTURE IN TYRRHENA

This image is from the Tyrrhena Terra region. There are well-exposed chloride-rich deposits (like NaCl, table salt) close to this site. Slightly purple colours in CaSSIS images are potential indicators of chlorides and there are some small patches to the right of this image. However, adjacent to the crater with the brownish ejecta is a circular feature that was probably a filled crater that has now been totally eroded. Note the valley to the bottom left indicating past water flow.

CRATÈRE ET STRUCTURE CIRCULAIRE DANS TYRRHÉNA

Cette image provient de la région de Tyrrhena Terra. On observe des dépôts riches en chlorures (tels que NaCl, le sel de table) bien exposés près de ce site. Les couleurs légèrement violettes dans les images CaSSIS sont des indicateurs potentiels de chlorures. On peut en noter quelques petits affleurements à droite de cette image. A gauche du petit cratère entouré d'éjectas brunâtres se trouve une structure circulaire qui était probablement un cratère comblé qui a maintenant été totalement érodé. Notez également la vallée en bas à gauche indiquant un ancien écoulement d'eau.

CRATERE E STRUTTURA CIRCOLARE IN TYRRHENA

Questa immagine della regione di Tyrrhena Terra. Ci sono depositi ricchi di cloruri (come NaCl, sale da cucina) ben esposti in superficie vicino a questo sito. I colori leggermente viola nelle immagini CaSSIS sono potenziali indicatori di cloruri e se ne possono osservare alcune esempi in piccole macchie sulla destra di questa immagine. Tuttavia, adiacente al cratere con l'ejecta marroncino si trova una forma circolare che era probabilmente un cratere riempito e che ora è stato completamente eroso. Si noti la valle in basso a sinistra che indica un antico flusso d'acqua.

КРАТЕР И КОЛЬЦЕВАЯ СТРУКТУРА НА ТИРРЕНСКОЙ ЗЕМЛЕ

Изображение района Тирренской земли. Рядом с этим местом находятся значимые обнажения пород, богатых хлоридами (такими как NaCl, поваренная соль). Пурпурные цвета в фильтрах CaSSIS являются потенциальным признаком наличия хлоридов, и справа на этом снимке есть несколько небольших пятен, которые отличаются от тёмно-фиолетовых участков. Рядом с кратером, окружённым коричневатыми выбросами, находится круглое образование, которое, вероятно, было другим заполненным кратером, но теперь полностью разрушено. Обратите внимание на долину в левом нижнем углу, указывающую на поток воды в прошлом.

3.11 CRATER AND CIRCULAR STRUCTURE IN TYRRHENA

ACQUISITION TIME: 2020-12-12T20:31:40.015
LONGITUDE: 91.0 E LATITUDE: 18.2 S
REGION: TYRRHENA

GEFÜLLTER KRATER IN ARCADIA

Dieser gefüllte Krater befindet sich in Arcadia Planitia. Arcadia Planitia ist eine glatte Ebene mit vulkanischen, glazialen oder Sedimentströmen, die erst relativ spät in der Geschichte des Mars (in der Amazonas-Epoche) entstanden sind. Der Rand des Kraters hier ist kaum sichtbar, und das Füllmaterial weist nur wenige kleine Krater auf, was auf ein sehr junges Alter hinweist. Andere Krater in der Nähe sind ebenfalls mit frischem Material gefüllt.

FILLED CRATER IN ARCADIA

This filled crater is in Arcadia Planitia. Arcadia Planitia is a smooth plain with volcanic, glacial, or sediment flows that have occurred relatively late in the history of Mars (the Amazonian epoch). The rim of the crater here is barely visible and the filling material shows few small-sized craters indicating a very young age. Other nearby craters are also filled with fresh material.

CRATÈRE COMBLÉ DANS LA RÉGION D'ARCADIA

Ce cratère comblé se trouve dans Arcadia Planitia, une plaine lisse avec des coulées volcaniques, glaciaires ou sédimentaires qui se sont produites relativement tard dans l'histoire de Mars (à l'époque amazonienne). Le bord du cratère est ici à peine visible et la formation le remplissant présente quelques cratères de petite taille indiquant un âge très jeune. D'autres cratères proches ont également été récemment comblés.

CRATERE RIEMPITO IN ARCADIA

Questo cratere riempito si trova in Arcadia Planitia. Arcadia Planitia è una pianura liscia con colate vulcaniche, glaciali o di sedimenti che si sono verificati relativamente tardi nella storia di Marte (periodo Amazzonico). Il bordo del cratere è appena visibile e il materiale di riempimento mostra pochi crateri di piccole dimensioni, che suggeriscono un'età molto giovane. Anche altri crateri vicini sono riempiti con materiale fresco.

ЗАПОЛНЕННЫЙ КРАТЕР НА РАВНИНЕ АРКАДИЯ

Этот заполненный кратер находится на равнине Аркадия. Равнина Аркадия – это гладкая равнина со следами неких потоков (вулканических, ледниковых или осадочных, а возможно, и какой-то их совокупности), возникших относительно поздно в истории Марса (в Амазонийский период). Вал кратера здесь едва заметен, а в материале заполнения видно только несколько кратеров небольшого размера, что указывает на его очень молодой возраст. Другие близлежащие кратеры также заполнены свежим материалом.

3.12 FILLED CRATER IN ARCADIA

ACQUISITION TIME: 2021-03-07T00:17:23.391
LONGITUDE: 187.3 E   LATITUDE: 50.7 N
REGION: ARCADIA

KLEINE KRATER MIT UNGEWÖHNLICHEM INNENLEBEN IN NILOSYRTIS MENSAE

Diese Krater befinden sich in der Region Nilosyrtis Mensae und zeigen geschichtete innere Ablagerungen. Die meisten kleinen Krater sind schüsselförmig, aber hier liegt das Material in der Mitte der Krater auf einer höheren Ebene. Dieses Bild wurde am späten Abend (16.56 Uhr Ortszeit) aufgenommen, wobei die Sonne von Westen (unten) scheint. Im Osten des Innenmaterials sind Schatten zu erkennen.

SMALL CRATERS WITH UNUSUAL INTERIORS IN NILOSYRTIS MENSAE

These craters are in the Nilosyrtis Mensae region and show layered interior deposits. Most small craters are bowl-shaped but here the material in the centre of the craters is at a higher elevation. This image was taken late in the day (16:56 local time) with the Sun coming the west (down). Shadows can be seen on the east of the interior material.

PETITS CRATÈRES AUX DÉPOTS INTÉRIEURS INHABITUELS DANS NILOSYRTIS MENSAE

Ces cratères se trouvent dans la région de Nilosyrtis Mensae et présentent des dépôts intérieurs stratifiés. La plupart des petits cratères ont la forme d'un bol, mais le matériau au centre des cratères se trouve ici à une altitude plus élevée. Cette image a été prise en fin de journée (16:56 heure locale) avec le soleil venant de l'ouest (vers le bas). Les ombres à l'est des dépôts intérieurs accentuent le relief de leur stratification.

PICCOLI CRATERI CON INTERNI INSOLITI IN NILOSYRTIS MENSAE

Questi crateri si trovano nella regione di Nilosyrtis Mensae e mostrano depositi interni stratificati. La maggior parte dei piccoli crateri sono a forma di ciotola, e il materiale al centro dei crateri è ad un'altezza maggiore. Questa immagine è stata scattata alla fine della giornata (16:56 ora locale) con il Sole che viene da ovest (in basso). Le ombre del materiale interno possono essere viste estendersi a est.

НЕОБЫЧНОЕ ЗАПОЛНЕНИЕ МАЛЫХ КРАТЕРОВ В СТОЛОВЫХ ГОРАХ НИЛОСИРТ

Эти кратеры находятся в области столовых гор Нилосирт и демонстрируют так называемые «слоистые внутренние отложения». Большинство небольших кратеров имеют чашеобразную форму, но здесь материал в центре кратеров находится на бóльшей высоте. Снимок сделан в конце дня (16:56 по местному времени), когда Солнце светило на запад (вниз). На внутренних восточных поверхностях кратеров видны тени.

3.13 SMALL CRATERS WITH UNUSUAL INTERIORS IN NILOSYRTIS MENSAE

ACQUISITION TIME: 2021-03-04T07:30:52.247
LONGITUDE: 69.3 E   LATITUDE: 38.5 N
REGION: NILOSYRTIS MENSAE

NAHEGELEGENE KRATER AN EINER STRUKTURGRENZE IN NOACHIS

Zwei Krater von etwa gleicher Grösse (2–3 km) auf den beiden Seiten einer Grenze in der Oberflächenstruktur. Dies ist Noachis Terra nördlich des Arkangelsky-Kraters. Die Veränderung der Struktur fällt mit einem unbenannten degradierten Kraterrand zusammen.

NEARBY CRATERS ACROSS A STRUCTURAL BOUNDARY IN NOACHIS

Two craters of roughly the same size (2–3 km) on either side of a boundary in surface structure. This is from Noachis Terra north of Arkangelsky Crater. The change in structure coincides with an unnamed degraded crater rim.

CRATÈRES PROCHES TRAVERSANT UNE LIMITE STRUCTURALE DANS NOACHIS

Deux cratères de taille à peu près identique (2–3 km) de part et d'autre d'une disparité de structure de la surface dans la région de Noachis Terra au nord du cratère Arkangelsky. Le changement de structure coïncide avec le bord d'un cratère sans nom, très dégradé.

CRATERI ADIACENTI LUNGO UN CONFINE STRUTTURALE IN NOACHIS

Due crateri di circa la stessa dimensione (2–3 km) su entrambi i lati di una linea di confine sulla struttura superficiale. Questa è Noachis Terra a nord del cratere Arkangelsky. Il cambiamento della topografia coincide con un margine degradato di un cratere senza nome.

КРАТЕРЫ НА ЗЕМЛЕ НОЯ, СОСЕДСТВУЮЩИЕ ЧЕРЕЗ СТРУКТУРНУЮ ГРАНИЦУ

Это изображение области в земле Ноя к северу от кратера Архангельский. По обе стороны границы между разными типами поверхности видны два кратера примерно одного размера (2–3 км). Изменение в структуре поверхности совпадает с валом разрушенного безымянного кратера.

3.14   NEARBY CRATERS ACROSS A STRUCTURAL BOUNDARY IN NOACHIS

ACQUISITION TIME: 2020-12-24T09:42:52.613
LONGITUDE: 333.5 E   LATITUDE: 36.8 S
REGION: NOACHIS

SCHICHT IN EINER KRATERWAND

Dieser gut erhaltene 6 km grosse Krater liegt nordöstlich des Kraters Isil in Tyrrhena Terra. An seiner Basis ist felsiges Material freigelegt, und ausserhalb des nordöstlichen Randes (oben links) gibt es eine Vielfalt an Farben. Noch auffälliger ist der Ring aus hellem Material in der Kraterwand und die verschiedenfarbigen Erdrutsche, die durch Erosion dieses Materials entstanden sind. Es wird angenommen, dass dieser Teil von Tyrrhena Terra eine grosse Vielfalt an durch Wasser verändertem Material aufweist.

LAYER IN A CRATER WALL

This well-preserved 6 km crater is just to the north-east of Isil Crater in Tyrrhena Terra. There is exposed rocky material at its base and there is diversity in colour outside the north-eastern (top left) rim. Even more noticeable is the ring of bright material in the crater wall and the different coloured landslides produced by erosion of this material. This part of Tyrrhena Terra is thought to have a wide variety of aqueously-altered materials.

STRATE DANS LA PAROI D'UN CRATÈRE

Ce cratère bien préservé de 6 km se trouve juste au nord-est du cratère Isil dans la région de Tyrrhena Terra. On observe des formations rocheuses bien exposées à sa base et une diversité de couleurs à l'extérieur du bord nord-est (en haut à gauche). L'anneau de matériau clair dans la paroi du cratère et les glissements de terrain de différentes couleurs produits par l'érosion de ce matériau sont encore plus remarquables. On pense que cette partie de Tyrrhena Terra présente une grande variété de roches altérées par l'eau.

STRATIFICAZIONE NELLA PARETE DI UN CRATERE

Questo cratere ben conservato di 6 km è appena a nord-est del cratere Isil in Tyrrhena Terra. Si può osservare materiale roccioso esposto alla sua base e sfumature di colore fuori dal bordo nord-orientale (in alto a sinistra). Ancora più evidente è l'anello di materiale chiaro nella parete del cratere e le diverse frane colorate prodotte dall'erosione di questo materiale. Si pensa che questa parte di Tyrrhena Terra abbia un'ampia varietà di minerali alterati dall'acqua.

СЛОЙ В СТЕНЕ КРАТЕРА

Этот хорошо сохранившийся 6-километровый кратер находится к северо-востоку от кратера Исиль в Тирренской земле. На его дне имеется обнаженный скальный материал, также видно разнообразие цветов за северо-восточным (верхним левым) краем кратера. Еще более заметным является кольцо яркого материала, идущее по внутренним склонам кратера, и разноцветные оползни, образовавшиеся в ходе эрозии этого кольца. Считается, что в данной части Тирренской земли имеется большое разнообразие пород, измененных водой.

3.15 LAYER IN A CRATER WALL

ACQUISITION TIME: 2021-01-28T03:32:40.416
LONGITUDE: 88.9 E   LATITUDE: 26.4 S
REGION: TYRRHENA

154

4 KRATERRÄNDER UND RINNEN

CRATER RIMS AND GULLIES

BORDURES DE CRATÈRES ET RAVINES

BORDI DI CRATERI E CALANCHI

ВАЛЫ КРАТЕРОВ И ОВРАГИ

VIELFARBIGER KRATERRAND IN NOACHIS

Dieser Krater befindet sich in Noachis Terra, etwas nördlich vom Roddenberry-Krater und nordwestlich vom Asimov-Krater. Der Krater ist viel grösser als das CaSSIS-Sichtfeld und hat noch keinen Namen. In der Mitte des Bildes ist eine kleine isolierte Reihe von Rinnen am östlichen Rand des Kraters zu sehen und darunter eine Reihe von hellen Dünen. Der Kraterrand selbst weist verschiedene Farben auf, die auf Gesteine mit unterschiedlicher Zusammensetzung hinweisen. Noachis ist eine sehr grosse Region mit altem Terrain und vielen Anzeichen für fluviale Aktivitäten.

VARIEGATED CRATER RIM IN NOACHIS

This crater is in Noachis Terra a little way to the north of Roddenberry Crater and northwest of Asimov Crater. The crater is much larger than the CaSSIS field of view and not yet named. A small isolated set of gullies can be seen in the centre of the image on the eastern rim of the crater and beneath them a series of light-toned dunes. The rim itself shows variegated colours indicating rocks with different compositions. Noachis is a very large region with ancient terrain and a lot of evidence of fluvial activity.

BORDURE DE CRATÈRE VARIÉE DANS NOACHIS

Ce cratère se trouve dans la région de Noachis Terra, un peu au nord du cratère Roddenberry et au nord-ouest du cratère Asimov. Ce cratère est beaucoup plus grand que le champ de vue de CaSSIS mais il n'a pas encore été nommé. On peut voir un petit ensemble isolé de ravines au centre de l'image sur la bordure est du cratère et, en dessous, une série de dunes claires. La bordure elle-même présente des couleurs variées indiquant des roches de compositions différentes. Noachis est une très grande et ancienne région avec de nombreuses preuves d'activité fluviale.

MARGINE VARIEGATO DI UN CRATERE IN NOACHIS

Questo cratere si trova in Noachis Terra un po' più a nord del cratere Roddenberry e a nord-ovest del cratere Asimov. Il cratere è molto più grande del campo visivo di CaSSIS e non ha ancora un nome. Una piccola serie isolata di calanchi può essere osservata al centro dell'immagine sul bordo orientale del cratere e sotto di essi si trova una serie di dune chiare. Il bordo stesso mostra colori variegati che indicano rocce con composizioni diverse. Noachis è una regione antica molto grande e mostra molte prove di attività fluviale.

ПЕСТРЫЙ ОБОД КРАТЕРА ЗЕМЛИ НОЯ

Этот кратер находится на земле Ноя немного севернее кратера Родденберри и северо-западнее кратера Азимов. Он намного больше, чем поле зрения CaSSIS, и пока не имеет названия. В центре снимка на восточном валу кратера виден небольшой изолированный комплекс оврагов, а под ним – серия дюн светлых тонов. Сам вал имеет пестрые цвета, указывающие на породы разного состава. Земля Ноя – очень большой регион с древним рельефом и множеством свидетельств флювиальной активности.

4.1　VARIEGATED CRATER RIM IN NOACHIS

ACQUISITION TIME: 2020-06-21T11:27:48.099
LONGITUDE: 355.2 E　LATITUDE: 41.9 S
REGION: NOACHIS

BUNTER KRATERRAND BEIM ARKHANGELSKY-KRATER

Der östliche Teil eines Kraterrands in Noachis Terra, südwestlich des Arkhangelsky-Kraters. Der Rand weist unterschiedliche Farben auf, die zeigen, dass das Grundgestein unterschiedlich zusammengesetzt ist. Bei Betrachtung mit der höchsten Auflösung scheint es wahrscheinlich, dass die Oberfläche geschichtet ist. Beachten Sie die Freilegung von hellerem Material oben links. Dieser Krater ist teilweise mit eisigen Ablagerungen gefüllt, welche von der Kraterwand abgebrochen sind. Noachis ist eine sehr grosse Region mit altem Terrain und vielen Anzeichen für frühere fluviale Aktivitäten.

VARIEGATED CRATER RIM NEAR ARKHANGELSKY CRATER

The eastern part of a crater rim in Noachis Terra, south-west of Arkhangelsky Crater. The rim has variegated colours revealing that the bedrock has contrasting compositions. When viewed at the highest resolution, it seems probable that the surface is stratified. Notice the exposure of brighter material to the top left. This crater can be partially filled by icy deposits in winter leading to cracks on the crater wall. Noachis is a very large region with ancient terrain and a lot of evidence of past fluvial activity.

BORDURE DE CRATÈRE VARIÉE PRÈS DU CRATÈRE ARKHANGELSKY

La partie orientale d'une bordure de cratère dans la région de Noachis Terra, au sud-ouest du cratère Arkhangelsky. Cette bordure montre des couleurs variées qui révèlent les contrastes de composition du substrat rocheux. Lorsqu'on l'observe à la plus haute résolution disponible, la surface semble stratifiée. Notez également l'affleurement de roches plus claires en haut à gauche. Le cratère est partiellement rempli de dépôts glacés, qui se sont fissurés sur sa paroi. Noachis Terra est une très grande et ancienne région avec de nombreuses preuves d'activité fluviale passée.

MARGINE VARIEGATO DI UN CRATERE VICINO AL CRATERE ARKHANGELSKY

La parte orientale del bordo di un cratere in Noachis Terra, a sud-ovest del cratere Arkhangelsky. Il bordo presenta colori variegati che rivelano le differenti litologie del fondo roccioso. Se visto alla massima risoluzione, sembra probabile che la superficie sia stratificata. Si noti l'esposizione di materiale più chiaro in alto a sinistra. Questo cratere è parzialmente riempito da depositi ghiacciati, che si sono crepati sulla parete del cratere. Noachis è una regione molto grande con un terreno antico e presenta molte prove della passata attività fluviale.

ПЕСТРЫЙ КРАТЕРНЫЙ ВАЛ В РАЙОНЕ КРАТЕРА АРХАНГЕЛЬСКИЙ

Восточная часть вала кратера на земле Ноя, к юго-западу от кратера Архангельский. Вал имеет пестрые цвета, свидетельствующие о том, что коренные породы имеют разнообразный состав. При просмотре с максимальным разрешением кажется, что поверхность стратифицирована. Обратите внимание на обнажение более яркого материала в левом верхнем углу. Сам кратер, вероятно, частично заполняется зимой льдистыми отложениями, приводящими к образованию трещин на его стене.

4.2   VARIEGATED CRATER RIM NEAR ARKHANGELSKY CRATER

ACQUISITION TIME: 2020-12-19T07:54:48.350
LONGITUDE: 331.7 E   LATITUDE: 42.9 S
REGION: NOACHIS

EINGESCHNITTENER KRATERRAND IM MORELLA-KRATER

Dieser Krater liegt innerhalb des Morella-Kraters mit einem Durchmesser von 77 km und direkt nordöstlich von Ganges Cavus in der Region Valles Marineris. Die Ostwand des Morella-Kraters wurde durchbrochen, was zur Erosion der Kraterwände dieses kleineren Kraters geführt haben könnte. Der Morella-Krater enthält kieselsäurearme, magnesiumreiche Lava und Asche, was wahrscheinlich das farbenfrohe Gelände rund um den Krater erklärt. Beachten Sie auch die Talstrukturen im Süden (rechts) des Kraters.

INCISED CRATER RIM WITHIN MORELLA CRATER

This crater is within the 77 km diameter Morella Crater and just to the north-east of Ganges Cavus in the Valles Marineris region. The east wall of Morella Crater was breached and this may have led to the erosion of the crater walls of this smaller crater within. Morella Crater contains low-silica, Mg-rich lava and ash which probably leads to the colourful terrain surrounding the crater. Note also the valley structures to the south (right) of the crater.

BORDURE DE CRATÈRE INCISÉE DANS LE CRATÈRE MORELLA

Ce petit cratère se trouve à l'intérieur du cratère Morella, d'un diamètre de 77 km, au nord-est de Ganges Cavus dans la région de Valles Marineris. La paroi est du cratère Morella a été percée et cela a pu aboutir à l'érosion des parois de ce petit cratère. Le cratère Morella contient des laves et des cendres riches en magnésium et à faible teneur en silice qui causent probablement la couleur du terrain qui entoure le cratère. Notez également les structures de la vallée au sud (à droite) du cratère.

ORLO DI UN CRATERE SFONDATO NEL CRATERE MORELLA

Questo cratere si trova all'interno del cratere Morella di 77 km di diametro e appena a nord-est di Ganges Cavus nella regione delle Valles Marineris. La parete est del cratere Morella è stata sfondata e questo può aver portato all'erosione delle pareti del cratere più piccolo al suo interno. Il cratere Morella contiene cenere e lava a basso contenuto di silice e ricca di magnesio che probabilmente produce i colori del terreno che circonda il cratere. Si notino anche le strutture della valle a sud (a destra) del cratere.

ВРЕЗАННЫЙ КРАТЕРНЫЙ ВАЛ ВНУТРИ КРАТЕРА МОРЕЛЛА

Этот кратер находится внутри 77-километрового кратера Морелла, что к северо-востоку от котловины Ганг в районе долин Маринер. Восточная стена кратера Морелла была разрушена, что, возможно, привело к эрозии стен меньшего кратера. Морелла содержит низкокремнистую, богатую магнием лаву и пепел, что, вероятно, обуславливает пестрые цвета вокруг него. Обратите внимание также на долинные структуры к югу (справа) от кратера.

4.3 INCISED CRATER RIM WITHIN MORELLA CRATER

ACQUISITION TIME: 2018-11-01T16:15:21.016
LONGITUDE: 308.9 E LATITUDE: 9.5 S
REGION: MARINERIS

GEFÄRBTER KRATERRAND UND RINNEN

Dieser Krater befindet sich in Aonia Terra südlich von Bosporos Planum. In den nahe gelegenen Ebenen wurde Olivin entdeckt. Der Rand des Kraters ist auf den CaSSIS-Bildern bemerkenswert farbenfroh, was auf verschiedene Mineralogien hinweist. Der Krater kann im Winter stellenweise mit Eisablagerungen gefüllt sein, was zu Rissen in der Kraterwand führt. Man beachte auch die Rinnen im äusseren Rand des kleinen Einschlagkraters, die den Rand im Nordosten durchbrochen haben (oben links). Das hangabwärts rutschende Material ist im Vergleich zur Umgebung sehr hell.

COLOURED CRATER RIM AND GULLIES

This crater is in Aonia Terra south of Bosporos Planum. Olivine has been detected in nearby plains. The rim of the crater in CaSSIS imagery is remarkably colourful indicating different mineralogies. The crater itself has been partially filled with icy material. Note also the gullies in the outer rim of small impact crater that has breached the rim to the north-east (top left). The material sliding downslope is very bright compared to the surroundings.

BORDURE DE CRATÈRE COLORÉE ET RAVINES

Ce cratère se trouve dans la région d'Aonia Terra au sud de Bosporos Planum. De l'olivine a été détectée dans les plaines voisines. La bordure du cratère apparaît remarquablement colorée, indiquant diverses minéralogies. Le cratère lui-même a été partiellement rempli de matériaux glacés. Notez également les ravines dans la bordure extérieure du petit cratère d'impact qui a ouvert une brèche dans la bordure au nord-est (en haut à gauche). Le matériau glissant vers le bas de la pente est très clair par rapport aux alentours.

ORLO DI UN CRATERE COLORATO CON CALANCHI

Questo cratere si trova in Aonia Terra a sud di Bosporos Planum. É stata rilevata dell'olivina nelle pianure vicine. Il bordo del cratere nelle immagini CaSSIS è notevolmente colorato e indica diverse mineralogie. Il cratere stesso è stato parzialmente riempito da materiale ghiacciato. Si notino anche i calanchi sul bordo esterno del piccolo cratere da impatto che interrompe il bordo a nord-est (in alto a sinistra). Il materiale che scivola verso il basso è molto chiaro rispetto ai dintorni.

ПЁСТРЫЙ ОБОД КРАТЕРА И ОВРАГИ

Этот кратер находится в земле Аонид к югу от Босфорского плато. На близлежащих равнинах был обнаружен оливин. Вал кратера на снимках CaSSIS отличается удивительной пестротой, указывающей на минеральное разнообразие. Сам кратер был частично заполнен льдистыми породами. Обратите внимание также на овраги во внешнем ободе маленького ударного кратера, пробившего большой вал на северо-востоке (вверху слева). Материал, сползающий вниз по склону, очень яркий, по сравнению с окружающей средой.

4.4   COLOURED CRATER RIM AND GULLIES

ACQUISITION TIME: 2020-09-06T14:05:53.997
LONGITUDE: 296.4 E   LATITUDE: 40.2 S
REGION: AONIA

EINSCHLAG UND RANDKOLLAPS

Dieser Krater befindet sich in Tyrrhena Terra nördlich des Talsystems Vichada Valles. Der Einsturz und das Absacken eines grossen Teils der Nordwand des Kraters hat dazu geführt, dass der Umriss des Kraters nicht mehr kreisförmig ist. Später brach ein weiterer, kleinerer Teil des Kraterrands ein und verursachte einen «zungenförmigen» Erdrutsch auf dem Kraterboden. Die lokale grüne Farbe an der Kraterwand deutet auf die Lage von Gesteinen anderer Zusammensetzung hin.

IMPACT AND RIM COLLAPSE

This crater is in Tyrrhena Terra north of the Vichada Valles valley system. Collapse and slumping of a large portion of the northern wall of the crater has lead to its non-circular outline. Later, another, smaller part of the rim of the crater has collapsed producing a "tongue"-shaped landslide on the crater floor. The locally green colour on the crater wall indicates the position of rocks of a different composition.

IMPACT ET EFFONDREMENT DE BORDURE DE CRATÈRE

Ce cratère est situé dans la région de Tyrrhena Terra, au nord du système de vallées de Vichada Valles. L'effondrement et l'affaissement d'une grande partie de la paroi nord du cratère a conduit à son contour non circulaire. Plus tard, une autre partie, plus petite, de la bordure du cratère s'est effondrée, produisant un glissement de terrain en forme de «langue» sur le fond du cratère. La couleur localement verte sur la paroi du cratère indique la présence de roches de composition différente.

COLLASSO DEL MARGINE DI UN CRATERE PER IMPATTO

Questo cratere si trova in Tyrrhena Terra a nord del sistema di valli Vichada Valles. Il crollo e successivo franamento di una grande porzione della parete settentrionale del cratere ha portato al suo profilo non più circolare. Più tardi, un'altra più piccola parte del bordo del cratere è crollata producendo una frana a forma di «lingua» sul fondo del cratere. Il colore verde della zona sulla parete del cratere indica la presenza di rocce di composizione diversa.

УДАР И ОБРУШЕНИЕ КРАТЕРНОГО ВАЛА

Этот кратер находится в Тирренской земле к северу от системы долин Вичада. Обрушение и оползание значительной части северной стены кратера сделало его очертания неправильными. Позже другая, меньшая часть вала кратера, обрушилась, образовав внутри кратера оползень в форме языка. Зеленый цвет на стене кратера указывает на наличие пород другого состава.

4.5   IMPACT AND RIM COLLAPSE

ACQUISITION TIME: 2021-03-30T18:32:45.144
LONGITUDE: 89.2 E   LATITUDE: 15.9 S
REGION: TYRRHENA

FARBENFROHE KRATERWÄNDE

Dies ist ein relativ frischer Krater auf dem Boden des Holden-Kraters. Die Innenwände und der äussere Rand des Kraters weisen stark variierende Farben auf, was auf eine grosse mineralogische Vielfalt hinweist. Der Holden-Krater liegt formal in Noachis Terra, ist aber Teil eines Komplexes von Kratern und Tälern, die nach Margaritifer Terra führen. Er beherbergt zahlreiche alluviale Fächerablagerungen.

COLOURFUL CRATER WALLS

This is a relatively fresh crater on the floor of Holden Crater. The interior walls and the external rim of the crater have highly variegated colours indicating significant mineralogical diversity. Holden Crater is formally in Noachis Terra but is part of a complex of craters and valleys leading into Margaritifer Terra. It hosts numerous alluvial fan deposits.

PAROIS COLORÉES DE CRATÈRE

Petit cratère relativement frais situé au fond du cratère Holden. Les parois intérieures et la bordure extérieure du cratère présentent des couleurs très variées indiquant une importante diversité minéralogique. Le cratère Holden se situe au sein de la région de Noachis Terra mais fait partie d'un complexe de cratères et de vallées menant à Margaritifer Terra. Il abrite de nombreux cônes de dépôts alluviaux.

PARETI COLORATE DI UN CRATERE

Questo è un cratere relativamente fresco sul fondo del cratere Holden. Le pareti interne e il bordo esterno del cratere hanno colori molto variegati che indicano una significativa diversità mineralogica. Il cratere Holden si trova formalmente in Noachis Terra, ma fa parte di un complesso di crateri e valli che si estendono a Margaritifer Terra. Ospita numerosi conoidi alluvionali.

РАЗНОЦВЕТНЫЕ СТЕНЫ КРАТЕРА

Это относительно свежий кратер на дне кратера Холден. Внутренние его стены и внешний край имеют очень пёстрые цвета, что указывает на значительное минералогическое разнообразие. Кратер Холден формально находится на земле Ноя, но является частью комплекса кратеров и долин, ведущих в Жемчужную землю. Здесь расположены многочисленные аллювиальные веера (отложения обломков в форме конуса или веера, вынесенные потоками).

4.6 COLOURFUL CRATER WALLS

ACQUISITION TIME: 2020-11-25T21:48:19.137
LONGITUDE: 326.5 E   LATITUDE: 25.7 S
REGION: NOACHIS

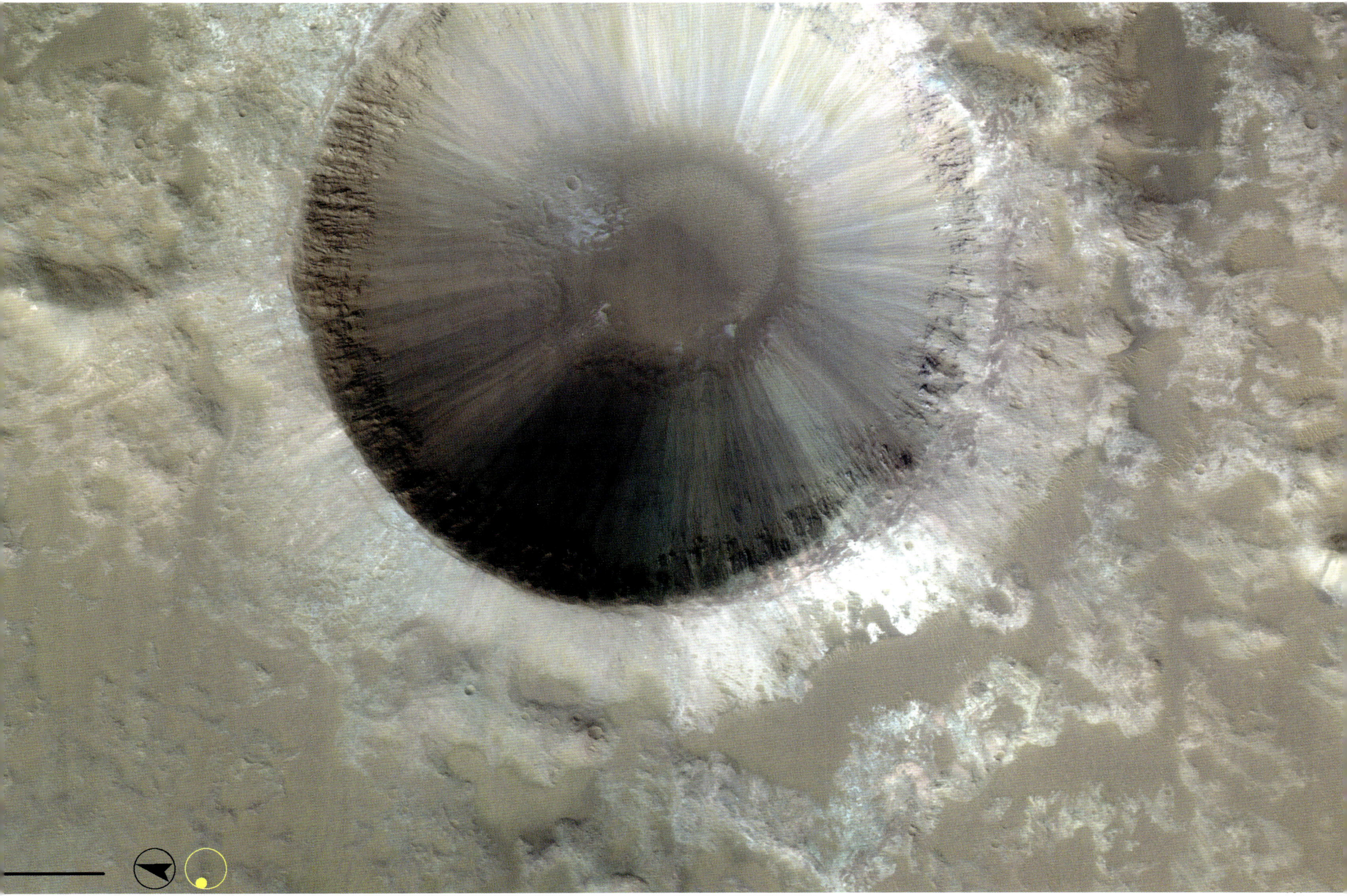

FARBENFROHE KRATERWAND IN DEN NEREIDUM MONTES

Dieses Bild einer Kraterwand im östlichen Randmaterial des Argyre-Einschlagbeckens südlich des Hale-Kraters zeigt eine breite Palette von subtilen Farben. Die Rinnen an der Wand dieses Kraters offenbaren das Grundgestein in diesem Gebiet, dessen kontrastreiche Farben auf eine vielfältige Mineralogie schliessen lassen. Die glatten Texturen innerhalb und ausserhalb des Kraters sind typisch für eisreiche Materialien.

COLOURFUL CRATER WALL IN NEREIDUM MONTES

This image is of a crater wall in the eastern rim materials of the Argyre impact basin south of Hale Crater and shows a wide range of subtle colours. The gullies on the wall of this crater are revealing the bedrock in this area, whose contrasting colours imply a diverse mineralogy. The smooth textures inside and outside the crater are typical of ice-rich materials.

PAROI DE CRATÈRE COLORÉE DANS NEREIDUM MONTES

Image de la paroi d'un cratère situé sur la bordure orientale du bassin d'impact Argyre au sud du cratère Hale. Les ravines sur la paroi de ce cratère révèlent le substrat rocheux de cette zone, dont les couleurs très diverses impliquent une minéralogie variée. Les textures lisses à l'intérieur et à l'extérieur du cratère sont typiques des matériaux riches en glace.

COLORATA PARETE DI CRATERE IN NEREIDUM MONTES

Questa immagine di una parete del cratere del bordo orientale del bacino d'impatto di Argyre a sud di Hale Crater mostra una vasta gamma di sfumature colorate nei materiali che lo compongono. I calanchi sulla parete di questo cratere rivelano il basamento roccioso di questa zona, i cui colori contrastanti implicano una mineralogia diversificata. Le strutture lisce all'interno e all'esterno del cratere sono tipiche dei materiali ricchi di ghiaccio.

РАЗНОЦВЕТНЫЕ СТЕНЫ КРАТЕРА В ГОРАХ НЕРЕИД

На этом снимке показаны стены кратера, расположенного в ударном бассейне Аргир к югу от кратера Хейл. Здесь представлен широкий диапазон тонких цветовых оттенков. Овраги на стене кратера обнажают коренные породы этой области, контрастные цвета которых указывают на минералогическое разнообразие. Гладкие текстуры внутри и снаружи кратера типичны для богатых льдом материалов.

4.7   COLOURFUL CRATER WALL IN NEREIDUM MONTES

ACQUISITION TIME: 2020-09-30T08:30:12.221
LONGITUDE: 325.4 E   LATITUDE: 39.3 S
REGION: ARGYRE

ZERKLÜFTETES, GESCHICHTETES MATERIAL IN ARABIA TERRA

Dies ist ein Bild eines kleinen Kraters innerhalb eines grösseren Kraters, der an den Jiji-Krater in Arabia Terra angrenzt. Der schüsselförmige Krater selbst ist nicht ungewöhnlich, aber die Umgebung zeigt geschichtete Sedimentablagerungen, die zerklüftet sind. In diesem Gebiet werden Schichten beobachtet (z. B. im Jiji-Krater), aber es ist nicht klar, ob der Sedimentationsprozess im Wasser oder durch Ablagerung aus der Luft (z. B. vulkanische Asche) erfolgte.

FRACTURED LAYERED MATERIAL IN ARABIA TERRA

This is an image of a small crater within a larger crater that is adjacent to Jiji Crater in Arabia Terra. The bowl-shaped crater itself is not unusual but the surroundings show layered sedimentary deposits that are fractured. Layers are observed in this area (e.g. in Jiji Crater) but it is not clear whether the sedimentation process was within water or as a result of deposition by airfall (e.g. volcanic ash).

DÉPOT STRATIFIÉ ET FRACTURÉ DANS LA RÉGION D'ARABIA TERRA

Image d'un petit cratère situé à l'intérieur d'un plus grand cratère dans la région d'Arabia Terra. La forme en bol du cratère est fréquemment observée et parfaitement préservée ici. Les environs présentent des dépôts sédimentaires stratifiés et fracturés, par exemple dans le cratère adjacent Jiji, mais on ne sait pas si le processus de sédimentation était aqueux ou éolien (dépôt de cendres volcaniques).

MATERIALE STRATIFICATO FRATTURATO IN ARABIA TERRA

Questa è un'immagine di un piccolo cratere all'interno di un cratere più grande che è adiacente al cratere Jiji in Arabia Terra. Il cratere a forma di ciotola in sé non è insolito, ma i dintorni mostrano depositi sedimentari stratificati che sono fratturati. Si osservano stratificazioni in quest'area (ad esempio nel cratere Jiji) ma non è chiaro se il processo di sedimentazione sia stato dovuto all'acqua o sia il risultato della deposizione per caduta aerea (ad esempio cenere vulcanica).

РАСТРЕСКАННЫЙ СЛОИСТЫЙ МАТЕРИАЛ НА ЗЕМЛЕ АРАВИЯ

Это изображение небольшого кратера внутри более крупного кратера, который находится рядом с кратером Джиджи в земле Аравия. Сам кратер чашеобразной формы не является необычным, но в его окрестностях видны слоистые осадочные отложения с трещинами. Подобные слои встречаются в этой области (например, в кратере Джиджи), но неясно, происходил ли процесс осаждения в воде или в воздухе (например при выпадении вулканического пепла).

4.8   FRACTURED LAYERED MATERIAL IN ARABIA TERRA

ACQUISITION TIME: 2021-05-13T06:57:36.554
LONGITUDE: 358.0 E   LATITUDE: 8.7 N
REGION: ARABIA

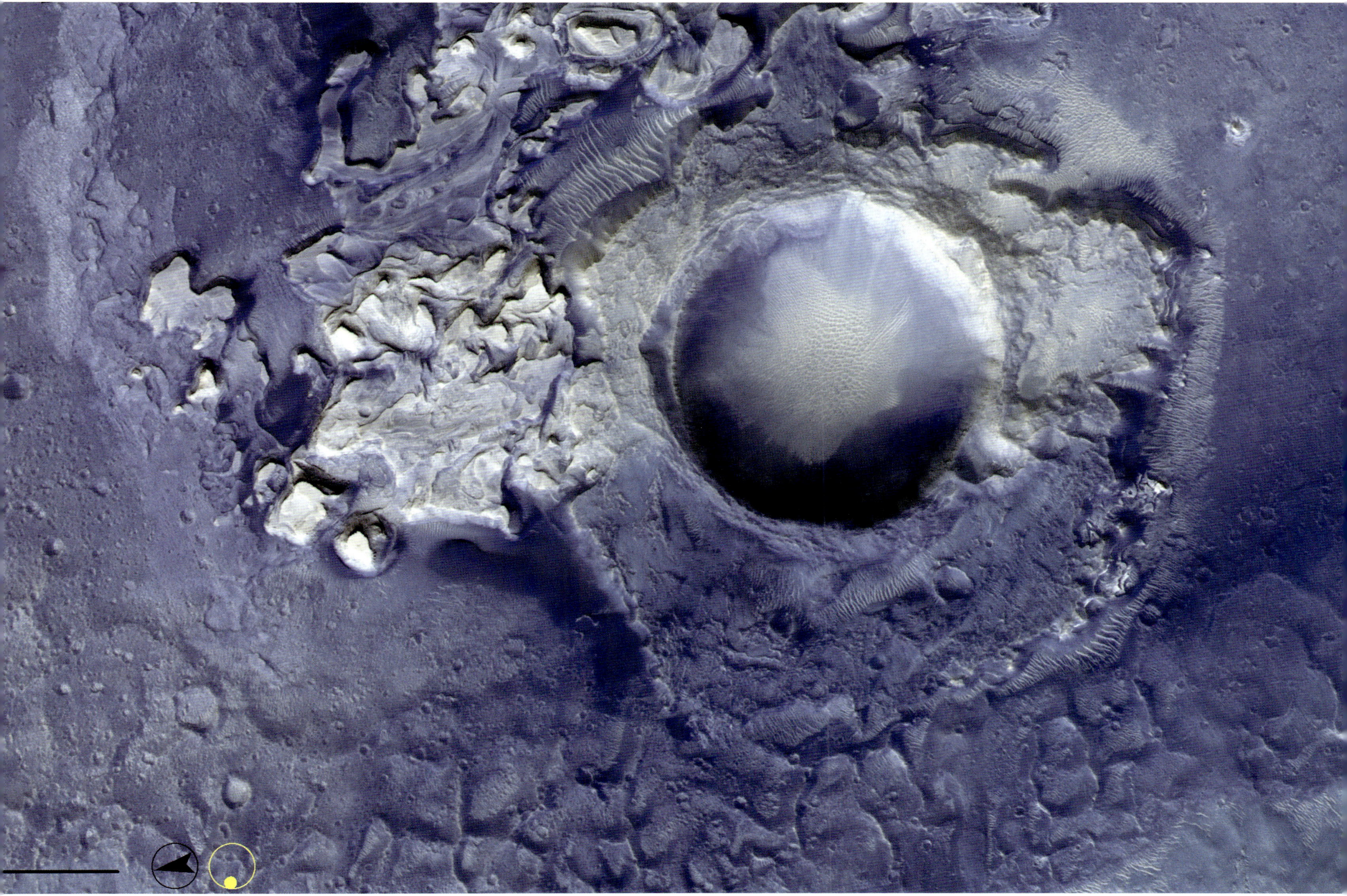

FREIGELEGTES HELLES MATERIAL IN EINER KRATERWAND

Die Kraterwand weist hier helles Material an der Oberfläche auf. Die Kraterwand sowie der Kraterboden weisen hell getöntes Material auf. Es gibt weitere Beobachtungen von Einschlägen kleiner Krater in Wänden nahe gelegener Krater, die helles Auswurfmaterial zeigen, was darauf hindeutet, dass hell getöntes Material hier relativ nah an der Oberfläche vorhanden sein könnte. Hell getönte Ablagerungen sind von Bedeutung, weil sie auf eine Veränderung der Zusammensetzung durch flüssiges Wasser hinweisen können.

EXPOSED LIGHT-TONED MATERIAL IN A CRATER WALL

The crater wall here shows light-toned material. The crater floor also shows exposures of light-toned material. There are other observations of small crater impacts into the walls of nearby craters that show bright ejecta suggesting that light-toned material might be pervasive here relatively close to the surface. Light-toned deposits have significance because they can indicate alteration of composition by liquid water

ROCHE EXPOSÉE DE COULEUR CLAIRE DANS LA PAROI D'UN CRATÈRE

La paroi du cratère expose ici des roches claires à la surface mais certaines sont également visibles dans des affleurements sur le fond du cratère. Il existe d'autres observations d'impacts de petits cratères dans les parois de cratères voisins qui montrent des éjectas clairs. Ces observations suggèrent que cette roche claire pourrait être omniprésente ici, relativement près de la surface. Les dépôts clairs sont importants car ils peuvent indiquer une altération de la composition de la roche par l'eau liquide.

MATERIALE CHIARO AFFIORANTE NELLA PARETE DI UN CRATERE

La parete di questo cratere espone superficialmente materiale dai toni chiari. Anche i materiali sul fondo del cratere hanno tonalità chiare. Sono stati osservati altri piccoli crateri nelle pareti di crateri vicini che mostrano ejecta chiari, suggerendo che in questa zona il materiale chiaro potrebbe essere pervasivo relativamente vicino alla superficie. I depositi dai toni chiari sono importanti perché possono indicare un'alterazione di composizione mineralogica da parte di acqua liquida.

ОБНАЖЕНИЕ СВЕТЛЫХ ПОРОД СТЕНЫ КРАТЕРА

Здесь, на стене и дне кратера обнажаются светлые породы. Наблюдения небольших ударных кратеров в стенах прочих близлежащих кратеров показали наличие яркого выброшенного материала. Это позволяет предположить, что подобная порода светлых тонов может быть очень распространена в этой местности и находится относительно близко к поверхности. Отложения светлых оттенков важны для исследователей, поскольку они могут указывать на изменение состава под воздействием жидкой воды.

4.9 EXPOSED LIGHT-TONED MATERIAL IN A CRATER WALL

ACQUISITION TIME: 2020-12-15T17:23:45.164
LONGITUDE: 162.1 E   LATITUDE: 33.9 S
REGION: CIMMERIA

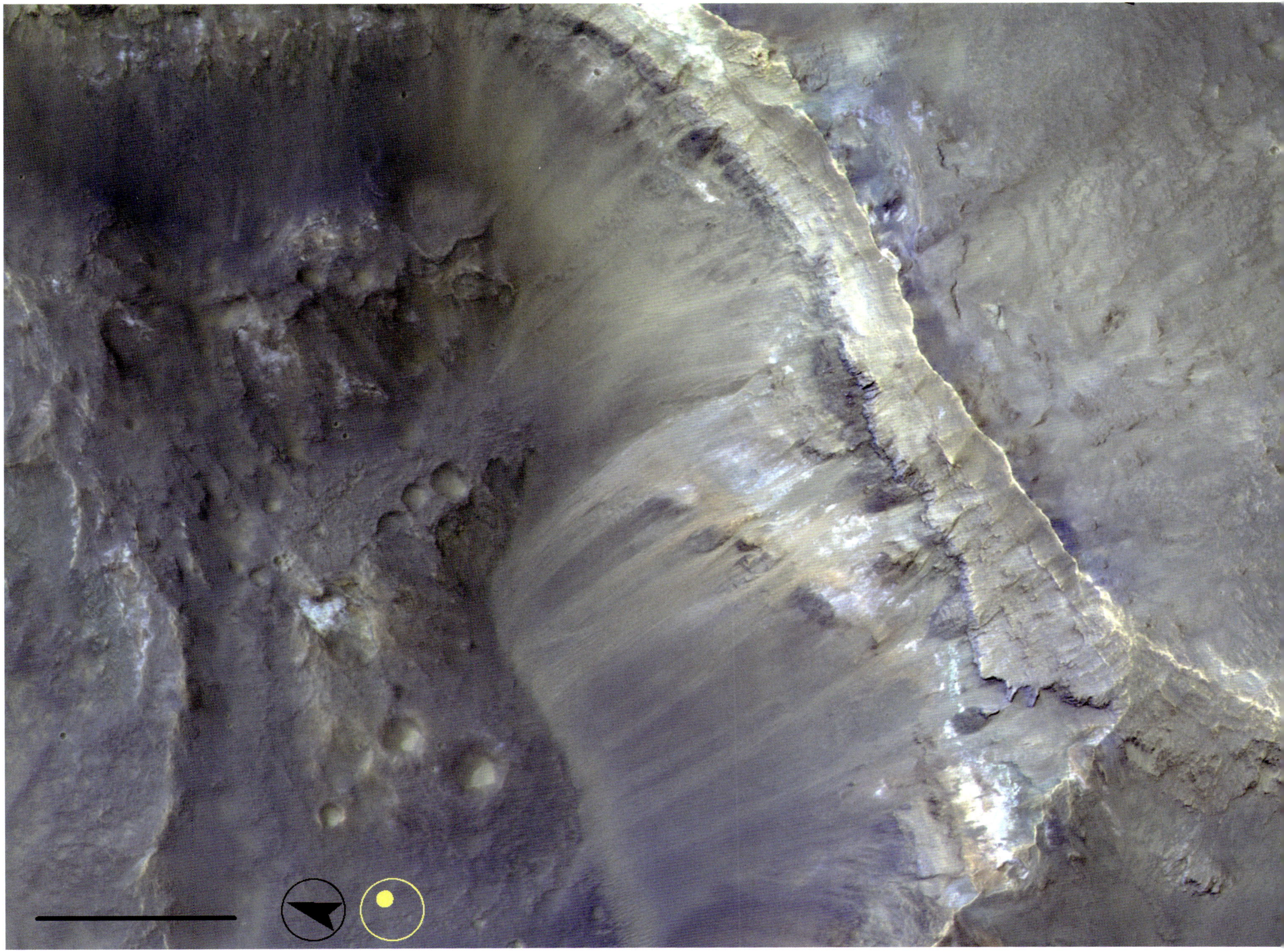

UNGEWÖHNLICHE AUSWURFDECKE

Unten in der Mitte dieses Bildes von Utopia Planitia nördlich von Astapus Colles ist der Rand eines gefüllten Einschlagkraters zu erkennen. Er ist von hellem, radial strukturiertem Auswurfmaterial umgeben, dessen lappiges Aussehen auf die Überdeckung durch eine dicke Schicht aus Eis und Staub zurückzuführen ist. Man beachte, dass diese Schicht eine polygonal gemusterte Oberfläche hat, die durch den dunklen Staub ganz rechts im Bild hervorgehoben wird und durch die thermische Kontraktion des Eises verursacht wird.

UNUSUAL EJECTA BLANKET

The edge of a filled impact crater can just be seen, in bottom centre of this image from Utopia Planitia just north of Astapus Colles. It is surrounded by bright radially structured ejecta material the lobate appearance of which is due to mantling by a thick layer of ice and dust. Note this layer has a polygonally patterned surface, which is picked out by dark dust to the far right of the image, and is caused by thermal contraction of the ice.

COUVERTURE D'ÉJECTAS INHABITUELLE

La bordure d'un cratère d'impact comblé est tout juste visible, en bas au centre de cette image d'Utopia Planitia juste au nord d'Astapus Colles. Le cratère est entouré d'éjectas clairs, structurés radialement, dont l'aspect lobé est causé par un manteau de glace et de poussière. Notez que cette couche présente une surface à motifs polygonaux, particulièrement mise en évidence par la poussière sombre à droite de l'image, et qui est causée par la contraction thermique de la glace.

INSOLITO MANTO DI EJECTA

In questa immagine da Utopia Planitia, appena a nord di Astapus Colles, si può a malapena osservare il bordo di un cratere da impatto riempito di materiale (basso al centro). È circondato dal materiale chiaro dell'ejecta strutturato a raggiera, il cui aspetto lobato è dovuto alla copertura di uno spesso strato di ghiaccio e polvere. Si noti che questo strato ha una superficie a forme poligonali, che è evidenziato dalla polvere scura all'estrema destra dell'immagine, ed è causata dalla contrazione termica del ghiaccio.

НЕОБЫЧНЫЙ ПОКРОВ ВЫБРОСОВ

На этом снимке в центре внизу виден вал заполненного ударного кратера участка равнины Утопия к северу от холмов Астап. Он окружен яркими радиальными выбросами, лопастная форма которых обусловлена покровом из толстого слоя льда и пыли. Обратите внимание на узорчатую полигональную поверхность, которая проявляется в правой части изображения скопившейся в трещинах темной пылью, и вызвана тепловым сжатием льда.

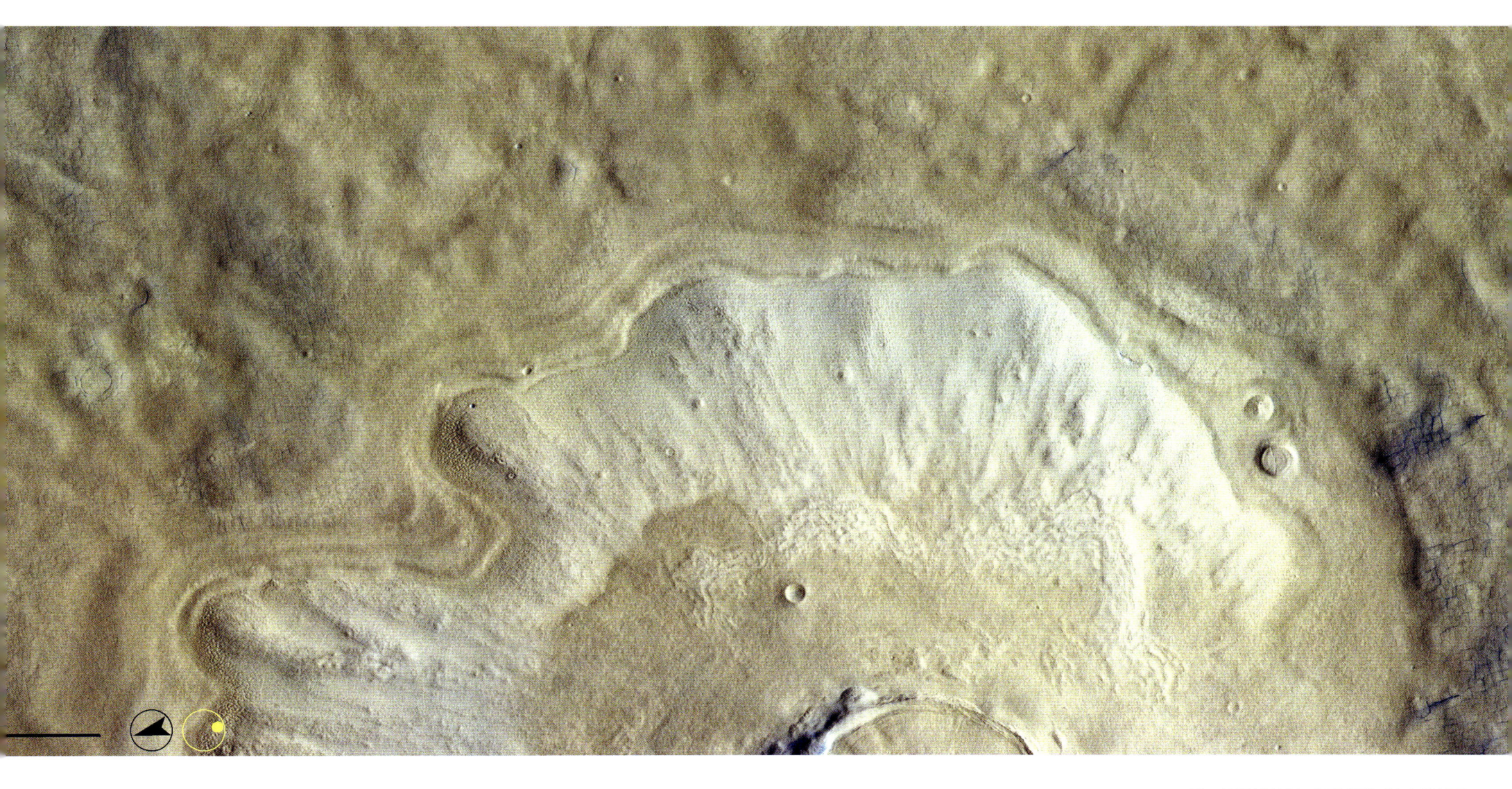

4.10 UNUSUAL EJECTA BLANKET

ACQUISITION TIME: 2021-02-05T08:27:22.838
LONGITUDE: 88.7 E   LATITUDE: 41.5 N
REGION: UTOPIA

SCHLAMMARTIGE AUSWURFDECKE BEI SUNGARI VALLIS

Dieses Bild zeigt die Auswurfdecke eines Kraters im Osten von Hellas Planitia. Zwei grosse Täler, Dao und Harmakhis Valles, münden in diesem Gebiet in das Hellas-Becken. Zwischen ihnen liegt das kleinere Sungari Vallis. Dieses Bild wurde in der Nähe dieses Tals aufgenommen. Die Talsysteme transportierten Wasser in das Hellas-Becken und enthalten nun schuttbedeckte Gletscher. Der Krater selbst ist mit schuttbedeckten Gletscherablagerungen gefüllt, und die Ejekta weist breite Risse auf, die durch den Verlust von Eis aus dem darunterliegenden Gletschermaterial verursacht worden sein könnten.

MUD-LIKE EJECTA BLANKET NEAR SUNGARI VALLIS

This image is of the ejecta blanket of a crater in east Hellas Planitia. Two large valleys, Dao and Harmakhis Valles, flow into Hellas Basin in this area. Between them is the smaller Sungari Vallis. This image was obtained close to this valley. The valley systems transported water into Hellas Basin and now contain debris covered glaciers. The crater itself is filled with debris covered glacial deposits and the ejecta has wide cracks which might be caused by loss of ice from the underlying glacial material.

COUVERTURE D'ÉJECTAS DE TYPE BOUEUX PRÈS DE SUNGARI VALLIS

Cette image montre la couverture d'éjectas d'un cratère dans l'est de Hellas Planitia. Dans cette zone, deux grandes vallées, Dao et Harmakhis, se jettent dans le bassin d'Hellas. Entre elles se trouve la plus petite vallée de Sungari près de laquelle cette image a été prise. Les systèmes de vallées ont transporté de l'eau dans le bassin d'Hellas et contiennent maintenant des glaciers couverts de débris rocheux. Le cratère lui-même est rempli de dépôts glaciaires couverts de débris et les éjectas présentent de larges fissures qui pourraient être causées par la perte de glace du matériau sous-jacent.

MANTO DI EJECTA SIMILE A FANGO VICINO A SUNGARI VALLIS

Questa immagine rappresenta una coltre di ejecta di un cratere nell'Hellas Planitia orientale. In questa zona due grandi valli, Dao e Harmakhis Valles, sfociano nel bacino di Hellas. Tra di loro c'è la più piccola Sungari Vallis. Questa immagine è stata ottenuta vicino a questa valle. I sistemi di valli hanno trasportato l'acqua nel bacino di Hellas e ora contengono ghiacciai coperti di detriti. Il cratere stesso è riempito da depositi glaciali coperti di detriti e l'ejecta presenta ampie fenditure che potrebbero essere causate dalla perdita di ghiaccio dal materiale glaciale sottostante.

ГРЯЗЕПОДОБНЫЙ ПОКРОВ ВЫБРОСОВ В РАЙОНЕ ДОЛИНЫ СУНГАРИ

На этом снимке показан покров выбросов (эжектитов) кратера на востоке равнины Эллада. В этом районе две большие долины, Дао и Хармахис, впадающие в бассейн Эллады. Между ними находится меньшая по размерам долина Сунгари, область снимка находится недалеко от нее. Эти системы долин переносили воду в бассейн Эллады, а теперь содержат покрытые обломками ледники. Сам кратер заполнен гляциальными отложениями, покрытыми разрушенной породой, а в покрове выбросов имеются широкие трещины, которые могут быть вызваны потерей льда в нижележащих ледниковых слоях.

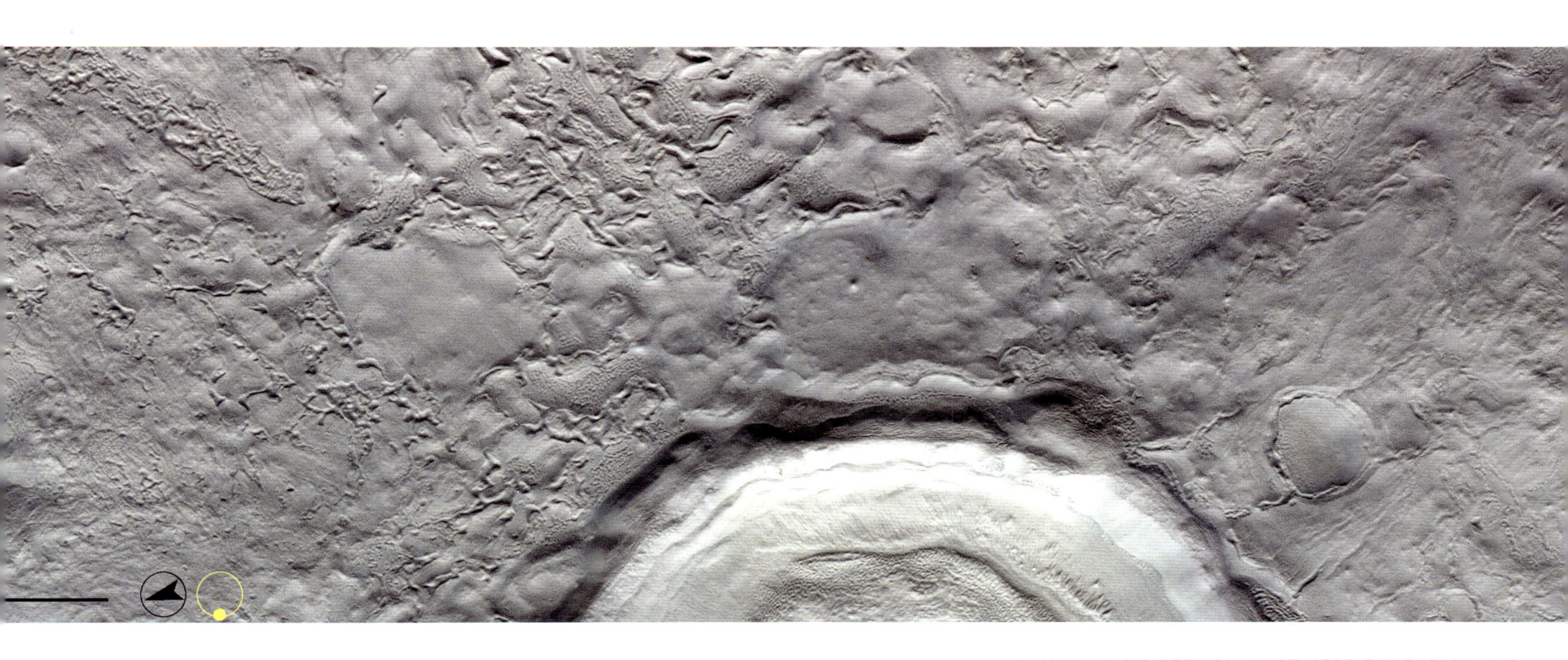

4.11 MUD-LIKE EJECTA BLANKET NEAR SUNGARI VALLIS

ACQUISITION TIME: 2021-01-09T19:16:08.463
LONGITUDE: 89.2 E LATITUDE: 41.1 S
REGION: HELLAS

DOPPELKRATER MIT INNEREM PERIGLAZIALEM GELÄNDE

Dieses Bild stammt aus dem nördlichsten Teil von Xanthe Dorsa nahe der Grenze zu Acidalia Planum und zeigt einen doppelten «Erdnuss»-Krater mit zahlreichen interessanten Merkmalen. Das Innere des Kraters ist von Rissen durchzogen, die typisch für schuttbedeckte Gletscheroberflächen sind. An den nach Süden gerichteten Kraterwänden gibt es einen kleinen Fleck mit dunklen Dünen und Rinnen sowie hell gefärbtes Grundgestein am Kraterrand.

DOUBLE CRATER WITH INTERNAL PERIGLACIAL TERRAIN

This image is from the northernmost part of Xanthe Dorsa close to the border with Acidalia Planum and shows a double "peanut" crater with numerous interesting features. The interior of the crater has aligned cracks in the surface which is typical of debris covered glacial surfaces. There is a small patch of dark dunes and gullies on the south-facing crater walls as well as bright coloured bedrock exposed on the crater rim.

DOUBLE CRATÈRE AU TERRAIN PÉRIGLACIAIRE INTERNE

Cette image provient de la partie la plus septentrionale de Xanthe Dorsa, près de la frontière avec Acidalia Planum, et montre un double cratère en forme de cacahuète avec de nombreuses caractéristiques intéressantes. L'intérieur du cratère présente des fissures alignées en surface, ce qui est typique des surfaces glaciaires recouvertes de débris. Il y a une petite zone de dunes sombres et de ravines sur les parois du cratère orientées vers le sud, ainsi qu'un substrat rocheux de couleur vive exposé sur le bord du cratère.

DOPPIO CRATERE CON TERRENO PERIGLACIALE ALL'INTERNO

Questa immagine proviene dalla parte più settentrionale di Xanthe Dorsa vicino al confine con Acidalia Planum e mostra un doppio cratere a forma di «arachide» con numerose caratteristiche interessanti. L'interno del cratere ha crepe allineate nella superficie, il che è tipico delle superfici glaciali coperte da detriti. C'è una piccola zona di dune scure e calanchi sulle pareti del cratere rivolte a sud, che contrasta con la roccia del pavimento roccioso di colore chiaro esposta sul bordo del cratere.

ДВОЙНОЙ КРАТЕР С ВНУТРЕННИМ ПЕРИГЛЯЦИАЛЬНЫМ РЕЛЬЕФОМ

На этом снимке, сделанном в самой северной части гряд Ксанфы, недалеко от границы с плато Ацидалия, изображен двойной «арахисовый» кратер с многочисленными интересными особенностями. Внутренняя часть кратера имеет согласованные трещины на поверхности, что типично для погребённых ледниковых поверхностей. На стенах кратера, обращенных к югу, есть небольшое пятно темных дюн и оврагов, а также яркие скальные породы, обнаженные на кромке кратера.

4.12 DOUBLE CRATER WITH INTERNAL PERIGLACIAL TERRAIN

ACQUISITION TIME: 2019-08-19T04:52:50.196
LONGITUDE: 326.4 E LATITUDE: 47.8 N
REGION: XANTHE

STREUUNG VON EJEKTA

Dieser frische Einschlagkrater in Terra Sirenum mit einem Durchmesser von etwas weniger als 1 km zeigt in diesem synthetischen RGB-Bild Strahlen aus dunklem Material, die sich nach aussen erstrecken und ältere Krater verdecken. In der Nähe des Kraters selbst ist die Ejektadecke recht asymmetrisch. Die frischen felsigen Schuttablagerungen zeigen sich als kontrastreiche helle und dunkle Streifen im Inneren des Kraters. Das Bild zeigt ein felsiges Gebiet in Terra Sirenum.

SPREAD OF EJECTA

This fresh impact crater in Terra Sirenum which is a bit less than 1 km in diameter shows rays of dark material in this synthetic RGB image, extending outwards and covering older craters. Close to the crater itself, the ejecta blanket is quite asymmetric. The fresh rocky talus deposits show up as contrasting light and dark streaks inside the crater. The image is of a rocky area in Terra Sirenum.

ÉTALEMENT DES ÉJECTAS

Ce cratère d'impact récent d'un peu moins de 1 km de diamètre est situé dans une zone rocheuse de la région de Terra Sirenum. Il a produit des éjectas radiaux plus sombres qui s'étendent vers l'extérieur et recouvrent des cratères plus anciens. Près du cratère lui-même, la couverture d'éjecta est assez irrégulière. Les dépôts d'éboulis rocheux frais apparaissent comme des bandes claires et sombres contrastées à l'intérieur du cratère.

MANTO DI EJECTA

In questa immagine composta RGB, un cratere da impatto fresco in Terra Sirenum, che ha un diametro di poco meno di 1 km, presenta dei raggi di materiale scuro che si estendono verso l'esterno e coprono i crateri più vecchi. Vicino al cratere stesso, la coltre di ejecta è piuttosto asimmetrica. I depositi freschi di ghiaia si delineano come striature contrastanti chiare e scure all'interno del cratere. L'immagine proviene da un'area rocciosa in Terra Sirenum.

РАСПРОСТРАНЕНИЕ ВЫБРОСОВ

На этом синтезированном RGB-изображении свежий ударный кратер, диаметром чуть менее 1 км, демонстрирует лучи темного материала, выходящие наружу и покрывающие более старые кратеры. Вблизи самого кратера покров выброса довольно асимметричен. Свежие скальные осыпи проявляются в виде контрастных светлых и темных полос внутри кратера. На снимке – скалистая область земли Сирен.

4.13 SPREAD OF EJECTA

ACQUISITION TIME: 2018-09-26T20:43:29.496
LONGITUDE: 187.3 E LATITUDE: 29.6 S
REGION: SIRENUM

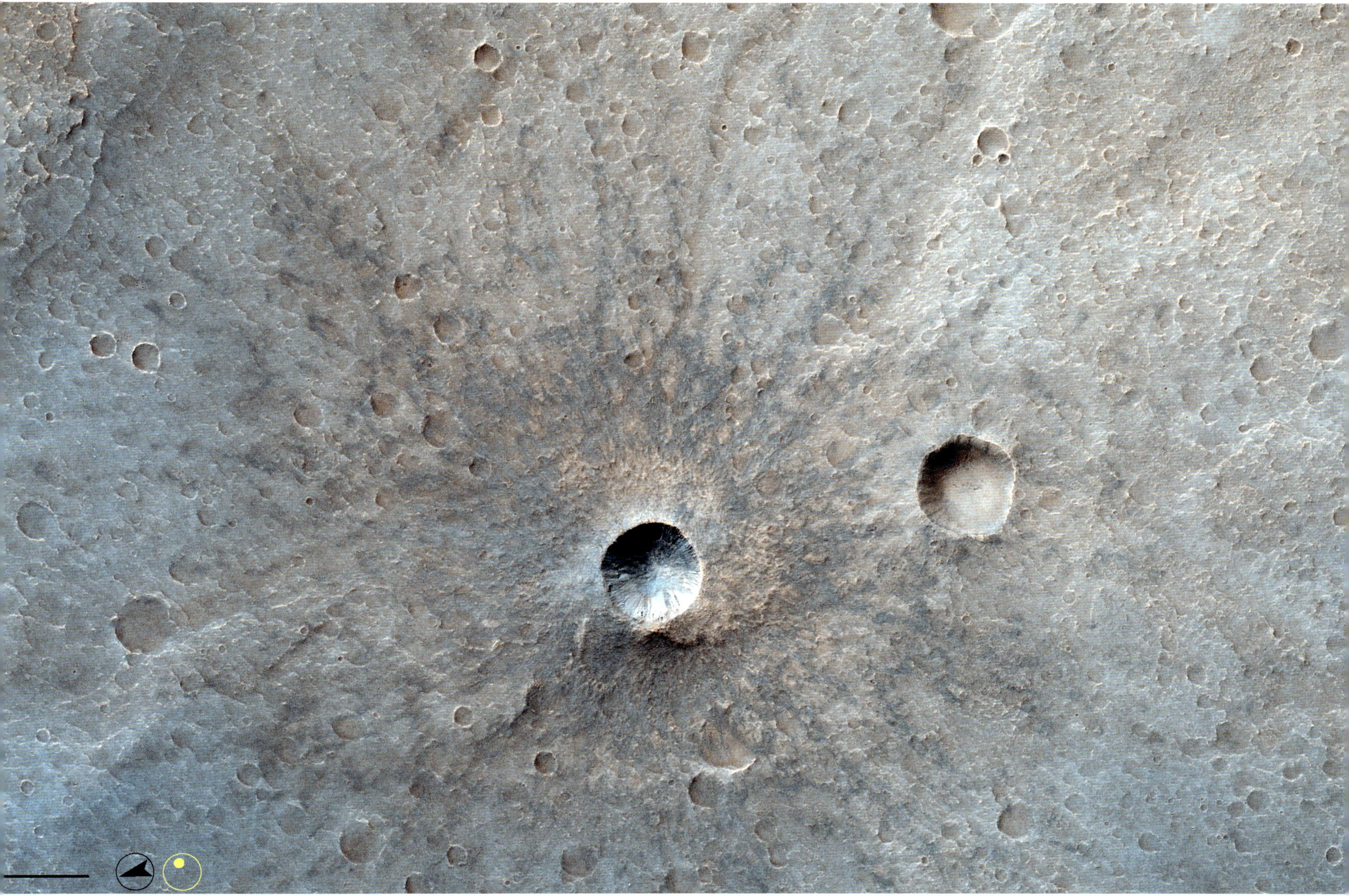

RINNEN, DIE VON EINEM BUNTEN KRATERRAND AUSGEHEN

Das Ziel dieser Beobachtung war die Untersuchung einer Reihe von Rinnen in Noachis Terra. Der Krater, in dem sie sich befinden, ist das Ergebnis eines Einschlags in den Rand eines älteren Kraters. Der Kraterboden weist markante gekrümmte Erhebungen auf, von denen eine gerade noch unten rechts am Rand des Sichtfelds zu sehen ist.

GULLIES EMANATING FROM A COLOURFUL CRATER RIM

The objective of this observation was to examine a set of gullies in Noachis Terra. The crater that they are in was the result of an impact into the rim of an older crater. The crater floor has prominent curved ridges one of which can just be seen bottom right at the edge of field of view.

RAVINES ÉMANANT D'UNE BORDURE DE CRATÈRE COLORÉE

L'objectif de cette observation était d'examiner un ensemble de ravines dans la région de Noachis Terra. Le cratère dans lequel elles se trouvent est le résultat d'un impact sur le bord d'un cratère plus ancien. Le fond du cratère présente de proéminentes crêtes incurvées dont l'une est visible en bas à droite, à la limite du champ de vue.

CALANCHI CHE SI ORIGINANO DA UN BORDO CRATERICO COLORATO

L'obiettivo di questa osservazione era quello di esaminare una serie di calanchi in Noachis Terra. Il cratere in cui si trovano è il risultato di un impatto sul bordo di un cratere più vecchio. Il pavimento del cratere ha prominenti creste curvate, una delle quali si può vedere a malapena in basso a destra al bordo del campo visivo.

ОВРАГИ НА КРАСОЧНОМ КРАТЕРНОМ ВАЛЕ

Целью данного наблюдения было изучение ряда оврагов в земле Ноя. Кратер, в котором они находятся, образовался в результате удара метеорита в обод более старого кратера. Его дно имеет заметные изогнутые гребни, один из которых виден внизу справа на самом краю поля зрения.

4.14 GULLIES EMANATING FROM A COLOURFUL CRATER RIM

ACQUISITION TIME: 2020-06-23T10:34:50.198
LONGITUDE: 16.4 E LATITUDE: 49.1 S
REGION: NOACHIS

RINNEN IN SISYPHI CAVI

In Sisyphi Planum gibt es ein Gebiet mit dem Namen Sisyphi Cavi, das zahlreiche tafelartige Strukturen aufweist, von denen aus sich Rinnen bilden. Hier ist ein hervorragendes Beispiel. Die Ränder der Tafelberge erodieren durch die Aktivität der Gullys zurück. In der Oberfläche des Materials, aus dem die Schutthalden der Tafelberge bestehen, sind lange, dünne Gullykanäle zu erkennen. Ein besonders gutes Beispiel befindet sich in der Nähe der Bildmitte. Dieses Bild wurde im Spätsommer aufgenommen, nachdem das saisonale CO_2-Eis sublimiert war.

GULLIES IN SISYPHI CAVI

There is an area within Sisyphi Planum called Sisyphi Cavi that has numerous mesa-like structures from which gullies extend. Here is an excellent example. The edges of the mesas are eroding back through gully activity. Long thin gully channels can be seen in the surface of the material comprising the talus slope of the mesas. There is a particularly good example close to the centre of the image. This image was taken in late-summer after the seasonal CO_2 ice has sublimed.

RAVINES DANS SISYPHI CAVI

Une zone de Sisyphi Planum, appelée Sisyphi Cavi, présente de nombreuses structures en forme de mésa d'où partent des ravines. En voici un excellent exemple. Les bords des mésas sont érodés par l'activité des ravines. On peut voir de longs et minces chenaux de ravinement à la surface du matériau qui constitue le talus des mésas près du centre de l'image. Cette image a été prise à la fin de l'été, après la sublimation de la glace saisonnière de CO_2.

CALANCHI IN SISYPHI CAVI

C'è un'area all'interno di Sisyphi Planum chiamata Sisyphi Cavi che ha numerose strutture simili a mesa da cui si estendono i calanchi. Questa immagine ne è un eccellente esempio. I bordi delle mesa si stanno erodendo a causa dell'attività dei calanchi. Si vedono lunghi e sottili canali di deflusso nel materiale superficiale che compone i ghiaioni sui pendii delle mesa. Un buon esempio di tale morfologia si trova vicino al centro dell'immagine. Questa immagine è stata scattata in tarda estate dopo la sublimazione di ghiaccio stagionale di CO_2.

ОВРАГИ КОТЛОВИН СИЗИФА

В пределах плато Сизифа есть область под названием котловины Сизифа с многочисленными столовыми горами, края которых эродированы оврагами. На изображении – отличный пример того, как она выглядит. Длинные тонкие русла оврагов видны на поверхности осыпных склонов, что особенно хорошо заметно вблизи центра снимка. Изображение было получено в конце лета, после того, как сублимировался сезонный лед CO_2.

4.15 GULLIES IN SISYPHI CAVI

ACQUISITION TIME: 2018-12-16T11:00:12.074
LONGITUDE: 358.5 E LATITUDE: 70.4 S
REGION: SISYPHI

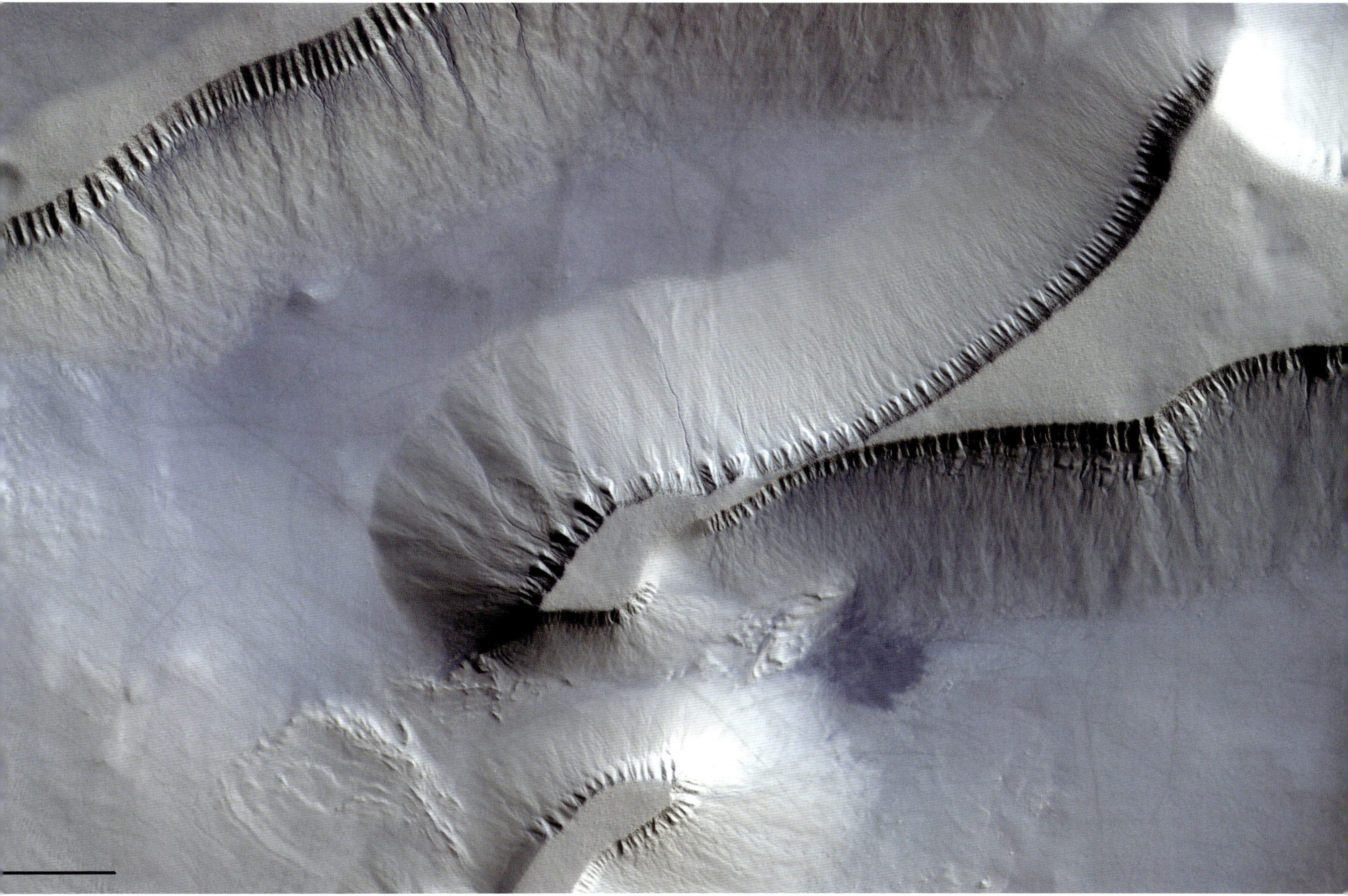

EINE FÜLLE VON VERSCHIEDENEN OBERFLÄCHENTYPEN

Die südliche Wand dieses Kraters (rechts) in der Nähe der Dichotomie-Grenze weist drei grosse Gruppen von Rinnen auf, die zur Ansammlung von Material an der Basis der Kraterwand beigetragen haben. Ein schuttbedeckter Gletscher erstreckt sich von den Rinnen hinunter zum Zentrum des Kraters, wo der Kraterboden durch ein rätselhaftes Bruchmuster strukturiert ist.

A PLETHORA OF DIFFERENT SURFACE TYPES

The southern wall of this crater (right) close to the dichotomy boundary has three large groups of gullies that have contributed to the build-up of material at the base of the crater wall. A debris covered glacier extends from the gullies down towards the centre of the crater where the crater floor is patterned by an enigmatic fracture pattern.

UNE PLÉTHORE DE DIFFÉRENTS TYPES DE SURFACE

La paroi sud de ce cratère (à droite), près de la limite de la dichotomie globale martienne, présente trois grands groupes de ravines qui ont contribué à l'accumulation de matériaux à la base de la paroi du cratère. Un glacier rocheux s'étend des ravines vers le centre du cratère dont le fond est modelé par un motif de fracturation à l'origine non encore élucidée.

UNA PLETORA DI DIVERSI TIPI DI SUPERFICIE

La parete meridionale di questo cratere (a destra) vicino al limite della dicotomia ha tre grandi gruppi di calanchi che hanno contribuito all'accumulo di materiale alla base della parete del cratere. Un ghiacciaio coperto di detriti si estende dai calanchi verso il centro del cratere, dove numerose fratture modellano il pavimento del cratere in uno disegno enigmatico.

МНОЖЕСТВО РАЗЛИЧНЫХ ТИПОВ ПОВЕРХНОСТЕЙ

На снимке изображён фрагмент южного вала и дна кратера, расположенного вблизи области Дихотомии. Выделяются три большие группы оврагов, выносящие обломочный материал к основанию стены. Погребённый под обломками ледник спускается от оврагов вниз, заполняя центр котловины, покрытый загадочным узором трещин.

4.16 A PLETHORA OF DIFFERENT SURFACE TYPES

ACQUISITION TIME: 2021-02-16T19:33:01.431
LONGITUDE: 354.0 E   LATITUDE: 37.3 N
REGION: ARABIA

RINNEN UND KRATERAUSWÜRFE

Dieses Bild zeigt eine ausgedehnte Reihe von Rinnen an der Wand dieses Kraters in Terra Sirenum, die über einer schuttbedeckten Gletscherablagerung liegen. Dieses Bild ist auch deshalb interessant, weil die radialen Rillen, die während des Einschlags in der Ejekta entstanden sind, noch nicht vollständig durch Oberflächenerosion oder Überdeckung durch andere Materialien verloren gegangen sind.

GULLIES AND CRATER EJECTA

This image shows an extensive series of gullies on the wall of this crater in Terra Sirenum, which are overlain on a debris covered glacial deposit. This is also interesting because of the radial grooves produced in the ejecta during the impact have not been completely lost by surface erosion or mantling by other materials.

RAVINES ET ÉJECTA DE CRATÈRE

Cette image montre une série étendue de ravines sur la paroi d'un cratère de Terra Sirenum, qui sont recouvertes d'un dépôt glaciaire lui-même couvert de débris rocheux. Cette image est également intéressante parce que les rainures radiales produites dans les éjectas pendant l'impact n'ont pas été complètement effacées par l'érosion de surface ou le recouvrement par d'autres matériaux.

CALANCHI ED EJECTA DI UN CRATERE

Questa immagine mostra un'estesa serie di calanchi sulla parete di questo cratere in Terra Sirenum, che sono sovrapposti a un deposito glaciale coperto di detriti. Questa immagine è interessante perché le scanalature radiali prodotte nell'ejecta durante l'impatto non sono state completamente perse dall'erosione superficiale o coperte da altri materiali.

ОВРАГИ И КРАТЕРНЫЕ ВЫБРОСЫ

На этом снимке показана обширная система оврагов, разрезающих внутреннюю стену вала кратера в земле Сирен. Материал, выносимый оврагами, наползает на покрытый обломками пород ледник, заполняющий центральную котловину. Это изображение интересно также тем, что радиальные борозды, образовавшиеся при метеоритном ударе, сохранились, не подвергшись воздействию эрозии или перекрытию другими материалами.

4.17 GULLIES AND CRATER EJECTA

ACQUISITION TIME: 2021-01-06T08:44:01.280
LONGITUDE: 225.6 E   LATITUDE: 40.6 S
REGION: SIRENUM

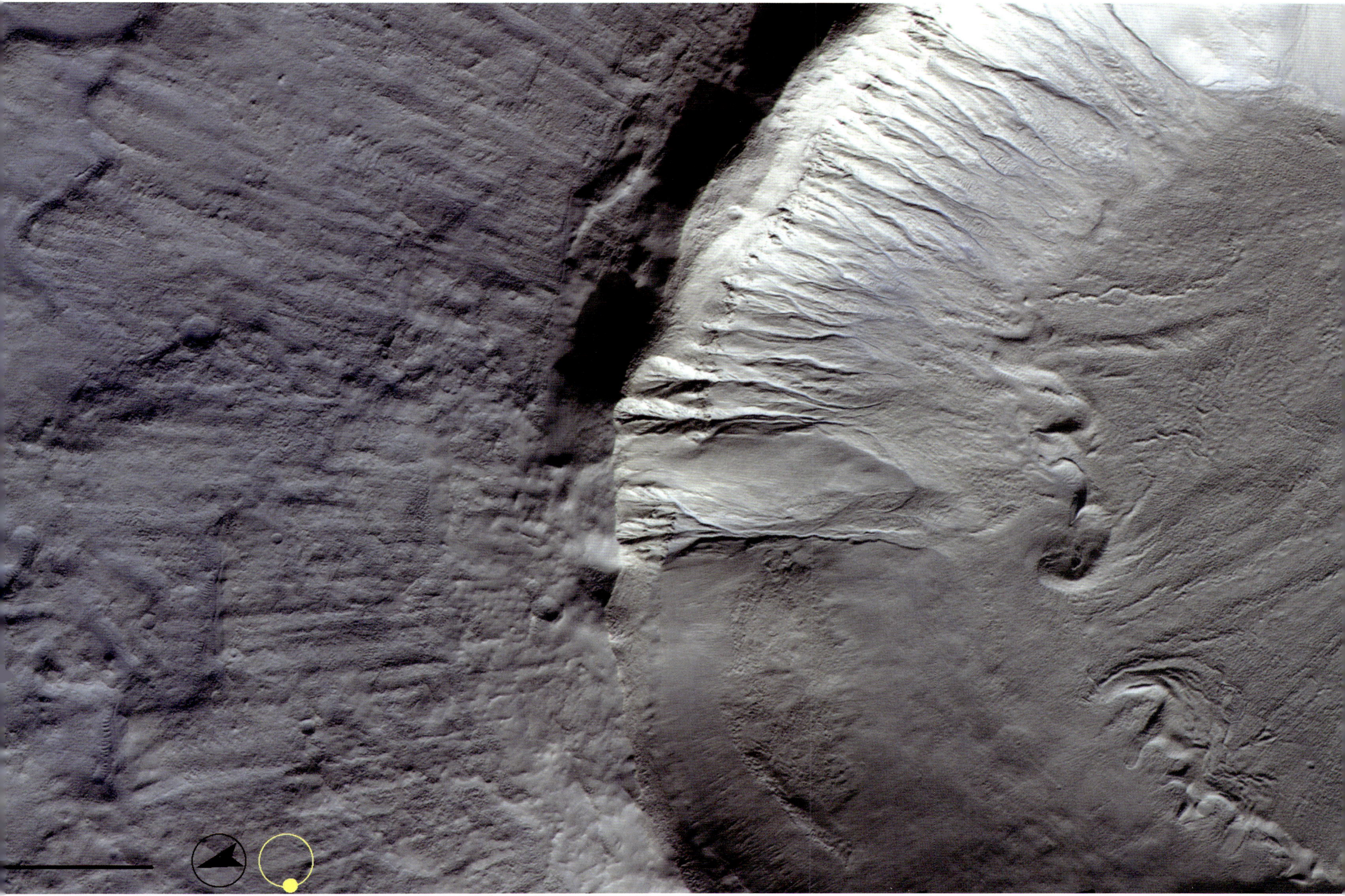

RINNEN UND DÜNEN

Dieses synthetische RGB-Bild zeigt Rinnen in einer Kraterwand südwestlich des Rabe-Kraters in Noachis Terra. Viele der Krater in diesem Gebiet enthalten Dünenfelder, wie das hier dunkelblau dargestellte. Das Material aus den Rinnen im Norden (links) interagiert mit dem Dünenmaterial an der Basis der Kraterwand.

GULLIES AND DUNES

This synthetic RGB image is of gullies in a crater wall southwest of Rabe Crater in Noachis Terra. Many of the craters in this area contain dune fields like the one seen in dark blue here. The material from the gullies to the north (left) interacts with the dune material at the base of the crater wall.

RAVINES ET DUNES

Cette image montre des ravines dans le mur d'un cratère au sud-ouest du cratère Rabe dans la région de Noachis Terra. Plusieurs des cratères de cette région contiennent des champs de dunes comme celui que l'on voit en bleu foncé ici. Les débris provenant des ravines au nord (à gauche) interagissent avec le sable des dunes à la base du mur du cratère.

CALANCHI E DUNE

Questa immagine composta nei filtri RGB raffigura dei calanchi in una parete del cratere a sud-ovest del cratere Rabe in Noachis Terra. Molti dei crateri in questa zona contengono campi di dune come quello in blu scuro qui. Il materiale dei canaloni a nord (a sinistra) interagisce con il materiale delle dune alla base della parete del cratere.

ОВРАГИ И ДЮНЫ

Это синтезированное RGB-изображение оврагов в стене кратера к юго-западу от кратера Рабе на земле Ноя. Многие кратеры в этой области содержат дюнные поля, подобные тому, что выделяется на снимке темно-синим цветом. Обломочный материал, поступающий из оврагов (слева) смешивается с материалом дюн у основания стены кратера.

4.19 GULLIES AND DUNES

ACQUISITION TIME: 2018-12-13T14:32:23.948
LONGITUDE: 33.8 E   LATITUDE: 45.7 S
REGION: NOACHIS

HELLE GULLY-SCHUTTFÄCHER

Dies ist ein Bild der nach Norden gerichteten Wand eines Kraters in der Vastitas-Borealis-Formation in der nördlichen Hemisphäre (direkt nördlich von Tempe Terra und dem Perepelkin-Krater). Der Grund, warum die Schuttfächer der Rinnen hell sind, ist nicht klar, aber es könnte daran liegen, dass vom Wind verwehter Staub an der rauen Oberfläche der Ablagerung hängen bleibt. Frischere Ablagerungen sind heller.

BRIGHT GULLY DEBRIS FANS

This is an image of the north-facing wall of a crater in the Vastitas Borealis Formation in the northern hemisphere (just north of Tempe Terra and Perepelkin Crater). The reason why the debris fans of the gullies are bright is not clear but it could be because of wind-blown dust being trapped on the rough surface of the deposit. Fresher deposits are brighter.

CÔNES DE DÉBRIS CLAIRS DE RAVINES

Image de la paroi nord d'un cratère de la formation de Vastitas Borealis dans l'hémisphère nord martien, juste au nord de Tempe Terra et du cratère Perepelkin. La raison pour laquelle les cônes de débris des ravines sont clairs n'est pas évidente mais cela pourrait être dû à la poussière soufflée par le vent qui est piégée sur la surface rugueuse du dépôt. Les dépôts les plus frais sont les plus clairs.

DEPOSITI DI DETRITI PRODOTTI DA CALANCHI CHIARI

Questa è un'immagine della parete di un cratere rivolta a nord nella Formazione Vastitas Borealis nell'emisfero settentrionale (appena a nord di Tempe Terra e del cratere Perepelkin). La ragione per cui i conoidi di deiezione dei calanchi sono chiari non è nota, ma potrebbe essere a causa della polvere trasportata dal vento che viene intrappolata sulla superficie ruvida del deposito. I depositi più recenti sono più chiari.

ЯРКИЕ КОНУСЫ ВЫНОСА ОВРАГОВ

Это изображение южной внутренней стены кратера на Великой Северной равнине (чуть севернее земли Темпе и кратера Перепёлкин). Возможная причина, по которой конусы выноса оврагов выделяются более светлым тоном, состоит в том, что пыль, переносимая ветром, лучше задерживается на шероховатой поверхности свежих осадочных отложений. Чем моложе отложения, тем они ярче.

4.20 BRIGHT GULLY DEBRIS FANS

ACQUISITION TIME: 2021-05-18T14:58:02.831
LONGITUDE: 293.6 E LATITUDE: 55.6 N
REGION: VBF

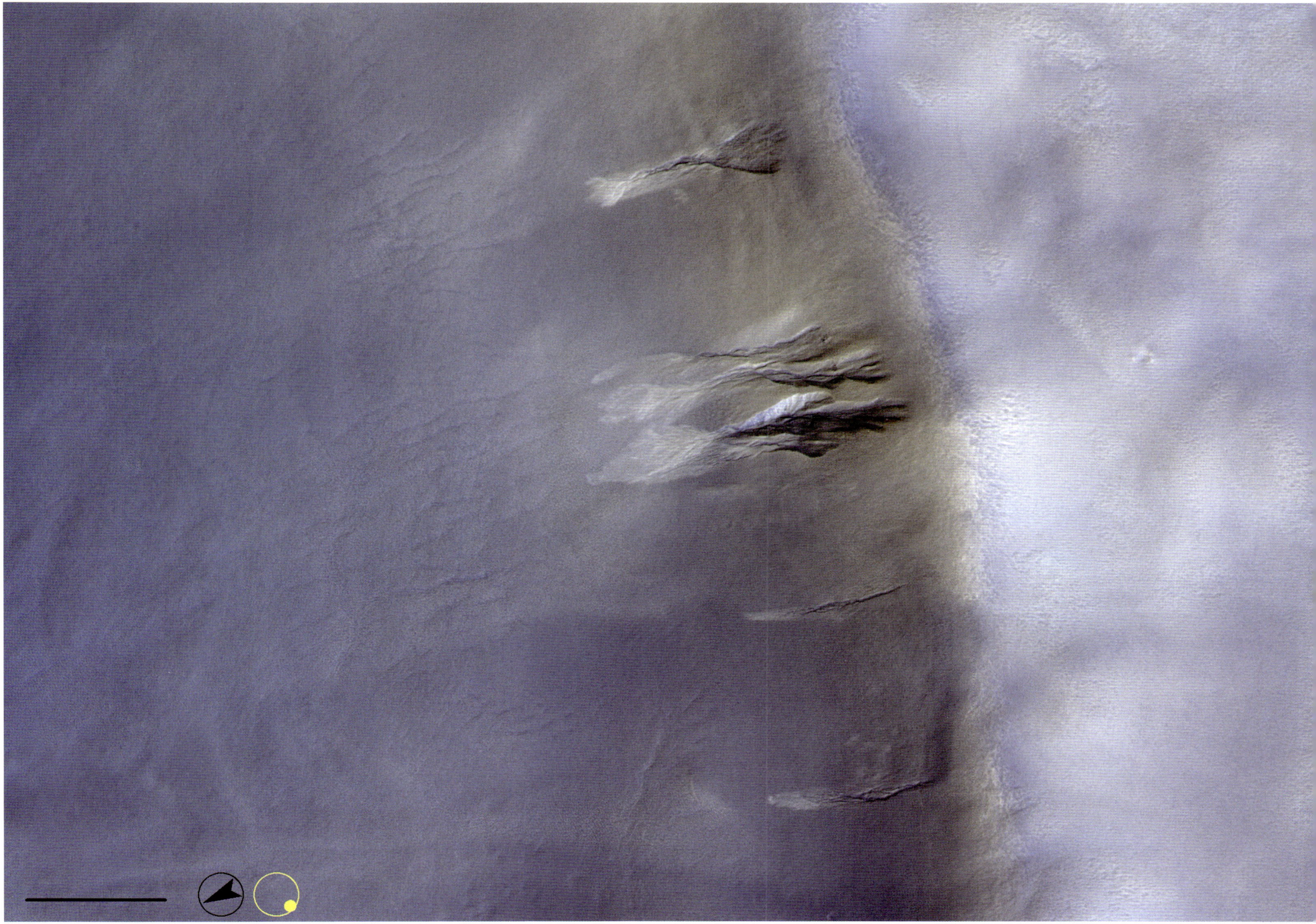

TRÜMMERFÄCHER UNTER EINER RINNE

Dies ist ein Bild der südlichen Wand eines unregelmässigen Kraters in der Nähe des südwestlichen Randes des Ptolemaeus-Kraters in Terra Sirenum. Andere Bilder von CaSSIS und HiRISE zeigen, dass die Wand dieses Kraters viele Rinnen enthält, aber dieses Bild ist besonders interessant wegen des Trümmerfächers, der an der Basis einer Rinne unten in der Mitte deutlich zu sehen ist.

DEBRIS FAN BENEATH A GULLY

This is an image of the southern wall of an irregular crater close to the southwest rim of Ptolemaeus Crater in Terra Sirenum. Other images from CaSSIS and HiRISE show that the wall of this crater contains many gullies but this particular image is interesting because of the debris fan brought out in relief at the base of a gully located bottom-centre.

CÔNE DE DÉBRIS SOUS UNE RAVINE

Image du mur sud d'un cratère irrégulier près de la bordure sud-ouest du cratère Ptolemaeus dans la région de Terra Sirenum. D'autres images de CaSSIS et HiRISE montrent que la paroi de ce cratère contient de nombreuses ravines, mais cette image particulière est intéressante en raison du cône de débris en relief à la base d'une ravine située en bas au centre.

DEPOSITO DI DETRITI SOTTO UN CALANCO

Questa è un'immagine della parete meridionale di un cratere irregolare vicino al bordo sud-ovest del cratere Ptolemaeus in Terra Sirenum. Altre immagini di CaSSIS e HiRISE mostrano che la parete di questo cratere contiene molti calanchi, ma questa particolare immagine è interessante per il conoide di deiezione messo in rilievo alla base di un calanco situato in basso al centro.

КОНУС ВЫНОСА ПОД ОВРАГОМ

Это изображение южной стены кратера неправильной формы, расположенного недалеко от юго-западного края кратера Птолемея в земле Сирен. Стена кратера, как это видно здесь и на многих других снимках CaSSIS и HiRISE, изрезана множеством радиальных оврагов. Данное изображение особенно интересно тем, что оно позволяет различить конус выноса одного из наиболее крупных оврагов (внизу слева).

4.21 DEBRIS FAN BENEATH A GULLY

ACQUISITION TIME: 2020-12-13T14:22:24.053
LONGITUDE: 200.5 E   LATITUDE: 46.1 S
REGION: SIRENUM

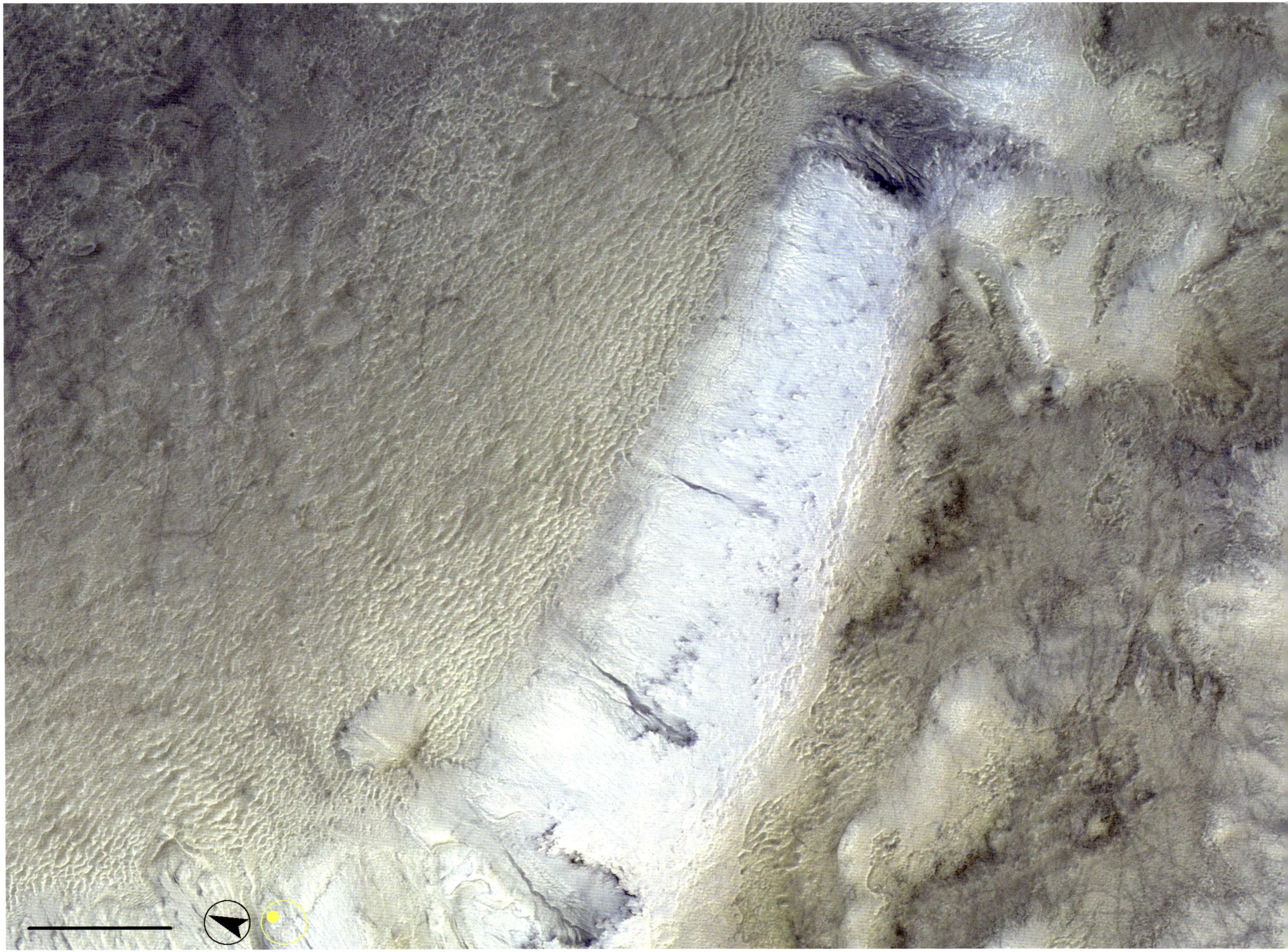

5 ERDHÜGEL, CHAOS, TAFELBERGE UND KONTAKTE

MOUNDS, CHAOS, MESAS,
AND CONTACTS

MONTICULES, CHAOS, MÉSAS
ET CONTACTS

TUMULI, CHAOS, MESAS
E ZONE DI CONTATTO

КУРГАНЫ, ХАОСЫ, СТОЛОВЫЕ ГОРЫ
И КОНТАКТЫ РЕЛЬЕФОВ

CHAOTISCHES TERRAIN

Dieses Bild zeigt Gorgonum Chaos in der Region Terra Cimmeria. Chaotische Geländeformen auf dem Mars sind durch unregelmässig angeordnete Tafelberge charakterisiert, die mehrere Kilometer breit und gelegentlich Hunderte von Metern hoch sind. Die oberen Flächen sind im Wesentlichen flach und weisen einige lokale Strukturen auf. In den Mulden dieser chaotischen Geländeformen wird erodiertes Material von den Flanken der Tafelberge aufgefangen. Man geht davon aus, dass sie durch gigantische Schmelzwasserfreisetzungen unter der Erdoberfläche entstehen, die zu einem Einsturz des Bodens führen. Es gibt mehrere ähnliche Beispiele für chaotische Geländeformen auf dem Mars.

CHAOTIC TERRAIN

This image features Gorgonum Chaos in Terra Cimmeria region. Chaotic terrains on Mars are characterised by irregularly organized mesas several kilometres across and occasionally hundreds of metres high. The upper most faces are essentially flat topped with some local structure. The troughs in these chaotic terrains are seen to trap eroded material from the flanks of the mesas. It is thought to form by gigantic releases of subsurface meltwater, which causes ground collapse. There are several similar examples of chaotic terrains on Mars.

TERRAIN CHAOTIQUE

Cette image montre Gorgonum Chaos dans la région de Terra Cimmeria. Les terrains chaotiques sur Mars sont caractérisés par des mésas organisées de façon irrégulière, de plusieurs kilomètres de diamètre et parfois de centaines de mètres de hauteur. Les faces supérieures sont essentiellement plates avec quelques structures locales. Les creux dans ces terrains chaotiques sont vus comme des pièges à matériaux érodés provenant des flancs des mésas. On pense qu'ils se sont formés par de gigantesques rejets d'eau de fonte souterraine, qui ont provoqué l'effondrement du sol. Il existe plusieurs exemples similaires de terrains chaotiques sur Mars.

TERRENO CAOTICO

Questa immagine mostra il Gorgonum Chaos nella regione di Terra Cimmeria. I terreni caotici su Marte sono caratterizzati da mesa organizzate in modo irregolare e larghe diversi chilometri e occasionalmente alte centinaia di metri. Le superfici più alte sono essenzialmente piatte con qualche struttura locale. I solchi in questi terreni caotici intrappolano il materiale eroso dai fianchi delle mesa. Si pensa che i terreni caotici si formino da giganteschi rilasci di acqua di fusione sotterranea, che causano il collasso del terreno. Ci sono diversi esempi simili a questo su Marte.

ХАОТИЧНЫЙ РЕЛЬЕФ

На этом снимке изображен хаос Горгоны на Киммерийской земле. Хаотичные рельефы на Марсе характеризуются нерегулярно организованными столовыми горами, имеющими несколько километров в поперечнике и иногда сотни метров в высоту. Самые верхние их грани, в основном, плоские, с локальной структурой. Желоба хаотичного рельефа (между возвышенностями), как видно на снимке, задерживают эродированный материал со склонов. Считается, что данные впадины образуются в результате гигантских выбросов подповерхностной талой воды, приводящих к обрушению грунта. На Марсе есть несколько подобных примеров хаотичного рельефа.

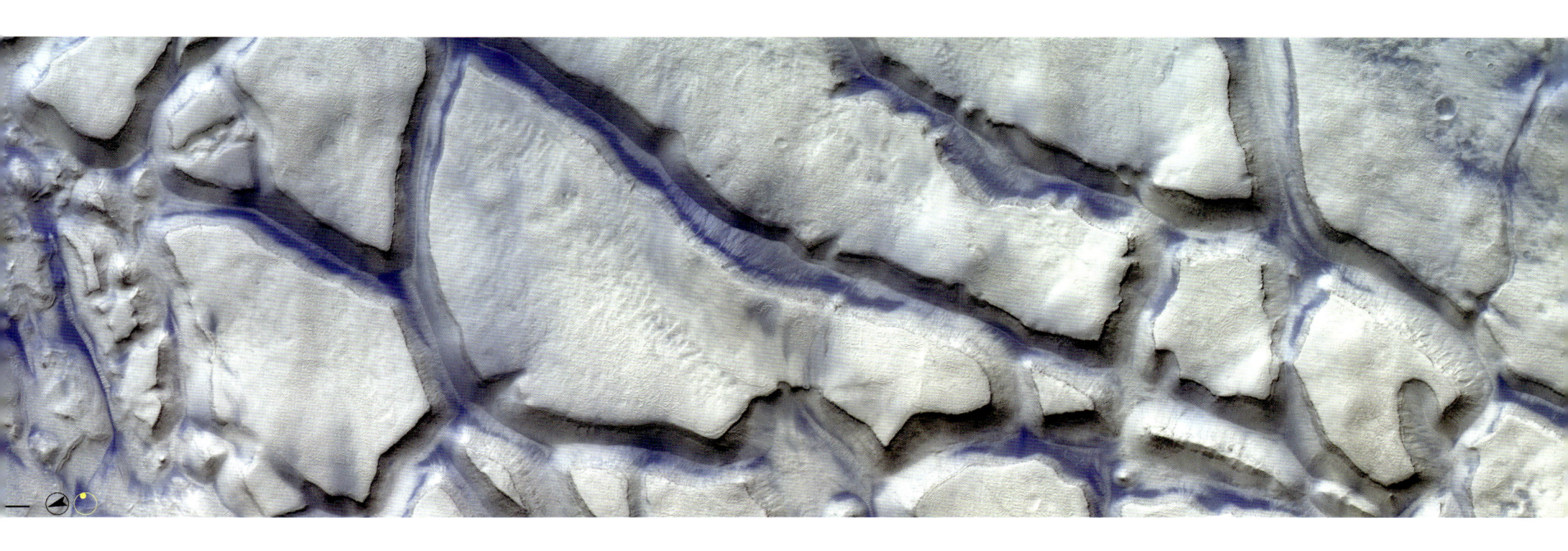

5.1 CHAOTIC TERRAIN

ACQUISITION TIME: 2018-10-01T22:27:25.711
LONGITUDE: 189.6 E   LATITUDE: 37.3 S
REGION: CIMMERIA

MINERALISCHE VIELFALT IN COPRATES CATENA

Coprates Catena besteht aus einer Reihe von Gruben und Mulden, die fast parallel zur Schlucht von Coprates Chasma verlaufen. Auf diesem Bild sehen wir die südliche Kante einer grossen Mulde in Coprates Catena. Dieses Gebiet ist ungewöhnlich wegen des grünlich gefärbten Materials, das den Rand und die Ebenen zwischen den Mulden bedeckt. Normalerweise werden farbige Ablagerungen auf den Canyonböden und nicht auf den Ebenen zwischen den Mulden beobachtet, weswegen ihr Ursprung unklar bleibt.

MINERAL DIVERSITY IN COPRATES CATENA

Coprates Catena comprises a set of pits and troughs that run almost parallel to the Coprates Chasma canyon. In this image, we are looking at the southern scarp of a large trough in Coprates Catena. This area is unusual because of the greenish-coloured material that caps the rim and the inter-trough plains. Usually coloured deposits are observed on the canyon floors and not on the inter-trough plains, therefore their origin remains unclear.

DIVERSITÉ MINÉRALE DANS COPRATES CATENA

Coprates Catena comprend un ensemble de fosses et de cuvettes qui sont presque parallèles au canyon principal de Coprates Chasma. Sur cette image, nous observons l'escarpement sud d'une grande cuvette de Coprates Catena. Cette zone est inhabituelle en raison du matériau de couleur verdâtre qui recouvre le rebord et les plaines inter-cuvettes. Habituellement, les dépôts colorés sont observés sur le fond des canyons et non sur les plaines inter-cuvettes, leur origine reste donc incertaine.

VARIETÀ MINERALI IN COPRATES CATENA

Il Coprates Catena comprende un insieme di fosse e solchi che corrono quasi parallelamente al canyon del Coprates Chasma. In questa immagine si può ammirare la scarpata meridionale di una grande depressione del Coprates Catena. Questa zona è inusuale a causa del materiale di colore verdastro che ricopre il bordo e le pianure tra i solchi. Di solito i depositi colorati si osservano sui pavimenti dei canyon e non sulle pianure inter-trogali, quindi la loro origine rimane poco chiara.

РАЗНООБРАЗИЕ ПРОЯВЛЕНИЙ МИНЕРАЛОВ В ЦЕПОЧКЕ КОПРАТ

Цепочка Копрат представляет собой набор ям и желобов, которые идут почти параллельно каньону Копрат. На этом снимке мы смотрим на южный склон большой впадины в цепочке Копрат. Эта область необычна из-за зеленоватого цвета материала, покрывающего край желоба и межложбинные равнины. Обычно цветные отложения наблюдаются на дне каньона, а не на равнинах между впадинами, поэтому в данном случае их происхождение остается неясным.

5.2 MINERAL DIVERSITY IN COPRATES CATENA

ACQUISITION TIME: 2021-04-16T11:25:38.105
LONGITUDE: 301.1 E LATITUDE: 15.0 S
REGION: VALLES MARINERIS

ARAM CHAOS

Dies ist ein Bild des nördlichen Teils von Aram Chaos. Der Boden von Aram Chaos enthält riesige Blöcke aus eingestürztem Material (Bildmitte), die als chaotisches Gelände bekannt sind. Der Einsturz scheint erfolgt zu sein, als Wasser oder Eis auf katastrophale Weise aus dem Untergrund entfernt wurde. Hell getönte Ablagerungen (z. B. ein kleiner Fleck auf der linken Seite) weisen auf eine mineralogische Vielfalt hin.

ARAM CHAOS

This is an image of the northern part of Aram Chaos. The floor of Aram Chaos contains huge blocks of collapsed material (image centre), which are known as chaotic terrains. The collapse appears to have occurred when water or ice was catastrophically removed from the sub-surface. Light-toned deposits (there is a small patch to the left, for example) indicate mineralogical diversity.

ARAM CHAOS

Image de la partie nord d'Aram Chaos. Le sol d'Aram Chaos contient d'énormes blocs de matériaux effondrés (centre de l'image), qui sont connus comme des terrains chaotiques. L'effondrement semble s'être produit lorsque l'eau ou la glace s'est retirée de manière catastrophique du sous-sol. Les dépôts de couleur claire (par exemple la petite unité à gauche) indiquent une diversité minéralogique.

ARAM CHAOS

Questa è un'immagine della parte settentrionale di Aram Chaos. Il pavimento di Aram Chaos contiene enormi blocchi di materiale collassato (immagine centrale), che sono chiamati terreni caotici. Il collasso sembra essere avvenuto quando l'acqua o il ghiaccio sono stati catastroficamente rimossi dal sottosuolo. I depositi di colore chiaro (c'è una piccola macchia a sinistra, per esempio) indicano una diversità mineralogica.

ХАОС АРАМ

Это изображение северной части хаоса Арам. На его поверхности находятся огромные блоки разрушенного материала (в центре изображения), известные как хаотичные рельефы. Обрушение, по-видимому, произошло, когда вода или лед резко ушли из подповерхностного слоя. Отложения светлых тонов (например, небольшой участок слева) свидетельствуют о минеральном разнообразии.

5.3   ARAM CHAOS

ACQUISITION TIME: 2021-01-28T11:34:16.433
LONGITUDE: 338.5 E   LATITUDE: 2.9 N
REGION: MARGARITIFER

TAFELBERG IN SYRTIS

Dies ist ein Bild des chaotischen Geländes an der Ostseite einer Mulde in den Nili Fossae. Die Region Nili Fossae ist für ihre mineralogische Vielfalt bekannt, aber es gibt keine hochauflösenden spektroskopischen Beobachtungen dieser speziellen Region des chaotischen Geländes. Die Tafelberge hier (oben in der Mitte) scheinen geschichtet zu sein, obwohl die Schichten nicht die starke Farbvielfalt aufweisen, die wir normalerweise mit unterschiedlichen mineralischen Lithologien in CaSSIS-Bildern assoziieren.

MESA IN SYRTIS

This is an image of chaotic terrain on the east side of a trough in Nili Fossae. The Nili Fossae region is well known for its mineralogical diversity but there are no high resolution spectroscopic observations of this particular region of chaotic terrain. The mesas here (centre top) appear to be layered although the layers do not show the strong colour diversity that we normally associate with differing mineral lithologies in CaSSIS images.

MÉSA DANS LA RÉGION DE SYRTIS

Image de terrain chaotique sur le côté est d'une dépression de Nili Fossae. La région de Nili Fossae est bien connue pour sa diversité minéralogique mais il n'y a pas d'observations spectroscopiques à haute résolution de cette région particulière. Les mésas ici (au centre en haut) semblent être stratifiées. Ces dernières ne montrent cependant pas la forte diversité de couleurs que nous associons normalement aux différentes lithologies et variabilités minérales dans les images CaSSIS.

MESA IN SYRTIS

Questa è un'immagine del terreno caotico sul lato est di una depressione nelle Nili Fossae. La regione delle Nili Fossae è ben nota per la sua diversità mineralogica ma non ci sono osservazioni spettroscopiche ad alta risoluzione di questa particolare regione caotica. Le mesa qui (in alto al centro) sembrano essere stratificate anche se gli strati non mostrano la forte diversità di colore che normalmente associamo alle diverse litologie nelle immagini di CaSSIS.

СТОЛОВАЯ ГОРА В СИРТЕ

Это изображение хаотичного рельефа на восточной стороне желоба в бороздах Нила. Данный регион известен своим минеральным разнообразием, но пока ещё нет спектроскопических наблюдений высокого разрешения данного участка с хаотичным рельефом. Столовые горы здесь (в центре вверху) кажутся слоистыми, хотя слои не демонстрируют заметного цветового разнообразия, которое на снимках CaSSIS мы обычно связываем с вариациями минерального состава пород.

5.4 **MESA IN SYRTIS**

ACQUISITION TIME: 2018-04-29T01:53:40.000
LONGITUDE: 77.2 E LATITUDE: 24.5 N
REGION: SYRTIS

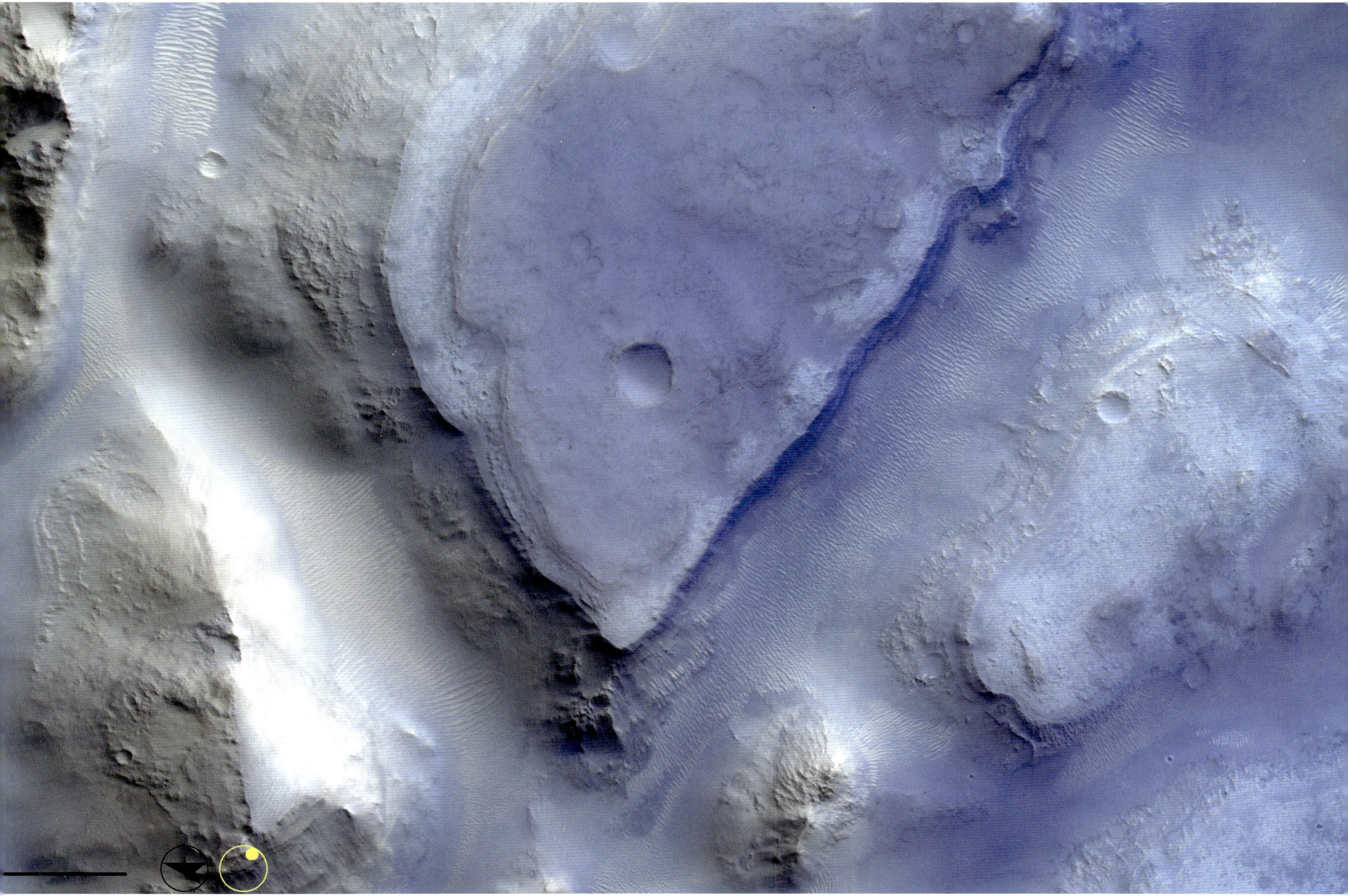

ERDHÜGEL IN GANGES CHASMA

Dies ist ein Bild von Hügeln an der Grenze zwischen Ganges Chasma und Aurorae Chaos. Es gibt erstaunlich wenig hochauflösende Daten aus diesem Gebiet. Die CaSSIS-Farbprodukte zeigen jedoch eine erhebliche Mineralienvielfalt in den Hügeln und Tafelbergen des chaotischen Geländes (Mitte und rechts). Links oben befindet sich ein mit hellbraunem Sand gefüllter Graben, der in das Ganges Chasma führt. Beachten Sie die Farbveränderungen auf jeder Seite des Grabens (ganz oben links), die zeigen, dass die lithologische Vielfalt nicht auf die Hügel in dieser Region beschränkt ist.

MOUNDS IN GANGES CHASMA

This is an image of mounds on the border between Ganges Chasma and Aurorae Chaos. There is surprisingly little high resolution data of this area. However, CaSSIS colour products show significant mineral diversity in the mounds and mesas of the chaotic terrain (centre and right). To the upper left, there is a graben filled with tan coloured sand that leads into Ganges Chasma. Note the changes in colour on each side of the graben (extreme upper left) indicating that lithological diversity is not restricted to the mounds in this region.

MONTICULES DANS GANGES CHASMA

Image de monticules à la frontière entre Ganges Chasma et Aurorae Chaos. Il y a étonnamment peu de données à haute résolution de cette zone. Les images couleur de CaSSIS montrent cependant une diversité minérale importante dans les monticules et les mésas du terrain chaotique (au centre et à droite). En haut à gauche, il y a un graben rempli de sable de couleur beige qui mène à Ganges Chasma. Notez les changements de couleur de chaque côté du graben (coin supérieur gauche) qui indiquent que la diversité lithologique ne se limite pas aux monticules dans cette région.

TUMULI NEL GANGES CHASMA

Questa è un'immagine di alcuni tumuli al confine tra Ganges Chasma e Aurorae Chaos. Ci sono sorprendentemente pochi dati ad alta risoluzione di quest'area. Tuttavia, i prodotti a colori di CaSSIS mostrano una significativa diversità di minerali nei tumuli e nelle mesa del terreno caotico (centro e destra). In alto a sinistra, c'è un graben pieno di sabbia di colore marrone che conduce al Ganges Chasma. Si notino le sfumature di colore su ogni lato del graben (in alto a sinistra) che indicano che la diversità litologica non è limitata ai tumuli in questa regione.

КУРГАНЫ В КАНЬОНЕ ГАНГ

Это изображение холмов, расположенных на границе между каньоном Ганг и хаосом Авроры. Снимков высокого разрешения этой территории на удивление мало. Однако данные CaSSIS показывают значительное разнообразие горных пород в холмах и столовых горах видимого хаотического рельефа (в центре и справа). Слева вверху находится грабен, заполненный песком бледно-коричневого цвета, ведущий в каньон Ганг. Обратите внимание на изменения в цвете на обеих сторонах грабена (крайний верхний левый угол), указывающие на то, что разнообразие горных пород в этом регионе не ограничивается холмами.

5.5   MOUNDS IN GANGES CHASMA

ACQUISITION TIME: 2018-12-20T23:31:29.275
LONGITUDE: 320.1 E   LATITUDE: 8.6 S
REGION: MARINERIS

GESCHICHTETE TAFELBERGE IN DEN HELLESPONTUS MONTES

Das Hauptziel dieser Aufnahme der Hellespontus Montes (westlich des Hellas-Beckens) war die Untersuchung von Barchan-Dünen, von denen einige unten rechts zu sehen sind. Diese Dünen bilden lange barchanoide Kämme, die in dieser Region durch unidirektionale Winde geformt wurden. Aber auch die geschichteten Tafelberge sind wegen des freigelegten hellen Materials interessant. Auch auf der linken Seite dieses Bildes ist eine subtile Farbvielfalt zu erkennen.

LAYERED MESAS IN HELLESPONTUS MONTES

The primary target of this image from Hellespontus Montes (west of Hellas Basin) was to study barchan dunes, some of which can be seen at the bottom right. These dunes form long barchanoid ridges that have been shaped by unidirectional winds in this region. However, the layered mesas are also interesting because of the exposed bright material. Some subtle colour diversity can be also seen at the left side of this image.

MÉSAS STRATIFIÉES D'HELLESPONTUS MONTES

Les cibles principales de cette image d'Hellespontus Montes (à l'ouest du bassin d'Hellas) étaient les dunes barchanoïdes, dont certaines sont visibles en bas à droite. Ces dunes forment de longues crêtes qui ont été façonnées par les vents unidirectionnels dans cette région. Cependant, les mésas stratifiées sont également intéressantes en raison des roches claires exposées. Une subtile diversité de couleurs est également visible sur le côté gauche de cette image.

MESA STRATIFICATE IN HELLESPONTUS MONTES

L'obiettivo primario di questa immagine di Hellespontus Montes (a ovest di Hellas Basin) era quello di studiare le dune barcanoidi, alcune delle quali possono essere viste in basso a destra. Queste dune formano lunghe creste barcanoidi che sono state modellate dai venti unidirezionali in questa regione. Tuttavia, le mesa stratificate sono interessanti anche per il materiale chiaro esposto. Si può vedere una sottile diversità di colore anche sul lato sinistro di questa immagine.

СЛОИСТЫЕ СТОЛОВЫЕ ГОРЫ В ГЕЛЛЕСПОНТСКИХ ГОРАХ

Основной целью этого снимка Геллеспонтских гор (к западу от бассейна Эллады) было изучение барханов, некоторые из которых видны внизу справа. Эти барханы образуют длинные гряды, которые сформировались здесь под воздействием ветров одного направления. Однако, видимые на этом снимке слоистые столовые горы также интересны благодаря обнажениям яркого материала. В левой части этого изображения также можно увидеть едва уловимые цветовые вариации.

5.6   LAYERED MESAS IN HELLESPONTUS MONTES

ACQUISITION TIME: 2021-04-12T03:25:26.879
LONGITUDE: 44.5 E   LATITUDE: 41.2 S
REGION: HELLESPONTUS

FARBENFROHE EROSIONSHÜGEL IN TERRA CIMMERIA

Dieses Bild zeigt ein kleines unbenanntes Gebiet in Terra Cimmeria, das dafür bekannt ist, chaotisches Terrain zu enthalten. Die Strukturen hier unterscheiden sich jedoch stark von denen, die man beispielsweise in Aram Chaos findet. Die Oberfläche scheint stark erodiert zu sein und zeigt helles, farbiges Material. Der Krater auf der rechten Seite zeigt freiliegendes helleres Grundgestein am Kraterrand und an der westlichen (unteren) Kraterwand.

COLOURFUL ERODED MOUNDS IN TERRA CIMMERIA

This image is of a small unnamed area in Terra Cimmeria that is known for containing chaotic terrain. However, the structures here are very different from those found in, for example, Aram Chaos. The surface appears to have been heavily eroded, revealing bright, colourful material. The crater to the right shows exposed brighter bedrock in the rim and the west (lower) wall.

MONTICULES ÉRODÉS COLORÉS DANS TERRA CIMMERIA

Cette image représente une petite région sans nom au sein de Terra Cimmeria, connue pour son terrain chaotique. Les structures visibles ici sont cependant très différentes de celles trouvées, par exemple, dans Aram Chaos. La surface semble avoir été fortement érodée, révélant des unités claires et colorées. Le cratère de droite montre un substrat rocheux plus clair sur le rebord et la paroi ouest (inférieure).

TUMULI EROSI COLORATI IN TERRA CIMMERIA

Questa immagine proviene da una piccola area senza nome in Terra Cimmeria che è nota per contenere una regione caotica. Tuttavia, le strutture qui sono molto diverse da quelle trovate, per esempio, in Aram Chaos. La superficie sembra essere stata pesantemente erosa, rivelando materiale chiaro e colorato. Il cratere a destra mostra una roccia più chiara esposta nel bordo e nella parete ovest (inferiore).

РАЗНОЦВЕТНЫЕ ЭРОДИРОВАННЫЕ ОСЫПИ В КИММЕРИЙСКОЙ ЗЕМЛЕ

Это изображение небольшой безымянной области в Киммерийской земле, которая известна своими хаотическими рельефами. Однако структуры здесь сильно отличаются от тех, что можно найти, например, в хаосе Арама. Поверхность, похоже, подверглась сильной эрозии, обнажив яркий, красочный материал. В ободе и западной (нижней) стене кратера, видимого справа, можно заметить обнажения более ярких коренных пород.

5.7 COLOURFUL ERODED MOUNDS IN TERRA CIMMERIA

ACQUISITION TIME: 2018-05-07T00:45:53.000
LONGITUDE: 164.0 E   LATITUDE: 36.9 S
REGION: CIMMERIA

ERDHÜGEL IN AUREUM CHAOS

Dieses Bild zeigt subtile Farbveränderungen in heller getönten Aufschlüssen in Aureum Chaos. Es gibt Hinweise auf wässrige Ablagerungen von hämatit- und sulfatreichem Material in dieser Region, was sich in CaSSIS-Bildern als heller gefärbtes Material zeigen würde. Die geringe Anzahl von Kratern deutet darauf hin, dass die Oberfläche hier erst relativ spät in der Marsgeschichte (spätes Hesperium) verändert wurde, obwohl Untersuchungen darauf hindeuten, dass wässrige Mineralien bereits vor dieser Oberflächenveränderung entstanden sind. Interessant sind auch die äolischen Kräuselungen, vor allem oben rechts, wo es wabenförmige Muster gibt.

MOUNDS IN AUREUM CHAOS

This image shows subtle colour changes in lighter-toned outcrops in Aureum Chaos. Evidence for aqueous deposition of hematite and sulphate-rich material in this region has been presented and this would show up in CaSSIS images as light-toned material. The low number of craters suggests that the surface here has been modified relatively late in Martian history (late Hesperian) although research suggests that aqueous minerals were produced before this re-surfacing. The aeolian ripples are also interesting here – particularly to the upper right where there are honeycomb patterns.

MONTICULES D'AUREUM CHAOS

Cette image montre des changements de couleurs subtils dans les affleurements d'Aureum Chaos. Des preuves de dépôts d'hématite et de sulfates ont été recueillies dans cette région. Les roches contenant ces minéraux apparaissent sur les images CaSSIS comme des formations de couleur claire. Le faible nombre de cratères suggère que la surface de cette région a été modifiée relativement tard dans l'histoire martienne (fin de l'Hespérien), bien que les recherches suggèrent que des minéraux aqueux ont été produits avant ce re-surfaçage. Les ondulations éoliennes sont également intéressantes ici – particulièrement en haut à droite où l'on trouve des motifs en nid d'abeille.

TUMULI NELL' AUREUM CHAOS

Questa immagine mostra deboli sfumature di colore negli affioramenti di colore più chiaro nell'Aureum Chaos. Ci sono prove di deposizione acquosa di ematite e di materiale ricco di solfato in questa regione, che si mostrerebbe nelle immagini CaSSIS come materiale dai toni chiari. Il basso numero di crateri suggerisce che la superficie è stata modificata relativamente tardi nella storia marziana (tardo Esperiano) anche se gli studi suggeriscono che i minerali acquosi sono stati prodotti prima di questa riemersione. Le increspature eoliche sono interessanti anche qui – in particolare in alto a destra dove ci sono strutture a nido d'ape.

КУРГАНЫ В ХАОСЕ ЗОЛОТОЙ РОГ

На этом снимке показаны тонкие цветовые вариации в светлых тонах обнажения в хаосе Золотой Рог. Для этого региона были представлены доказательства водного осаждения гематита и сульфат-содержащего материала, которые должны проявляться на снимках CaSSIS как участки светлых тонов. Малое количество кратеров говорит о том, что поверхность здесь была изменена относительно поздно в марсианской истории (поздний гесперийский период), хотя исследования показывают, что водные минералы сформировались до этого. Также здесь интересна эоловая рябь – особенно справа вверху, где есть узоры в виде сот.

5.8   MOUNDS IN AUREUM CHAOS

ACQUISITION TIME: 2021-01-14T05:35:29.695
LONGITUDE: 335.1 E   LATITUDE: 5.4 S
REGION: MARGARITIFER

ERDHÜGEL IN DER NÄHE DES LYOT-KRATERS

Der Lyot-Krater ist ein relativ junger (frühamazonischer) Krater mit 221 km Durchmesser am südwestlichen Rand der Vastitas-Borealis-Formation. Der Boden des Kraters besteht aus Hügeln, die gelegentlich zerklüftet sind und eine subtile Farbvielfalt aufweisen, wie auf diesem Bild zu sehen ist. Wir können hier auch einen mäandrierenden Kanal sehen, der den Kraterboden durchschneidet (von links nach rechts) und Teil eines grösseren fluvialen Systems in dieser Region ist. Es wird vermutet, dass das Schmelzen des oberflächennahen Eises und der Wasserabfluss eine wichtige Rolle bei der Gestaltung dieser fluvialen Merkmale gespielt haben.

MOUNDS NEAR LYOT CRATER

Lyot Crater is a relatively young (early Amazonian) 221 km diameter crater on the south-western edge of the Vastitas Borealis Formation. The floor of the crater contains mounds that are occasionally fractured and with subtle colour diversity as shown in this image. We can also see a meandering channel that dissects the crater floor here (left to right) that is part of a larger fluvial system in this region. It has been suggested that melting of near-surface ice and water runoff played a significant role in shaping these fluvial features.

MONTICULES PRÈS DU CRATÈRE LYOT

Le cratère Lyot est un cratère relativement jeune (début de l'Amazonien) de 221 km de diamètre sur le bord sud-ouest de Vastitas Borealis. Le plancher du cratère contient des monticules occasionnellement fracturés et présentent une subtile diversité de couleurs comme le montre cette image. On peut également voir ici (de gauche à droite) un chenal sinueux qui traverse le fond du cratère et qui fait partie d'un système fluvial plus large dans cette région. Il a été suggéré que la fonte de glace proche de la surface et le ruissellement de l'eau ont joué un rôle important dans la formation de ces caractéristiques fluviales.

TUMULI VICINO AL CRATERE LYOT

Il cratere Lyot è un cratere relativamente giovane (primo periodo Amazzonico) di 221 km di diametro sul bordo sud-occidentale della Formazione Vastitas Borealis. Il pavimento del cratere contiene tumuli che sono occasionalmente fratturati e con una leggera diversità di colore come mostrato in questa immagine. Possiamo anche vedere un canale serpeggiante che taglia il pavimento del cratere (da sinistra a destra) che fa parte di un sistema fluviale più grande in questa regione. È stato suggerito che lo scioglimento del ghiaccio vicino alla superficie e il deflusso dell'acqua abbiano giocato un ruolo significativo nel modellare queste conformazioni fluviali.

КУРГАНЫ ВОЗЛЕ КРАТЕРА ЛИО

Кратер Лио – относительно молодой кратер (ранний Амазонийский период) диаметром 221 км на юго-западном краю Великой Северной равнины. На дне кратера видны курганы, некоторые из которых разрушены и имеют едва уловимую цветовую неоднородность. Также можно видеть извилистый канал, прорезающий дно кратера (слева направо), который является частью более крупной флювиальной системы этого региона. Предполагается, что таяние приповерхностного льда и водный сток сыграли значительную роль в формировании этих флювиальных структур.

5.9 MOUNDS NEAR LYOT CRATER

ACQUISITION TIME: 2021-03-02T23:35:07.176
LONGITUDE: 30.6 E LATITUDE: 49.3 N
REGION: VASTITAS BOREALIS

6 WOLKEN, EIS, POLYGONE UND MUSCHELN

CLOUDS, ICE, POLYGONS
AND SCALLOPS

NUAGES, GLACE, POLYGONES
ET PÉTONCLES

NUVOLE, GHIACCIO, POLIGONI
E SCALLOPS

ОБЛАКА, ЛЕД, ПОЛИГОНЫ И ФЕСТОНЫ

WOLKENFRONT IM SÜDLICHEN AONIA TERRA

Die Viking-Orbiter entdeckten vor fast 50 Jahren Wolken in der Marsatmosphäre. Wolken auf dem Mars können aus Wasser oder aus CO2 bestehen. Viele Wolkenstrukturen sind mit der Topografie verbunden (z. B. auf der Leeseite der grossen Vulkane in Tharsis). Sie unterscheiden sich von atmosphärischen Trübungen, die durch Staubstürme entstehen. Wolken sind häufig in den Polarregionen zu sehen. Dieses Bild wurde in hohen südlichen Breitengraden in der Nähe des Charlier-Kraters im südlichen Frühling aufgenommen (Ls = 217,3°) und zeigt eine ungewöhnlich scharfe Kante der Wolkenstruktur.

CLOUD FRONT IN SOUTHERN AONIA TERRA

The Viking orbiters detected clouds in the atmosphere of Mars nearly 50 years ago. Clouds on Mars can consist of water or of CO_2. Many cloud structures are associated with topography (e.g. on the leeward side of the large volcanoes in Tharsis). They are distinct from atmospheric opacity produced by dust storms. Clouds are often seen in polar regions. This image was taken at high southern latitude near Charlier Crater in southern spring (Ls = 217.3°) and shows an unusually sharp edge to the cloud structure.

FRONT DE NUAGES DANS LE SUD D'AONIA TERRA

Les sondes Viking ont détecté des nuages dans l'atmosphère de Mars il y a près de 50 ans. Les nuages sur cette planète peuvent être constitués d'eau ou de CO_2. De nombreux nuages se forment grâce à la topographie (par exemple, sur le bord des grands volcans de Tharsis). On les distingue facilement des tempêtes de poussière. Les nuages sont aussi souvent observés dans les régions polaires. Cette image a été prise à une latitude méridionale élevée, près du cratère Charlier, au printemps austral (Ls = 217,3°) et montre un bord inhabituellement net de la structure nuageuse.

FRONTE NUVOLOSO IN AONIA TERRA MERIDIONALE

Gli orbiter Viking hanno rilevato delle nuvole nell'atmosfera di Marte quasi 50 anni fa. Le nuvole su Marte possono essere costituite da acqua o da CO_2. Molte strutture nuvolose sono associate alla topografia (ad esempio sul lato sottovento dei grandi vulcani di Tharsis). Sono differenti dall'opacità atmosferica tipica prodotta dalle tempeste di polvere. Le nuvole si vedono spesso nelle regioni polari. Questa immagine è stata scattata ad alta latitudine meridionale vicino al cratere Charlier in primavera (Ls = 217,3°) e mostra un confine insolitamente netto della struttura della nube.

ОБЛАЧНЫЙ ФРОНТ НА ЮГЕ ЗЕМЛИ АОНИД

Почти 50 лет назад во время орбитальных миссий «Викинг» 1 и 2 были обнаружены облака в атмосфере Марса. Они могут состоять из воды или углекислого газа (CO_2). Форма многих облачных структур связана с топографией местности, как например, на подветренной стороне крупных вулканов провинции Фарсида. Подобные структуры отличаются от непрозрачности атмосферы, вызываемой пылевыми бурями. Облака на Марсе часто видны в полярных регионах. Данное изображение было получено южной весной на высокой южной широте недалеко от кратера Шарлье (217.3°) и демонстрирует необычный для структуры облака резкий край.

6.1 **CLOUD FRONT IN SOUTHERN AONIA TERRA**

ACQUISITION TIME: 2020-06-10T15:54:50.556
LONGITUDE: 195.30 E   LATITUDE: 66.7 S
REGION: AONIA

WOLKEN IN MITTLEREN BREITENGRADEN IN ACIDALIA

Dieses Bild aus dem nördlichen Flachland zu Beginn des nördlichen Frühlings zeigt zarte Wolken in grosser Höhe, die wahrscheinlich aus Kohlendioxid bestehen, das während der Nacht kondensiert ist. Das Bild wurde kurz vor 08.00 Uhr morgens Ortszeit aufgenommen. Bei genauem Hinsehen erkennt man, dass sich die Farbe der Wolken verändert. Dies könnte darauf zurückzuführen sein, dass sich die Wolken während der Aufnahme bewegten oder dass sie sich so hoch über der Oberfläche befinden, dass das Timing der Kamera ungenau geworden ist. (Die Bildaufnahme wird in der Regel für Oberflächenaufnahmen vorberechnet, nicht für Aufnahmen in grosser Höhe.)

MID-LATITUDE CLOUDS IN ACIDALIA

This image from the northern plains in early northern spring shows a beautiful set of wispy high altitude clouds probably composed of carbon dioxide that condensed during the night. The image was taken just before 08:00 in the morning local time. Close inspection will give the impression that there are changes in colour around the clouds. This could be the result of the clouds moving during imaging or because they are so high above the surface that the timing of camera has become inaccurate. (The image acquisition is ususally pre-calculated for surface imaging, not high altitude imaging.)

NUAGES DE MOYENNES LATITUDES DANS ACIDALIA

Cette image prise dans les plaines Nord au début du printemps montre un bel ensemble de nuages de haute altitude, probablement composés de dioxyde de carbone condensé durant la nuit. L'image a été prise le matin avant 08:00, heure locale. En regardant précisement l'image, on pourrait penser qu'il y a des changements de couleur autour des nuages. Cela peut être dû au fait que les nuages se sont déplacés pendant la prise d'image ou parce qu'ils sont si hauts que la synchronisation de la caméra est devenue imprécise. L'acquisition de l'image est généralement pré-calculée pour l'imagerie de surface, pas pour l'imagerie à haute altitude.

NUBI DELLE MEDIE LATITUDINI IN ACIDALIA

Questa immagine dalle pianure settentrionali all'inizio della primavera mostra una bella serie di nubi vaporose ad alta quota, probabilmente composte da anidride carbonica che si è condensata durante la notte. L'immagine è stata scattata poco prima delle 08:00 del mattino ora locale. Da un esame ravvicinato si ha l'impressione che ci siano dei cambiamenti di colore intorno alle nuvole. Questo potrebbe essere il risultato del movimento delle nuvole durante l'acquisizione dell'immagine o perché sono così in alto sopra la superficie che la misurazione nel tempo della fotocamera è diventata imprecisa (l'acquisizione dell'immagine è di solito pre-calcolata per le immagini di superficie, non per quelle ad alta quota).

ОБЛАКА В СРЕДНИХ ШИРОТАХ НА РАВНИНЕ АЦИДАЛИЯ

Этот снимок северных равнин, сделанный ранней северной весной, демонстрирует прекрасный набор тонких высотных облаков, вероятно, состоящих из углекислого газа, сконденсировавшегося в течение ночи. Снимок сделан незадолго до 08:00 утра по местному времени. При внимательном рассмотрении создаётся впечатление, что цвет вокруг облаков варьируется. Это может быть вызвано движением облаков во время съёмки, или фактом их нахождения настолько высоко над поверхностью, что временные настройки камеры стали неточными. (Получение изображения обычно предварительно просчитывается для визуализации поверхности, а не для съёмки на большой высоте).

6.2   MID-LATITUDE CLOUDS IN ACIDALIA

ACQUISITION TIME: 2021-02-26T23:16:03.961
LONGITUDE: 7.8 E   LATITUDE: 44.3 N
REGION: ACIDALIA

WOLKEN ÜBER DEN SÜDPOLAREN SCHICHTABLAGERUNGEN

Dieses Bild zeigt Wolken über den südpolaren Schichtablagerungen (SPLD) in der mittleren Phase des südlichen Frühlings. Das Bild wurde in der Nähe von Thyles Rupes aufgenommen. Die saisonale Kohlendioxid-Polkappe ist hier bereits sublimiert und hinterlässt die kahle Oberfläche der SPLD. Ein unregelmässiger Wolkenkomplex treibt über die Schichten.

CLOUDS OVER THE SOUTH POLAR LAYERED DEPOSITS

This image shows clouds over the South Polar Layered Deposits (SPLD) in the middle of southern spring. The image was obtained near Thyles Rupes. The seasonal carbon dioxide polar cap has already sublimed here leaving the bare surface of the SPLD. An irregular set of clouds is drifting over the layers.

NUAGES AU-DESSUS DES DÉPÔTS STRATIFIÉS DU PÔLE SUD

Cette image, prise au milieu du printemps austral, montre des nuages au-dessus des dépôts stratifiés du pôle Sud (SPLD). L'image a été obtenue près de Thyles Rupes. La calotte polaire saisonnière de dioxyde de carbone s'est déjà sublimée ici, laissant à nu la surface du SPLD. On peut voir un ensemble irrégulier de nuages dériver au-dessus des couches.

NUBI SOPRA I DEPOSITI STRATIFICATI DEL POLO SUD

Questa immagine mostra delle nuvole sopra i depositi stratificati del Polo Sud (SPLD) nel mezzo della primavera meridionale. L'immagine è stata ottenuta vicino a Thyles Rupes. Qui la calotta polare stagionale di anidride carbonica è già sublimata lasciando la superficie nuda del SPLD. Un insieme irregolare di nuvole sta andando alla deriva sopra i depositi stratificati.

ОБЛАКА НАД ЮЖНЫМИ ПОЛЯРНЫМИ СЛОИСТЫМИ ОТЛОЖЕНИЯМИ

На этом изображении показаны облака над слоистыми отложениями Южного полюса (SPLD) в середине южной весны. Изображение было получено недалеко от уступа Тайлс. Сезонная полярная шапка углекислого льда здесь уже сублимировалась, обнажив поверхность SPLD. Над слоями дрейфует набор облаков несимметричной формы.

6.3   CLOUDS OVER THE SOUTH POLAR LAYERED DEPOSITS

ACQUISITION TIME: 2018-08-21T02:29:35.000
LONGITUDE: 133.9 E   LATITUDE: 72.8 S
REGION: PROMETHEI

WOLKEN ÜBER DEM OLYMPUS MONS

Die grossen Vulkane des Mars führen zur Bildung von orografischen Wolken. Diese werden von CaSSIS nur selten beobachtet (das Sichtfeld ist normalerweise zu klein, um zur richtigen Zeit am richtigen Ort zu sein). Hier haben wir jedoch ein bemerkenswertes Beispiel für die Caldera des Olympus Mons (links), die in unregelmässige Wolken gehüllt ist. Die Wolken sind relativ hoch und bewegen sich. Daher kommt es zu einer gewissen Farbfehlregistrierung. Der Effekt ist aber dennoch recht spektakulär.

CLOUDS OVER OLYMPUS MONS

The large volcanoes of Mars lead to the production of orographic clouds. CaSSIS rarely observes these (the field of view is usually too small to be in the right place at the right time). However, here we have a remarkable example of the caldera of Olympus Mons (to the left) blanketed in irregular clouds. The clouds are relatively high and moving. Therefore, there is some colour misregistration. However, the effect is still quite spectacular.

NUAGES AU-DESSUS D'OLYMPUS MONS

Les grands volcans de Mars causent la formation de nuages dus à leur topographie. CaSSIS observe rarement les nuages (le champ de vue est généralement trop petit pour se trouver au bon endroit au bon moment). Cependant, nous avons ici un exemple remarquable de la caldeira d'Olympus Mons (à gauche) recouverte de nuages irréguliers. Les nuages sont relativement hauts et en mouvement. Par conséquent, on peut observer certains défauts d'acquisition des couleurs. L'effet reste cependant assez spectaculaire.

NUBI SOPRA L'OLYMPUS MONS

I grandi vulcani di Marte causano la produzione di nubi orografiche. CaSSIS le osserva raramente (il campo visivo è di solito troppo piccolo per essere nel posto giusto al momento giusto). Tuttavia, qui abbiamo un notevole esempio della caldera di Olympus Mons (a sinistra) coperta da nubi irregolari. Le nuvole sono relativamente alte e in movimento, pertanto la misurazione dei colori è alterata. Tuttavia, l'effetto è comunque piuttosto spettacolare.

ОБЛАКА НАД ГОРОЙ ОЛИМП

Крупные вулканы Марса приводят к образованию орографической облачности. CaSSIS редко её наблюдает (его поле зрения обычно слишком мало, чтобы оказаться в нужном месте в нужное время). Однако, здесь мы имеем замечательный пример кальдеры (котловины вулкана) горы Олимп (слева), покрытой нерегулярными облаками. Данные облака относительно высокие и находятся в движении. Следовательно, наблюдается некоторое, несовпадение цветов. Однако эффект всё равно остается впечатляющим.

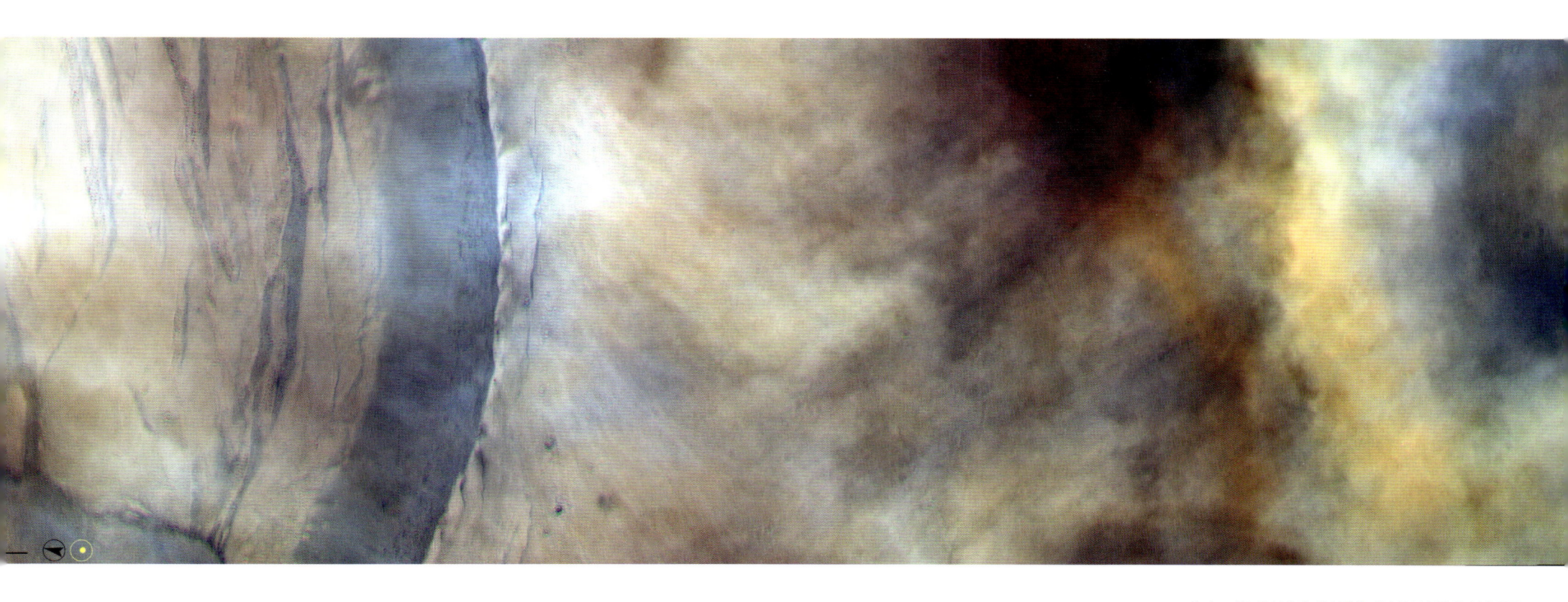

6.4   CLOUDS OVER OLYMPUS MONS

ACQUISITION TIME: 2021-03-30T08:32:23.122
LONGITUDE: 226.9 E   LATITUDE: 18.1 N
REGION: THARSIS

WELLENWOLKEN

Dieses Bild wurde auf einem hohen nördlichen Breitengrad zu Beginn des nördlichen Frühlings aufgenommen. Es zeigt regelmässige Wellenwolken über der Oberfläche. Wellenwolken bilden sich, wenn stabile Luft über erhöhtes Land strömt und eine atmosphärische Schwerkraftwelle erzeugt wird. Bei solchen Wellen bilden sich Wolken in der Region, in welcher Luft aufsteigt, während die Region, in welcher Luft sinkt, einen klaren Himmel begünstigt. Sie sind auf der Erde weit verbreitet. Auf dem Mars werden die Wolken wahrscheinlich durch Kohlendioxid-Eispartikel erzeugt und befinden sich vermutlich in relativ grossen Höhen (20–50 km).

WAVE CLOUDS

This image was acquired at high northern latitude just at the start of northern spring. It shows regular wave clouds over the surface. Wave clouds form when stable air flows over a raised land feature and an atmospheric gravity wave is created. In such waves, clouds form in the upward moving region, whereas the sinking region is favourable for clear skies. They are common on Earth. On Mars, the clouds are probably produced by carbon dioxide ice particles and are likely to be at relatively high altitudes (20–50 km).

NUAGES ONDULÉS

Cette image a été acquise à haute latitude dans l'hémisphère nord Martien, juste au début du printemps boréal. Elle montre des nuages ondulant régulièrement au-dessus de la surface. Les nuages ondulés se forment lorsque de l'air stable circule au-dessus d'un relief et qu'une onde de gravité atmosphérique est créée. Les nuages se forment dans la région ascendante, tandis que la région descendante est propice à un ciel clair. Ce processus est courant sur Terre. Sur Mars, les nuages sont probablement produits par des particules de glace de dioxyde de carbone et sont susceptibles de se trouver à des altitudes relativement élevées (20–50 km).

NUVOLE D'ONDA

Questa immagine è stata acquisita ad alta latitudine settentrionale proprio all'inizio della primavera settentrionale. Mostra nuvole a onda regolari sulla superficie. Le nuvole a onda si formano quando l'aria stabile scorre su un elemento terrestre sollevato e si crea un'onda di gravità atmosferica. In tali onde, le nuvole si formano nella regione che si muove verso l'alto, mentre la regione che affonda favorisce cieli limpidi. Sono piuttosto comuni sulla Terra. Su Marte, le nuvole sono probabilmente prodotte da particelle di ghiaccio di anidride carbonica e si trovano probabilmente ad altitudini relativamente alte (20–50 km).

ОРОГРАФИЧЕСКИЕ ОБЛАКА

Это изображение было получено на высоких северных широтах как раз в начале северной весны. На нём видны регулярные орографические (волновые) облака над поверхностью планеты. Такие облака образуются тогда, когда устойчивый поток воздуха проходит над возвышенностью, и создаётся «атмосферная гравитационная волна». В таких волнах облака образуются в более тёплой восходящей области при её охлаждении, тогда как холодная нисходящая область создаёт ясный участок. Это обычное явление на Земле. На Марсе облака, вероятно, образованы частицами углекислого льда и, похоже, находятся на относительно больших высотах (20–50 км).

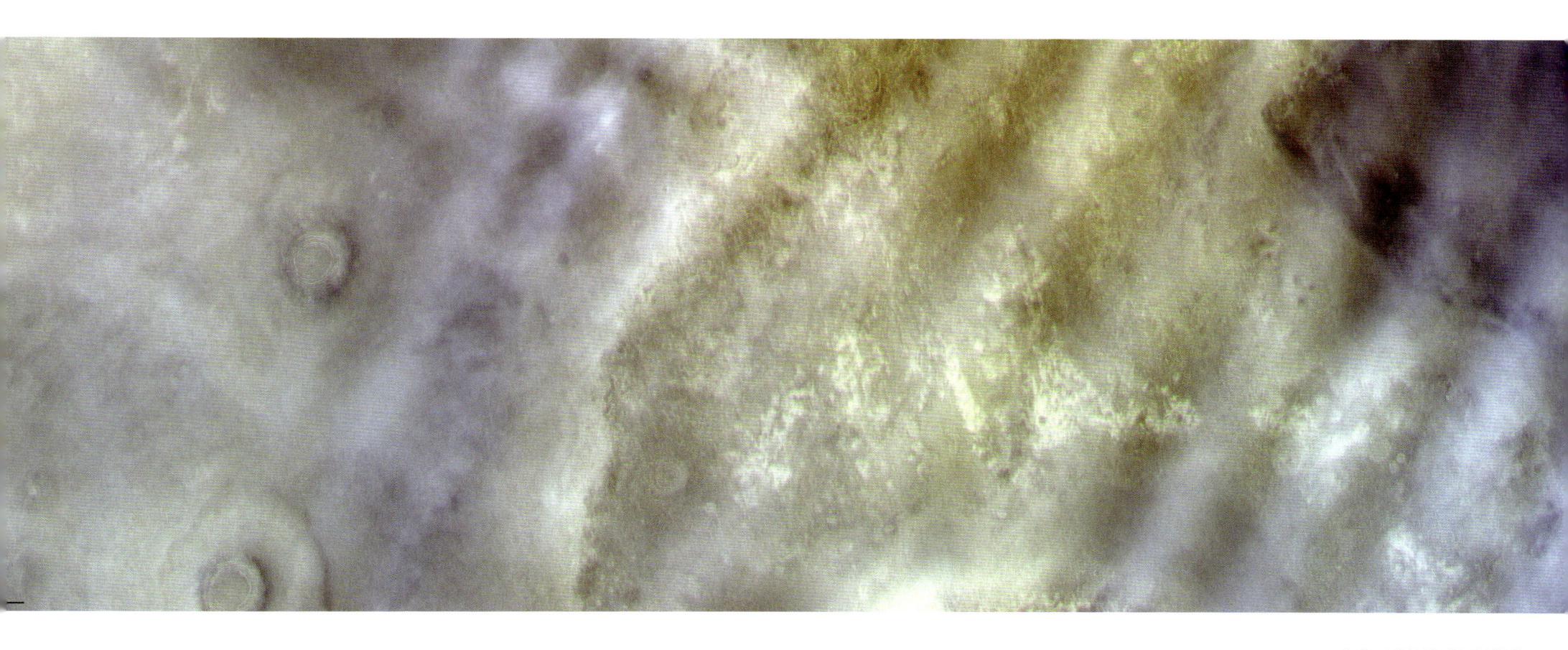

6.5 WAVE CLOUDS

ACQUISITION TIME: 2021-03-08T03:43:15.451
LONGITUDE: 128.9 E LATITUDE: 63.3 N
REGION: VASTITAS BOREALIS FORMATION

RESTEIS AUF DER NORDHEMISPHÄRE

CaSSIS kann nur bis zu einem Breitengrad von etwa 75° N abbilden. Der Grund dafür ist die feste Neigung der Umlaufbahn des Spurengas-Orbiters (Trace Gas Orbiter). Dieses Bild wurde nahe des maximalen nördlichen Breitengrads, den CaSSIS erreichen kann, in einem Gebiet südwestlich des Korolev-Kraters aufgenommen. Es wurde gegen Ende des nördlichen Frühlings aufgezeichnet und zeigt oberflächliche Eisablagerungen.

RESIDUAL ICES IN THE NORTHERN HEMISPHERE

CaSSIS can only image as far as about 75N in latitude. This is because of the inclination of the orbit of the Trace Gas Orbiter which is fixed. This image was acquired close to the maximum northern latitude that CaSSIS can reach from an area to the southwest of Korolev Crater. It was obtained towards the end of northern spring and shows surficial ice deposits.

GLACES RÉSIDUELLES DANS L'HÉMISPHÈRE NORD

CaSSIS ne prend pas d'images au delà de 75°N de latitude. Ceci est dû à l'inclinaison fixe de l'orbite de la sonde Trace Gas Orbiter. Cette image a été acquise près de la latitude nord maximale dans une zone située au sud-ouest du cratère Korolev. Prise à la fin du printemps boréal, cette image montre des dépôts de glace superficiels.

GHIACCI RESIDUI NELL'EMISFERO NORD

CaSSIS può acquisire immagini solo fino a circa 75°N di latitudine. Questo è dovuto all'inclinazione dell'orbita del Trace Gas Orbiter che è fissa. Questa immagine è stata acquisita vicino alla massima latitudine settentrionale che CaSSIS può raggiungere, su una zona a sud-ovest del cratere Korolev. È stata ottenuta verso la fine della primavera settentrionale e mostra depositi superficiali di ghiaccio.

ОСТАТОЧНЫЕ ЛЬДЫ В СЕВЕРНОМ ПОЛУШАРИИ

CaSSIS может делать снимки не далее 75° северной широты. Это происходит из-за фиксированного наклона орбиты TGO - Орбитального аппарата для исследования малых составляющих атмосферы. Данное изображение было получено вблизи максимальной северной широты, которую может достичь CaSSIS из области к юго-западу от кратера Королёв. Снимок был сделан ближе к концу северной весны и демонстрирует поверхностные ледяные отложения.

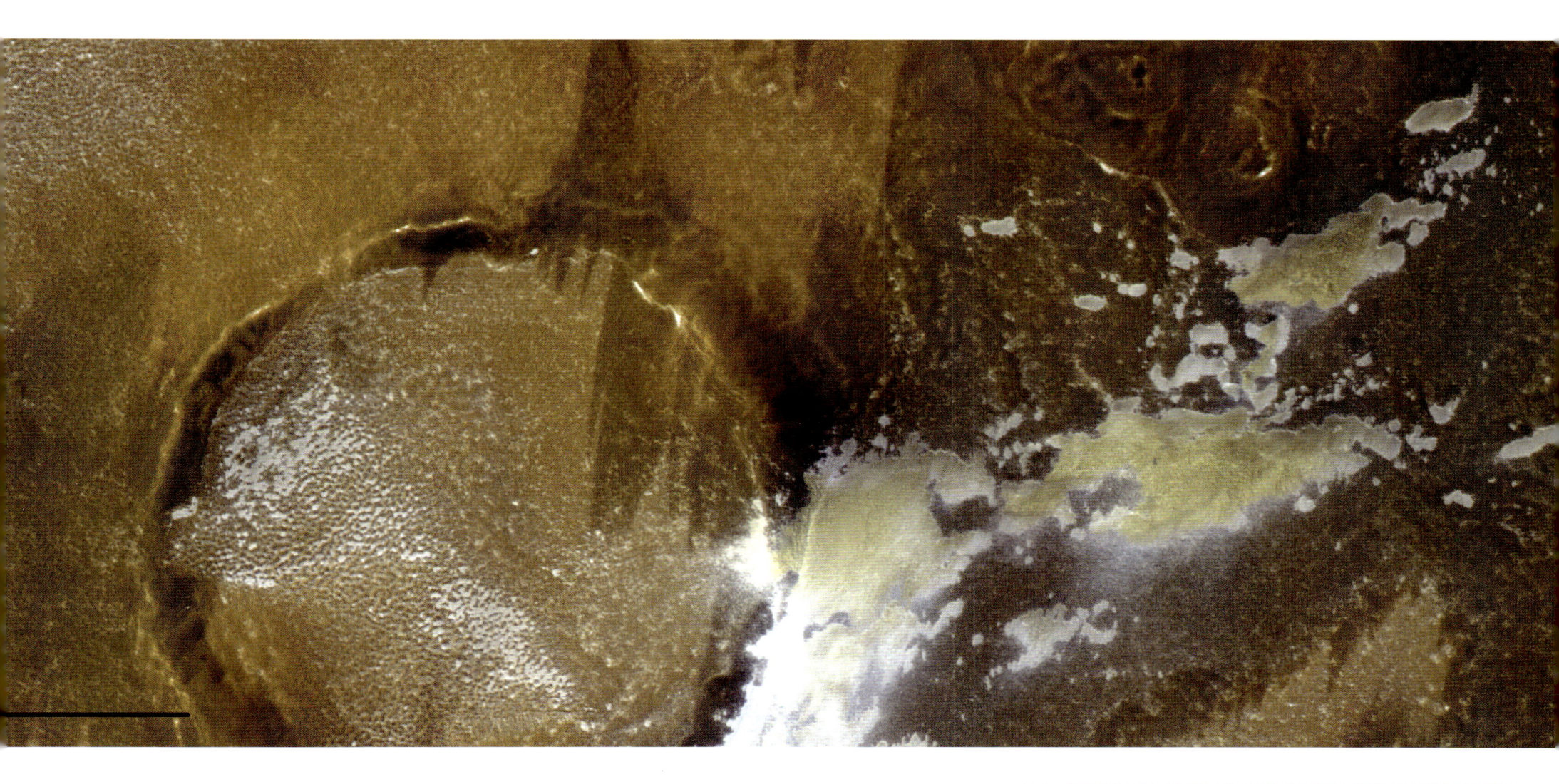

6.6   RESIDUAL ICES IN THE NORTHERN HEMISPHERE

ACQUISITION TIME: 2019-09-15T09:33:41.440
LONGITUDE: 167.7 E   LATITUDE: 73.7 N
REGION: VBF

FROST IN RINNEN UND AN EINEM KRATERRAND

Dieses Bild zeigt eine Reihe von Rinnen an einer inneren Kraterwand in Aonia Terra mit Reflexionen von hellem Material. Bei dem hellen Material handelt es sich wahrscheinlich um Kohlendioxid-Eis oder -Frost, der einige Wochen nach Beginn des südlichen Frühlings noch nicht sublimiert ist. Man beachte auch die hellen, diffusen Frostflecken an der nördlichen (linken) Kraterwand.

FROST IN GULLIES AND A CRATER RIM

This image shows a set of gullies on an interior crater wall in Aonia Terra with reflections from bright material within them. The bright material is probably carbon dioxide ice or frost that has yet to sublime a few weeks after the beginning of southern spring. Notice also the bright diffuse patches of frost on the northern (left) crater wall.

GIVRE DANS DES RAVINES ET SUR LA BORDURE D'UN CRATÈRE

Cette image prise quelques semaines après le début du printemps austral, montre un ensemble de ravines sur le bord d'un cratère dans la région d'Aonia Terra. Ces ravines présentent des reflets brillants qui correspondent probablement à des dépôts de givre de dioxyde de carbone qui n'ont pas encore été sublimés. Remarquez également les taches de givre brillantes sur la paroi à gauche du cratère.

BRINA SU CALANCHI E SUL BORDO DI UN CRATERE

Questa immagine mostra una serie di calanchi su una parete interna del cratere di Aonia Terra con riflessi di materiale chiaro al loro interno. Il materiale chiaro è probabilmente ghiaccio di anidride carbonica o brina che deve ancora sublimare a poche settimane dopo l'inizio della primavera meridionale. Si notino anche le macchie chiare diffuse di brina sulla parete settentrionale (sinistra) del cratere.

ИНЕЙ В ОВРАГАХ И НА ОБОДЕ КРАТЕРА

На этом изображении показаны овраги на внутренней стене кратера в земле Аонид с отражениями света от яркого материала внутри них. Яркий материал, вероятно, представляет собой углекислый лёд или иней, которому ещё предстоит сублимироваться через несколько недель после начала южной весны. Обратите внимание также на яркие размытые пятна изморози на северной (левой) стене кратера.

6.7 FROST IN GULLIES AND A CRATER RIM

ACQUISITION TIME: 2020-05-17T17:50:25.354
LONGITUDE: 257.8 E   LATITUDE: 56.9 S
REGION: AONIA

EIS AN DER NORDWAND EINES KRATERS

Kraterwände in mittleren und hohen Breitengraden, die polwärts gerichtet sind, verlieren ihre Eis- und Frostschichten im Frühjahr in der Regel am langsamsten. Hier ist ein Beispiel aus der nördlichen Hemisphäre auf einem hohen Breitengrad (Arcadia Dorsa). Die südliche Kraterwand (rechts) ist nach Norden ausgerichtet und wird von der Sonne relativ wenig erwärmt. Zu diesem Zeitpunkt (etwa vier Wochen nach dem Beginn des nördlichen Frühlings) begannen die im nördlichen Winter aus der Atmosphäre kondensierten, flüchtigen Stoffe zu sublimieren.

ICE ON THE NORTH-FACING WALL OF A CRATER

Crater walls at mid and high latitudes that face poleward are usually slowest to lose their ice and frost layers in spring. Here is an example from the northern hemisphere at high latitude (Arcadia Dorsa). The southern crater wall (right) faces north and receives relatively little warmth from the Sun. Condensed atmospheric volatiles in northern winter were beginning to sublime at this time (around 4 weeks after the start of northern spring).

GLACE SUR LA PAROI NORD D'UN CRATÈRE

Les parois des cratères orientées face au pôle et situées aux moyennes/hautes latitudes dégivrent généralement plus lentement que les parois orientées face à l'équateur lors du printemps. En voici un exemple dans la région d'Arcadia Dorsa dans l'hémisphère nord. La paroi sud du cratère (à droite) est orientée face vers le nord et reçoit relativement peu de chaleur. Les glaces condensées en hiver à la surface commencent à se sublimer à cette période (environ 4 semaines après le début du printemps).

GHIACCIO SULLA PARETE NORD DI UN CRATERE

Le pareti dei crateri alle medie e alte latitudini che sono rivolte verso il polo sono di solito le più lente a perdere i loro strati di ghiaccio e brina in primavera. Ecco un esempio dall'emisfero nord ad alta latitudine (Arcadia Dorsa). La parete meridionale del cratere (a destra) è rivolta verso nord e riceve relativamente poco calore dal Sole. I gas atmosferici, condensati nell'inverno settentrionale, stanno cominciando a sublimare in questo momento (circa 4 settimane dopo l'inizio della primavera settentrionale).

ЛЕД НА ОБРАЩЁННОЙ К СЕВЕРУ СТЕНЕ КРАТЕРА

Стены кратеров на средних и высоких широтах, обращённые к полюсу, весной обычно медленнее всего теряют свои покровы льда и инея. На снимке приведён пример из высоких широт северного полушария (гряды Аркадии). Внутренняя южная стена кратера (справа) обращена на север и получает относительно мало тепла от Солнца. Замёрзшие северной зимой атмосферные летучие вещества начинают сублимироваться примерно через 4 недели после начала северной весны (время создания снимка).

6.8 ICE ON THE NORTH-FACING WALL OF A CRATER

ACQUISITION TIME: 2021-03-13T23:15:40.755
LONGITUDE: 235.8 E   LATITUDE: 60.6 N
REGION: ARCADIA

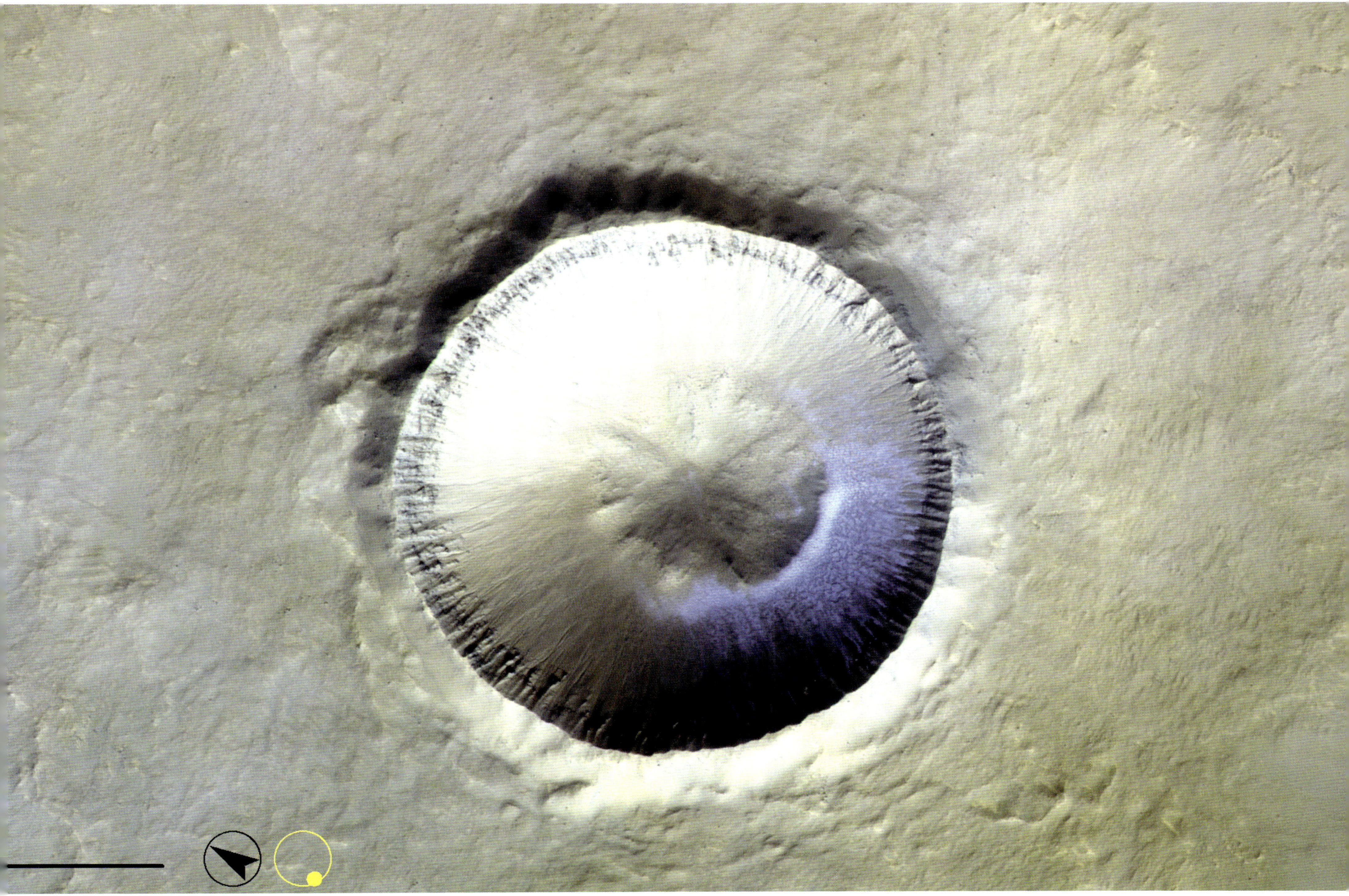

FROST AUF DÜNEN

Dieses eher ungewöhnliche Bild zeigt ein Dünenfeld in den mittleren südlichen Breitengraden zu Beginn des südlichen Frühlings. Das Dünenfeld befindet sich in einem unbenannten Krater in Terra Sirenum südwestlich des Liu-Hsin-Kraters. Der Streifeneffekt ist eine Folge regelmässiger Rippelmuster auf der Dünenoberfläche, die die Ausrichtung der Oberfläche zur Sonne leicht verändern. Dies führt zu einem quasi-regelmässigen Muster von Veränderungen in der Wärmezufuhr zur Eisoberfläche, die dann zur Sublimation genutzt wird. Man beachte auch das Eis an der nach Süden gerichteten Kraterwand links im Bild.

FROSTS ON DUNES

This rather unusual image is of a southern mid-latitude dune field at the beginning of southern spring. The dune field is located within an unnamed crater in Terra Sirenum southwest of Liu Hsin Crater. The striped effect is a consequence of regular ripple patterns on the surfaces of the dunes that slightly change the orientation of the surface with respect to the Sun. This leads to a quasi-regular pattern of changes in the heat given to the icy surface that is then used to provoke sublimation. Notice also the ice on the south facing crater wall to the left of the image.

GIVRE SUR LES DUNES

Cette image montre un champ de dunes aux moyennes latitudes sud, au début du printemps austral. Le champ de dunes est situé dans un cratère de Terra Sirenum, au sud-ouest du cratère Liu Hsin. L'aspect rayé des dépôts de givre, plutôt inhabituel, est une conséquence de motifs d'ondulation réguliers à la surface des dunes. Ils modifient légèrement l'orientation de la surface par rapport au soleil et cela conduit à une modulation locale dans la répartition de la chaleur apportée à la surface. La sublimation varie donc spatialement à de très courtes échelles. Remarquez également la glace sur la paroi du cratère orientée vers le sud, à gauche de l'image.

BRINA SULLE DUNE

Questa immagine piuttosto insolita rappresenta di un campo di dune delle medie latitudini all'inizio della primavera meridionale. Il campo di dune si trova all'interno di un cratere senza nome in Terra Sirenum a sud-ovest del cratere Liu Hsin. L'effetto a strisce è una conseguenza della struttura delle increspature sulla superficie delle dune che cambiano leggermente l'orientamento della superficie rispetto al Sole. Questo porta ad un disegno quasi regolare causato dai cambiamenti nel calore ceduto alla superficie ghiacciata, che viene poi utilizzato per la sublimazione. Si noti anche il ghiaccio esposto in superficie sulla parete del cratere a sud a sinistra dell'immagine.

ИНЕЙ НА ДЮНАХ

Это довольно необычное изображение дюнного поля в южных средних широтах в начале южной весны. Дюнное поле расположено в пределах безымянного кратера в земле Сирен к юго-западу от кратера Лю Син. Эффект полос является следствием регулярной ряби на поверхности дюн, которая слегка изменяет экспозицию поверхности по отношению к Солнцу. Это приводит к тому, что получение тепла ледяной поверхностью носит квазирегулярный характер, что провоцирует сублимацию льда на более освещённой стороне. Обратите внимание также на лёд на южной стене кратера в левой части снимка.

6.9 FROSTS ON DUNES

ACQUISITION TIME: 2020-05-16T22:11:08.313
LONGITUDE: 186.1 E LATITUDE: 56.3 S
REGION: SIRENUM

MIT SAISONALEM EIS GEFÜLLTER KRATER

Dieser gut erhaltene Krater in den nördlichen Hochebenen ist ein regelmässiges Ziel für CaSSIS. Dieses Bild, das im Hochsommer auf der Nordhalbkugel aufgenommen wurde, zeigt verbliebene saisonale Ablagerungen von eisigem Material sowohl innerhalb des Kraters als auch an seinem Rand. Andere Krater auf diesem Breitengrad weisen permanente Eisablagerungen auf, die Überreste einer früher grösseren Polkappe sind.

CRATER FILLED WITH SEASONAL ICE

This well-preserved crater in the high-latitude northern plains has been a regular target for CaSSIS. This image, taken in mid-summer in the northern hemisphere, shows residual seasonal deposits of icy material both within the crater and around its rim. Other craters at that latitude contain permanent ice deposits which are remnants of a formerly more extended polar cap.

CRATÈRE REMPLI DE GLACE SAISONNIÈRE

Ce cratère bien préservé dans les plaines des hautes latitudes de l'hémisphère nord a été une cible régulière de la caméra CaSSIS. Cette image, prise au milieu de l'été, montre des dépôts saisonniers résiduels de glace à l'intérieur et autour du cratère. D'autres cratères à cette latitude contiennent des dépôts de glace permanents qui sont les vestiges d'une calotte polaire autrefois plus étendue.

CRATERE RIEMPITO DA GHIACCIO STAGIONALE

Questo cratere ben conservato nelle pianure settentrionali ad alta latitudine è stato un obiettivo regolare per CaSSIS. Questa immagine, scattata a metà estate nell'emisfero settentrionale, mostra residui di depositi stagionali di materiale ghiacciato sia all'interno del cratere che intorno al suo bordo. Altri crateri a quella latitudine contengono depositi permanenti di ghiaccio che sono resti di una precedente calotta polare più estesa.

КРАТЕР, ЗАПОЛНЕННЫЙ СЕЗОННЫМ ЛЬДОМ

Этот хорошо сохранившийся кратер на высокоширотных северных равнинах находился под регулярным наблюдением CaSSIS. На этом изображении, сделанном в середине северного лета, виден остаточный сезонный покров ледяного материала как внутри кратера, так и вокруг его края. Другие кратеры на этой широте содержат постоянные ледяные отложения, которые являются остатками более протяженной полярной шапки, существовавшей когда-то.

6.10   CRATER FILLED WITH SEASONAL ICE

ACQUISITION TIME: 2019-12-26T23:50:28.292
LONGITUDE: 230.5 E   LATITUDE: 73.9 N
REGION: VBF

NÖRDLICHE DÜNEN AUF EINEM BEMERKENSWERTEN UNTERGRUND

Eine Dünenlandschaft in hohen nördlichen Breitengraden abgebildet kurz nach Beginn des nördlichen Frühlings. Die Dünen sind zu diesem Zeitpunkt grösstenteils mit Kohlendioxid-Eis bedeckt. Auf den Dünen sind einige Flecken- und Fächerablagerungen zu sehen, die wahrscheinlich durch den «Kieffer»-Mechanismus für die Erzeugung von Geysiren durch basale Sublimation einer Eisplatte entstehen. Besonders erwähnenswert sind hier die sehr fein ausgerichteten Rippen (unten links), die das Substrat in dieser Region bilden. Der abgebildete Ort befindet sich an der Grenze von Siton Undae in der VBF.

NORTHERN DUNES ON A REMARKABLE SUBSTRATE

A set of dunes at high northern latitude seen shortly after the start of northern spring. The dunes are mostly covered with carbon dioxide ice at this time. Some spots and fan deposits are seen on the dunes that probably arise from the "Kieffer" mechanism for the production of geysers from basal sublimation of an ice slab. Of particular note here, are the very fine aligned ridges (lower left) that form the substrate in this region. The site is on the boundary of Siton Undae in the VBF.

DUNES NORDIQUES SUR UN SUBSTRAT REMARQUABLE

Ensemble de dunes à haute latitude dans l'hémisphère nord, observées ici peu de temps après le début du printemps boréal. À cette période de l'année, les dunes sont principalement recouvertes de glace de dioxyde de carbone. On peut voir sur les dunes quelques taches et dépôts en éventail qui proviennent probablement du mécanisme de «Kieffer» pour la production de geysers à partir de la sublimation basale de la couche de glace. Les crêtes alignées très fines (en bas à gauche) qui forment le substrat de cette région sont particulièrement remarquables. Le site se trouve à la limite de la région de Siton Undae dans la formation Vastitas Borealis.

DUNE SETTENTRIONALI SU UN SUBSTRATO PARTICOLARE

Un insieme di dune ad alta latitudine settentrionale osservate poco dopo l'inizio della primavera settentrionale. Le dune sono per lo più coperte da ghiaccio di anidride carbonica in questo momento. Si vedono alcune macchie e depositi a ventaglio sulle dune che probabilmente derivano dal meccanismo «Kieffer» per la produzione di geyser dalla sublimazione basale di una lastra di ghiaccio. Di particolare nota qui, sono le creste allineate molto fini (in basso a sinistra) che formano il substrato in questa regione. Il sito è sul confine di Siton Undae nel VBF.

СЕВЕРНЫЕ ДЮНЫ НА ПРИМЕЧАТЕЛЬНОМ ФУНДАМЕНТЕ

Дюнное поле на высоких северных широтах вскоре после начала северной весны. В это время дюны в основном покрыты льдом углекислого газа. На их поверхности видны пятна и веерные отложения, которые, вероятно, возникают из-за так называемого механизма Киффера, создающего гейзеры в ходе сублимации нижних (контактирующих с основанием) слоёв льда. Особо следует отметить очень точно выровненные гребни (снизу слева), которые образуют фундамент этой области. Участок находится на границе дюн Ситон в Великой Северной равнине (VBF).

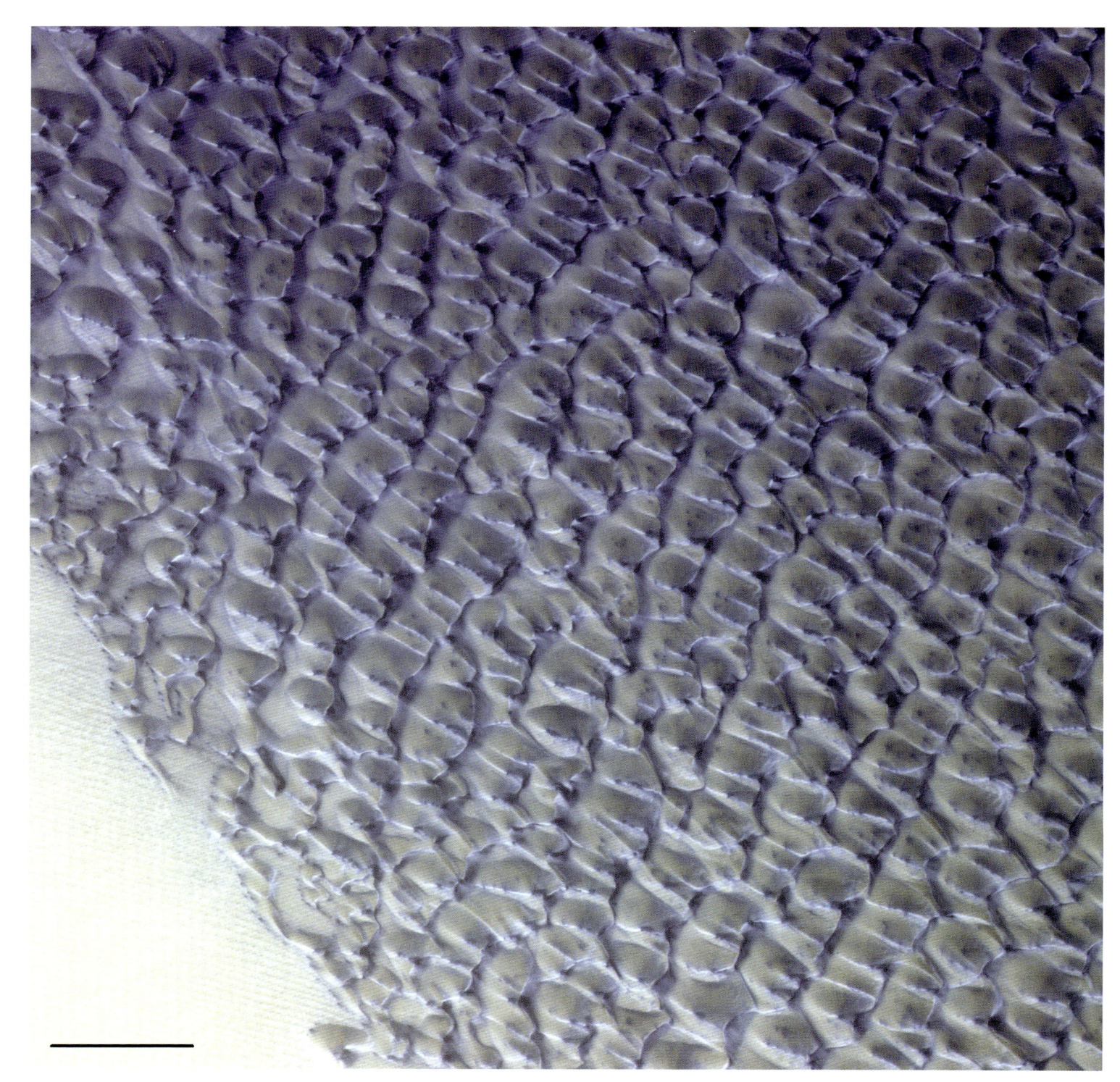

6.11 NORTHERN DUNES ON A
REMARKABLE SUBSTRATE

ACQUISITION TIME:
2021-02-27T09:19:47.982
LONGITUDE: 291.4 E
LATITUDE: 75.0 N
REGION: VBF

HELLES MATERIAL UM DUNKLE FLECKEN

Dieses Bild von Argentea Planum in der Nähe des Mellish-Kraters wurde nur vier Tage nach dem Beginn des südlichen Frühlings aufgenommen und zeigt, dass der Geysirmechanismus zur Erzeugung von Flecken und Fächern auf der saisonalen Kohlendioxid-Eiskappe bereits eingesetzt hatte. Dieses optimierte, synthetische RGB-Bild zeigt, dass viele der dunklen Flecken von relativ hellblauem Material umgeben sind. Es ist möglich, dass es sich bei diesem Material um Kohlendioxid-Schnee handelt, der kondensiert und ausfällt, wenn sich das Gas in den Geysiren schnell ausdehnt. Dies ist jedoch zu schwierig zu modellieren und bleibt Gegenstand der Forschung.

BRIGHT MATERIAL SURROUNDING DARK SPOTS

This image from Argentea Planum near Mellish Crater was taken just 4 days after the start of southern spring and shows that the geyser mechanism to produce spots and fans on the seasonal carbon dioxide ice cap had already started. This enhanced synthetic RGB image shows that many of the dark spots are surrounded by material that is relatively light blue in colour. It is conceivable that this material is carbon dioxide snow that condenses and precipitates as the gas in the geyser rapidly expands. This is however rather too difficult to model and remains the subject of research.

DÉPÔTS CLAIRS ENTOURANT DES TACHES SOMBRES

Cette image d'Argentea Planum près du cratère Mellish a été prise seulement 4 jours après le début du printemps austral. On voit que la production des taches sombres et des dépôts en éventail par le mécanisme de geyser sur la couche saisonnière de glace de dioxyde de carbone a déjà commencé. L'image montre que la plupart de ces taches sombres sont entourées d'un matériau relativement clair et de couleur bleue. Il est possible que ce matériau soit de la glace carbonique qui se condense et se précipite lorsque le gaz du geyser se dilate rapidement. Cette hypothèse est cependant difficile à prouver et fait encore l'objet d'études.

MATERIALE CHIARO INTORNO A MACCHIE SCURE

Questa immagine da Argentea Planum vicino al cratere Mellish è stata scattata appena 4 giorni dopo l'inizio della primavera australe, e mostra che il meccanismo dei geyser che produce macchie e ventagli sulla calotta di ghiaccio stagionale di anidride carbonica era già iniziato. Questa immagine composta nei filtri RGB, mostra che molte delle macchie scure sono circondate da materiale di colore blu relativamente chiaro. È ipotizzabile che questo materiale sia neve di anidride carbonica che si condensa e precipita quando il gas del geyser si espande rapidamente. Questo meccanismo è comunque piuttosto difficile da modellare e rimane oggetto di ricerca.

СВЕТЛЫЙ МАТЕРИАЛ ВОКРУГ ТЕМНЫХ ПЯТЕН

Этот снимок плато Серебряное возле кратера Меллиш был сделан всего через 4 дня после начала южной весны, когда стали активны гейзеры, создающие пятна и вееры на сезонной шапке углекислого льда. Данное RGB-изображение показывает, что многие темные пятна окружены материалом относительно светло-голубого цвета. Вполне возможно, что это снег из углекислого газа, конденсирующийся и оседающий по мере быстрого расширения и остывания газа в гейзере. Однако, смоделировать этот эффект довольно сложно, и он всё ещё остается предметом исследования.

6.12 BRIGHT MATERIAL SURROUNDING DARK SPOTS

ACQUISITION TIME: 2018-05-26T04:41:52.000
LONGITUDE: 334.6 E   LATITUDE: 70.5 S
REGION: ARGENTEA

GEYSIRE UND LOKALE WINDE

Dieses Bild stammt aus dem südlichsten Teil von Terra Cimmeria am Westrand des Richardson-Kraters. Es zeigt die typischen Merkmale der Geysiraktivität im südlichen Frühling. Das Bild wurde etwa einen Monat nach der Wintersonnenwende auf der südlichen Hemisphäre aufgenommen. Interessant sind hier die Ausläufer der dunklen Flecken, die darauf hinweisen, dass der vom Geysir ausgestoßene Staub vom Wind nach unten getragen wurde. Im Allgemeinen geht der Wind zwar grob von links oben nach rechts unten, aber es gibt lokale Schwankungen.

GEYSERS AND LOCAL WINDS

This image is from the southernmost part of Terra Cimmeria on the west rim of Richardson Crater. It shows the typical characteristics of geyser activity in southern spring. The image was taken around one month after the winter solstice in the southern hemisphere. Of interest here, are the tails of the dark spots that indicate that dust ejected by the geyser has been carried downwind. While the wind direction is roughly top-left to bottom-right, there are local variations.

GEYSERS ET VENTS LOCAUX

Cette image provient de la partie la plus au sud de Terra Cimmeria, sur la bordure ouest du cratère Richardson. Elle montre les caractéristiques typiques de l'activité des geysers au printemps austral. L'image a été prise quelques jours avant l'équinoxe de printemps dans l'hémisphère sud. Notez les queues des taches sombres qui indiquent que la poussière éjectée par le geyser a été transportée par le vent. Bien que la direction prédominante du vent soit approximativement de haut en bas et de gauche à droite, il existe des variations locales.

GEYSER E VENTI LOCALI

Questa immagine proviene dalla parte più meridionale della Terra Cimmeria sul bordo occidentale del Richardson Crater. Mostra le caratteristiche tipiche dell'attività dei geyser nella primavera meridionale. L'immagine è stata scattata circa un mese dopo il solstizio d'inverno nell'emisfero meridionale. Di particolare interesse sono le code delle macchie scure che indicano che la polvere espulsa dal geyser è stata trasportata sottovento. Anche se la direzione del vento è approssimativamente dall'alto a sinistra in basso a destra, ci sono variazioni locali.

ГЕЙЗЕРЫ И МЕСТНЫЕ ВЕТРА

Это изображение западного вала кратера Ричардсон, что на самом юге Киммерийской земли. Оно демонстрирует типичные характеристики гейзерной активности южной весной. Снимок был сделан примерно через месяц после зимнего солнцестояния в южном полушарии. Здесь примечательны хвосты тёмных пятен, которые указывают на то, что пыль, выброшенная гейзером, уносится ветром. Несмотря на то, что ориентация ветра примерно соответствует направлению от верхнего левого края изображения к правому нижнему, существуют локальные вариации.

6.13 GEYSERS AND LOCAL WINDS

ACQUISITION TIME: 2020-05-18T01:49:10.371
LONGITUDE: 177.6 E   LATITUDE: 72.3 S
REGION: CIMMERIA

NARBEN, VERTIEFUNGEN UND LINEARE STRUKTUREN IN DER PITYUSA PATERA

Pityusa Patera ist offiziell Teil von Noachis Terra, obwohl einige Autoren diese Region als Teil von Malea Planum bezeichnen. Sie wurde als alte Caldera und daher vulkanischen Ursprungs interpretiert. Dieses Bild, das in der Mitte des Frühlings auf der Südhalbkugel aufgenommen wurde, zeigt den nördlichen Rand eines grossen Tafelbergs. Man beachte, dass Norden ungefähr nach unten zeigt. Die obere Fläche des Tafelbergs weist eine parallele, quasi-periodische Struktur auf, die möglicherweise durch die vorherrschenden Hangwinde verursacht wurde, die nach Nordosten in das Hellas-Becken wehen. Bei näherer Betrachtung zeigt sich, dass die Oberfläche eine polygonale Struktur aufweist (hier besonders deutlich zwischen der Senke und der Steilwand links).

SCARPS, DEPRESSIONS AND LINEAR STRUCTURES IN PITYUSA PATERA

Pityusa Patera is formally part of Noachis Terra although some authors refer to it as being in Malea Planum. It has been interpreted as an old caldera and therefore volcanic in origin. This image, which was taken in mid-spring in the southern hemisphere, is from the northern edge of a large mesa. Note that north is roughly downwards. The upper surface of the mesa shows a parallel quasi-periodic structure which may have been caused by prevailing slope winds blowing to the northeast into the Hellas basin. Close inspection shows that the surface has a polygonal structure.

ESCARPEMENTS, DÉPRESSIONS ET STRUCTURES LINÉAIRES DE PITYUSA PATERA

Pityusa Patera fait officiellement partie de Noachis Terra, bien que certains auteurs la situent dans la région de Malea Planum. D'origine probablement volcanique, elle est considérée comme étant une ancienne caldeira. Cette image, qui a été prise au milieu du printemps dans l'hémisphère sud, provient du bord nord d'une grande mésa. Le nord se trouve approximativement vers le bas de l'image. La surface supérieure de la mésa présente une structure régulière quasi-périodique qui peut avoir été causée par les vents dominants soufflant vers le nord-est dans le bassin d'Hellas. Une observation plus minutieuse montre que la surface montre une structure polygonale.

SCARPATE, DEPRESSIONI E STRUTTURE LINEARI A PITYUSA PATERA

Pityusa Patera fa formalmente parte di Noachis Terra anche se alcuni autori si riferiscono ad essa come a Malea Planum. È stata interpretata come una vecchia caldera e quindi di origine vulcanica. Questa immagine, che è stata presa a metà primavera nell'emisfero meridionale, proviene dal bordo settentrionale di una grande mesa. Si noti che il nord è approssimativamente verso il basso. La superficie superiore della mesa mostra una struttura parallela quasi-periodica che può essere stata causata dai predominanti venti che soffiano verso nord-est nel bacino di Hellas. Un'ispezione ravvicinata mostra che la superficie ha una struttura poligonale (particolarmente visibile tra la depressione e la scarpata sulla sinistra).

УСТУПЫ, ВПАДИНЫ И ЛИНЕЙНЫЕ СТРУКТУРЫ В ПАТЕРЕ ПИТИУСА

Патера Питиуса формально является частью земли Ноя, хотя некоторые авторы относят её к плато Малея. Данный объект был интерпретирован как старая кальдера (котловина вулкана) и, следовательно, имеет вулканическое происхождение. На снимке, сделанном в середине весны в южном полушарии, виден край большой столовой горы. Обратите внимание, что север находится примерно внизу снимка. Вершина столовой горы демонстрирует параллельную квазипериодическую структуру, которая, возможно, была создана преобладающими склоновыми ветрами, направленными на северо-восток в бассейн Эллады. При внимательном рассмотрении видно, что поверхность имеет полигональную структуру (особенно это заметно между впадиной и уступом в левой части снимка).

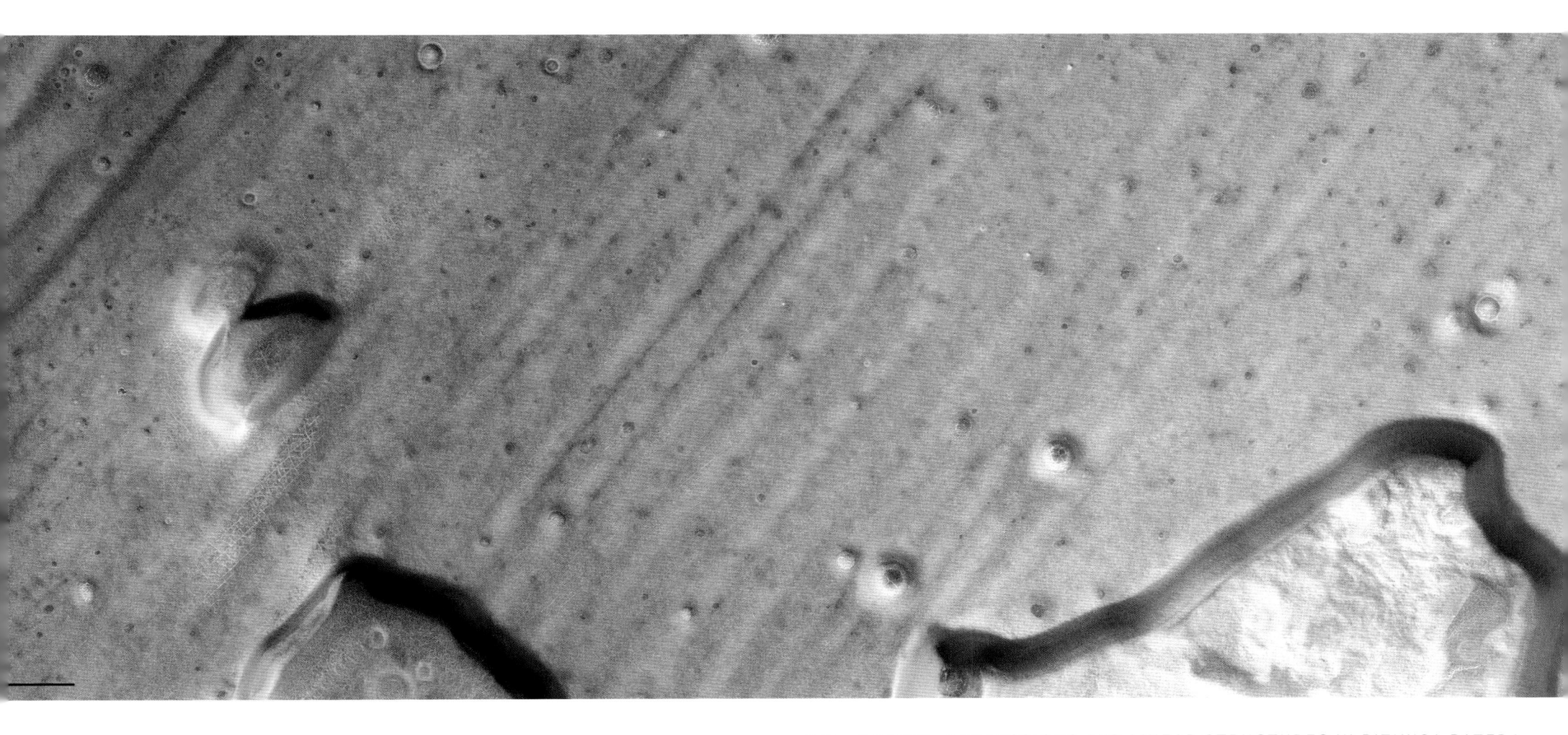

6.14 SCARPS, DEPRESSIONS AND LINEAR STRUCTURES IN PITYUSA PATERA

ACQUISITION TIME: 2020-06-26T09:12:35.346
LONGITUDE: 33.8E LATITUDE: 67.1S
REGION: PITYUSA

SAISONALES EIS AUF DEN SÜDPOLAREN SCHICHTABLAGERUNGEN

Dies ist ein Bild der südpolaren Schichtablagerungen (SPLD) im Planum Australe südlich von Terra Cimmeria, etwa 12 Wochen nach der Tagundnachtgleiche aufgenommen. Die saisonale Kohlendioxid-Eiskappe sublimiert und hinterlässt spinnenartige Strukturen. Es gibt immer noch Anzeichen für eine anhaltende Geysiraktivität (unten links). Die Schichten der SPLD sind ebenfalls deutlich sichtbar. Anhand dieser Schichten lassen sich möglicherweise die natürlichen Veränderungen des Marsklimas in den letzten Millionen Jahren untersuchen.

SEASONAL ICE ON THE SOUTH POLAR LAYERED DEPOSITS

This is an image of the South Polar Layered Deposits in Planum Australe south of Terra Cimmeria taken around 12 weeks after the equinox. The seasonal carbon dioxide ice cap is subliming and leaving behind spider-like structures. There is still evidence of continued geyser activity (bottom left). The layers of the SPLD are also clearly visible. These layers may eventually allow studies of natural changes in the Martian climate over the past several million years.

GLACE SAISONNIÈRE SUR LES DÉPÔTS STRATIFIÉS DU PÔLE SUD

Image des dépôts stratifiés du pôle Sud (SPLD) dans Australe Planum au sud de Terra Cimmeria. Prise environ 12 semaines après l'équinoxe, cette image montre des dépôts saisonniers de dioxyde de carbone en train de se sublimer en laissant des traces de structures radiales en forme d'araignée. Il y a des preuves d'une activité continue des geysers (en bas à gauche). Les couches du SPLD sont clairement visibles. Ces couches pourraient éventuellement permettre d'étudier les changements du climat martien au cours des derniers millions d'années.

GHIACCIO STAGIONALE SUI DEPOSITI STRATIFICATI DEL POLO SUD

Questa è un'immagine dei depositi stratificati polari meridionali nel Planum Australe a sud della Terra Cimmeria, scattata circa 12 settimane dopo l'equinozio. La calotta di ghiaccio stagionale di anidride carbonica sta sublimando e lascia dietro di sé strutture simili a ragni. C'è ancora evidenza di una continua attività dei geyser (in basso a sinistra). Anche gli strati dell'SPLD sono chiaramente visibili. Questi strati potrebbero alla fine permettere di studiare i cambiamenti naturali del clima marziano negli ultimi milioni di anni.

СЕЗОННЫЙ ЛЁД НА ЮЖНО-ПОЛЯРНЫХ СЛОИСТЫХ ОТЛОЖЕНИЯХ

Это снимок слоистых отложений (SPLD) на Южном плато, расположенном к полюсу от Киммерийской земли, сделанный примерно через двенадцать недель после равноденствия. Сезонная ледяная шапка из углекислого газа сублимируется и оставляет после себя «паукообразные» структуры. На снимке присутствуют свидетельства продолжающейся гейзерной активности (нижняя левая часть снимка). Также отчётливо видны слои SPLD. Эти ледяные отложения могут пролить свет на ход изменений марсианского климата в последние несколько миллионов лет.

6.15　SEASONAL ICE ON THE SOUTH POLAR LAYERED DEPOSITS

ACQUISITION TIME: 2020-06-23T20:12:18.218
LONGITUDE: 167.0 E　LATITUDE: 74.8 S
REGION: AUSTRALE

SPINNEN UND KLÜFTE IN DER NÄHE DES JEANS-KRATERS

Dies ist ein Bild aus dem südlichsten Teil von Terra Cimmeria und unmittelbar westlich des Jeans-Kraters. Es wurde kurz nach der Sommersonnenwende in der südlichen Hemisphäre aufgenommen (Beginn des südlichen Sommers). Es sind zwei grosse Brüche in der Oberflächenschicht zu erkennen. Die lokale Oberflächenrauigkeit ändert sich mit der Entfernung von der grösseren Bruchstelle.

SPIDERS AND FRACTURES NEAR JEANS CRATER

This is an image from the southernmost part of Terra Cimmeria and immediately to the west of Jeans Crater. It was acquired just after the summer solstice in the southern hemisphere (beginning of southern summer). Two large fractures in the surface layer can be seen. The local surface roughness changes with distance from the larger fracture.

ARAIGNÉES ET FRACTURES PRÈS DU CRATÈRE JEANS

Image prise à l'ouest du cratère Jeans, dans la partie la plus méridionale de Terra Cimmeria. Elle a été acquise juste après le solstice d'été dans l'hémisphère sud (début de l'été austral). On peut voir deux grandes fractures dans la couche de surface. La rugosité locale de la surface change avec la distance à la plus grande fracture.

RAGNI E FRATTURE VICINO AL CRATERE JEANS

Questa è un'immagine dalla parte più meridionale della Terra Cimmeria e immediatamente ad ovest del Jeans Crater. È stata acquisita subito dopo il solstizio d'estate nell'emisfero meridionale (inizio dell'estate australe). Si possono vedere due grandi fratture nello strato superficiale e l'asperità della superficie cambia con la distanza dalla frattura più grande.

ПАУКИ И РАЗЛОМЫ В РАЙОНЕ КРАТЕРА ДЖИНС

Это снимок самой южной части Киммерийской земли, к западу от кратера Джинс. Он был получен сразу после летнего солнцестояния в южном полушарии (начало южного лета). Отчётливо видны две большие трещины в поверхностном слое. Локальная неровность поверхности изменяется с удалением от более крупной трещины.

6.16 SPIDERS AND FRACTURES NEAR JEANS CRATER

ACQUISITION TIME: 2018-11-05T07:02:07.170
LONGITUDE: 154.2 E LATITUDE: 68.6 S
REGION: CIMMERIA

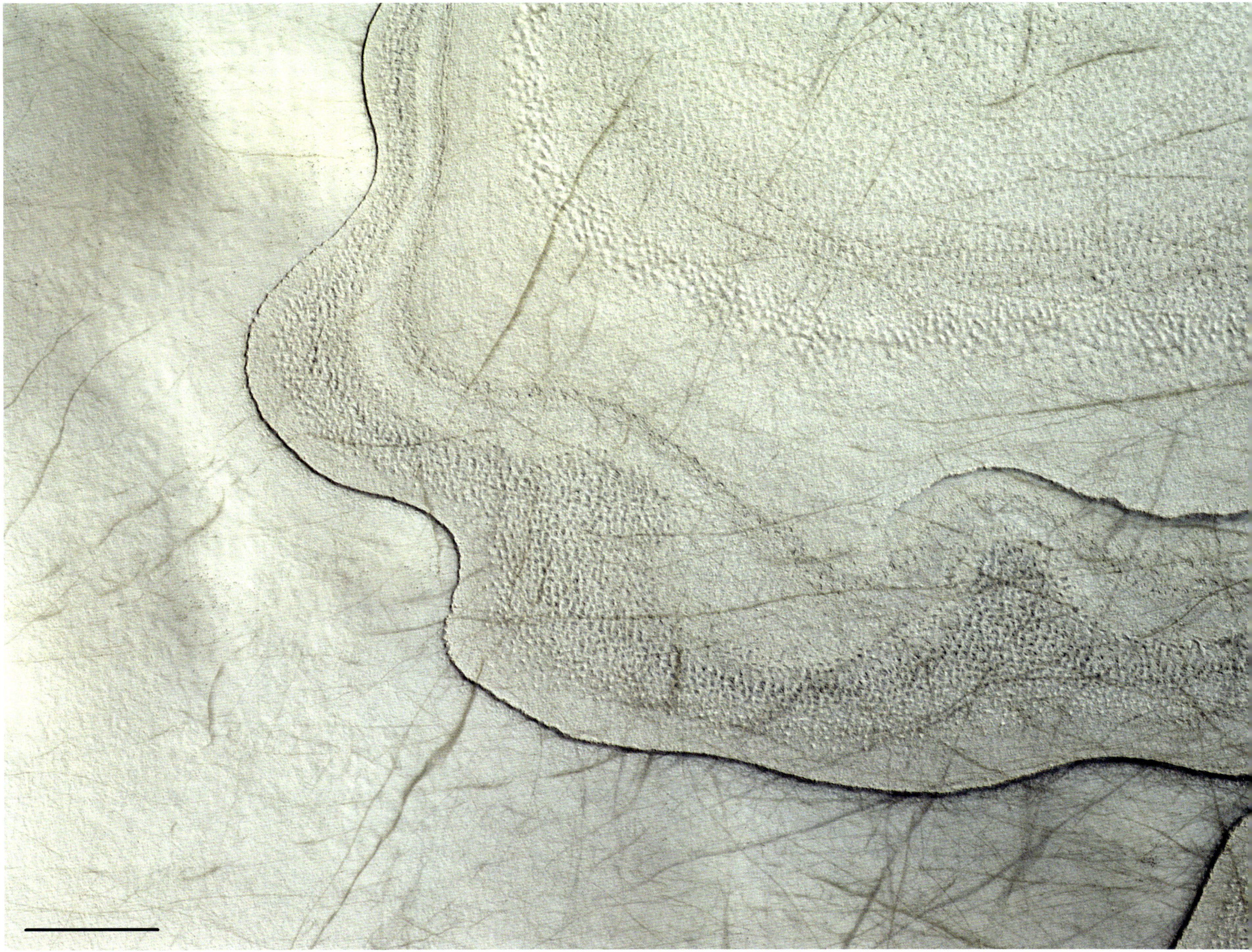

POLYGONE IN DER PITYUSA PATERA

Dieses Bild zeigt ein bemerkenswertes polygonales Terrain in der Region Pityusa Patera. Das Bild wurde in der Mitte des Frühlings auf der Südhalbkugel aufgenommen. Die Mulden des gemusterten Bodens enthalten immer noch Kohlendioxid-Eis, wodurch das polygonale Muster im Bild hervorsticht. Man beachte das Fehlen polygonaler Muster in den Kratern (Mitte rechts), obwohl der Boden im oberen Krater zerklüftet ist. Die Grösse und Form solcher Polygone spiegeln oft die Dicke der Bruchschicht und die lokalen Spannungsbedingungen wider, die möglicherweise durch Neigungen bestimmt werden.

POLYGONS IN PITYUSA PATERA

This image shows some remarkable polygonal terrain in the Pityusa Patera region. The image was taken in mid-spring in the southern hemisphere. The troughs of the patterned ground still contain carbon dioxide ice making the polygonal pattern stand out in the image. Notice the absence of polygonal patterns within the craters (centre-right) although there is fracturing of the floor in the upper one. The size and shape of such polygons often reflect the thickness of the fractured layer and the local stress conditions that may be controlled by slopes.

POLYGONES DANS LA RÉGION DE PITYUSA PATERA

Cette image montre un terrain polygonal remarquable dans la région de Pityusa Patera. L'image a été prise au milieu du printemps dans l'hémisphère sud. Les creux visibles contiennent encore de la glace de dioxyde de carbone, ce qui fait ressortir le motif polygonal sur l'image. Remarquez l'absence de motifs polygonaux à l'intérieur des cratères (centre-droite), bien que le sol soit fracturé dans le cratère supérieur. La taille et la forme de ces polygones reflètent souvent l'épaisseur de la couche fracturée et les conditions de contraintes mécaniques locales qui peuvent être contrôlées par les pentes.

POLIGONI NELLA PITYUSA PATERA

Questa immagine mostra un notevole terreno poligonale nella regione di Pityusa Patera. L'immagine è stata scattata a metà primavera nell'emisfero meridionale. Le depressioni del terreno a motivi geometrici contengono ancora ghiaccio di anidride carbonica, che li fa risaltare nell'immagine. Si noti l'assenza di strutture poligonali all'interno dei crateri (centro-destra) anche se c'è una frattura del pavimento in quello superiore. La dimensione e la forma di tali poligoni spesso riflettono lo spessore dello strato fratturato e le condizioni di stress locali che possono essere controllate dai pendii.

ПОЛИГОНЫ В ПАТЕРЕ ПИТИУСА

На этом снимке показан примечательный полигональный рельеф в районе патеры Питиуса. Снимок был сделан в середине весны в южном полушарии. Желоба между ровными участками всё ещё содержат лёд углекислого газа, что выделяет полигональный узор на изображении. Обратите внимание на отсутствие подобных узоров внутри кратеров (в центре справа), хотя в верхнем из них есть трещины на дне. Размер и форма многоугольников между трещинами часто отражают толщину их слоя и условия локальных напряжений, которые могут определяться уклонами поверхности.

6.17 POLYGONS IN PITYUSA PATERA

ACQUISITION TIME: 2020-06-22T06:57:12.140
LONGITUDE: 40.2 E   LATITUDE: 67.4 S
REGION: PITYUSA

POLYGONAL GEMUSTERTER BODEN IN MITTLEREN BREITEN

Utopia Rupes liegt auf der westlichen Seite von Utopia Planitia. Ein Grossteil der Region zeigt Anzeichen für periglaziale Prozesse. Die beobachteten Landformen sind möglicherweise die Folge tiefer Temperaturen unter dem Gefrierpunkt und des Expansions-/Kontraktionszyklus von Feuchtigkeit und Sedimenten in den oberen Schichten der Oberfläche, wenn diese gefrieren und auftauen. Dieses Bild zeigt einen polygonal gemusterten Boden, der ein Gebiet mit Kratern und Hügeln umgibt. Interessanterweise weisen einige der Hügel schwache Anzeichen des polygonalen Musters auf.

POLYGONAL PATTERNED GROUND AT MID-LATITUDES

Utopia Rupes is on the western side of Utopia Planitia. Much of the region shows evidence of periglacial processes. The observed landforms may be the consequence of cold sub-freezing temperatures and the expansion/contraction cycle of moisture and sediments in the upper layers of the surface as they freeze and thaw. This image shows polygonal patterned ground surrounding an area of craters and mounds. Interestingly, a couple of the mounds show some weak evidence of the polygonal pattern.

TERRAIN À MOTIFS POLYGONAUX AUX LATITUDES MOYENNES

Utopia Rupes se trouve dans la partie ouest d'Utopia Planitia. Une grande partie de la région présente des signes de processus périglaciaires. Les formes de terrain observées peuvent être la conséquence de températures froides sous le point de condensation ainsi que du cycle d'expansion/contraction de l'humidité et des sédiments dans les couches supérieures de la surface lorsqu'elles gèlent et dégèlent. Cette image montre un sol à motifs polygonaux entourant une zone de cratères et de monticules. Il est intéressant de noter que quelques monticules présentent de faibles traces du motif polygonal.

TERRENO POLIGONALE ALLE MEDIE LATITUDINI

Utopia Rupes si trova sul lato occidentale di Utopia Planitia. Gran parte della regione mostra prove di processi periglaciali. Le conformazioni del terreno possono essere la conseguenza di temperature fredde sotto lo zero e del ciclo di espansione/contrazione causato dall'umidità nei sedimenti sugli strati superiori della superficie mentre si congelano e si scongelano. Questa immagine mostra un terreno poligonale che circonda un'area di crateri e tumuli. È interessante notare che un paio di tumuli mostrano una debole evidenza di strutture poligonali.

ПОЛИГОНАЛЬНЫЙ УЗОРЧАТЫЙ ГРУНТ В СРЕДНИХ ШИРОТАХ

Уступ Утопия находится на западной стороне равнины Утопия. На большей части региона наблюдаются перигляциальные процессы. Эти формы рельефа могут быть следствием низких температур, а также цикла расширения/сжатия влаги и отложений в верхних слоях поверхности по мере их замерзания и оттаивания. На этом снимке показан полигональный узорчатый грунт, окружающий область кратеров и возвышений. Интересно, что пара возвышенностей здесь демонстрирует едва заметные признаки наличия полигонального рисунка.

6.18 POLYGONAL PATTERNED GROUND AT MID-LATITUDES

ACQUISITION TIME: 2021-05-17T03:30:46.754
LONGITUDE: 87.0 E LATITUDE: 42.1 N
REGION: UTOPIA

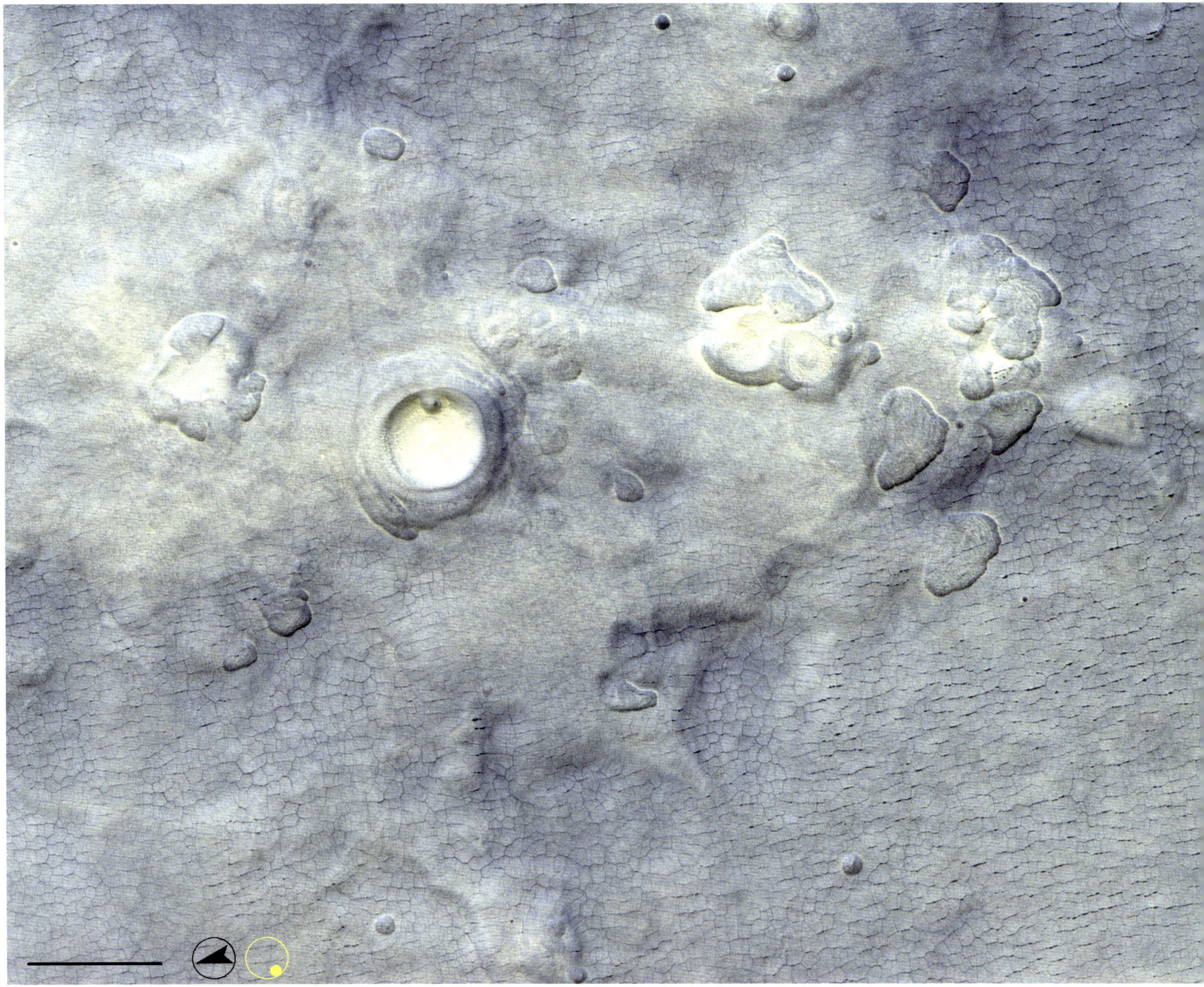

7 VULKANE, LAVEN, GRÄBEN, GRUBEN UND GÄNGE

VOLCANOES, LAVAS, GRABEN, PITS AND DIKES

VOLCANS, LAVES, GRABEN, FOSSES ET DYKES

VULCANI, COLATE LAVICHE, GRABEN, FOSSE E DICCHI

ВУЛКАНЫ, ЛАВЫ, ГРАБЕНЫ, ЯМЫ И ДАЙКИ

DIE MEHRFACHE CALDERA VON HECATES THOLUS

Hecates Tholus liegt in Elysium Planitia, und dieses Bild zeigt die Ostseite der Calderen (es gibt mindestens fünf). Die vulkanische Aktivität in der Region hat erst vor relativ kurzer Zeit aufgehört – vielleicht erst vor 100 Millionen Jahren. Man kann auch talähnliche Strukturen erkennen, die möglicherweise durch das Abschmelzen von Gletschern entstanden sind. Diese haben sich wahrscheinlich nach dem Ende der vulkanischen Aktivität gebildet und sind daher geologisch gesehen sehr jung. Sie könnten eine Folge der periodischen Änderungen der Schiefe des Mars sein.

THE MULTIPLE CALDERA OF HECATES THOLUS

Hecates Tholus is in Elysium Planitia and this image shows the east side of the calderas (there are at least 5). Volcanic activity in the region ceased relatively recently – perhaps as little as 100 million years ago. One can also see valley-like structures that may have resulted from melting of glaciers. These probably formed after volcanic activity stopped and are therefore geologically very recent indeed. They may be a consequence of the periodic changes in the obliquity of Mars.

LA CALDEIRA MULTIPLE D'HECATES THOLUS

Hecates Tholus se situe dans la région d'Elysium Planitia et cette image montre le côté est des caldeiras (il y en a au moins 5). L'activité volcanique dans la région a cessé relativement récemment – il y a peut-être seulement 100 millions d'années. On peut également voir des structures ressemblant à des vallées qui pourraient être le résultat de la fonte des glaciers. Ces structures se sont probablement formées après l'arrêt de l'activité volcanique et sont donc géologiquement très récentes. Elles peuvent être une conséquence des changements périodiques de l'obliquité de Mars.

LA CALDERA MULTIPLA DI HECATES THOLUS

Hecates Tholus si trova in Elysium Planitia e questa immagine mostra il lato est delle caldere (ce ne sono almeno 5). L'attività vulcanica nella regione è cessata relativamente di recente – forse appena 100 milioni di anni fa. Si possono anche vedere strutture simili a valli che possono essere il risultato dello scioglimento dei ghiacciai. Queste probabilmente si sono formate dopo la cessazione dell'attività vulcanica e sono quindi geologicamente molto recenti. Possono essere una conseguenza dei cambiamenti periodici dell'obliquità di Marte.

МНОГОЧИСЛЕННЫЕ КАЛЬДЕРЫ КУПОЛА ГЕКАТЫ

На этом снимке показана восточная сторона кальдер (их, как минимум, 5), расположенных в пределах Купола Гекаты на равнине Элизий. Вулканическая активность в этом районе прекратилась относительно недавно – возможно, не более 100 миллионов лет назад. Также видны структуры, похожие на долины, которые могли возникнуть в результате таяния ледников. Они, вероятно, образовались после прекращения вулканической активности и, следовательно, геологически очень молоды. Они могут быть следствием периодических изменений наклона оси Марса.

7.1 THE MULTIPLE CALDERA OF HECATES THOLUS

ACQUISITION TIME: 2018-06-09T03:13:13.000
LONGITUDE: 150.1 E LATITUDE: 31.5 N
REGION: ELYSIUM

EINE GRUBE UND EIN KRATER

Dieses Bild stammt von der Südostflanke des Alba Mons am Rande der Ceraunius Fossae. Es zeigt mehrere Merkmale, die in den vulkanischen Landschaften des Mars zu finden sind. Rechts oben ist ein gefüllter Einschlagkrater mit flachem Boden zu sehen. Daneben (oben in der Mitte) befindet sich eine Grube, die wahrscheinlich durch den Einsturz eines Hohlraums unter der Oberfläche entstanden ist, wobei dieser Hohlraum möglicherweise ein Lavatunnel oder ein durch Verwerfungen entstandener Hohlraum ist. In diese Grube führt von links (Norden) ein kleiner Kanal, und ein ähnlicher Kanal erstreckt sich rechts (Süden) der Grube. Der Ursprung dieses Kanals ist unbekannt, aber er ähnelt vulkanischen Kanälen in Lavaströmen auf der Erde und dem Mars. Das breite, flache Tal, das die untere rechte Seite des Bildes durchschneidet, ist wahrscheinlich ein Graben, der auf Dehnungsbrüche hinweist.

A PIT AND A CRATER

This image is from the southeast flank of Alba Mons on the edge of Ceraunius Fossae. It shows several features evident in the volcanic terrains of Mars. To the upper right is a filled impact crater with a flat floor. Adjacent to it (top centre) is a pit that was probably produced by the collapse of a sub-surface void, possibly a lava tube or one created by faulting. Leading into this pit from the left (north) is a small channel and a similar channel extends to the right (south) of the pit. The origin of this channel is unknown but it is similar in appearance to volcanic channels in lava flows on Earth and Mars. The broad, shallow valley cutting across the lower right of this image is probably a graben indicating extensional faulting.

UNE FOSSE ET UN CRATÈRE

Image du flanc sud-est d'Alba Mons, au bord de Ceraunius Fossae. Elle montre plusieurs caractéristiques fréquemment observées dans les terrains volcaniques de Mars. En haut à droite, on voit un cratère d'impact comblé avec un fond plat. À côté (en haut au centre) se trouve une fosse qui a probablement été produite par l'effondrement d'une cavité souterraine, peut-être un tube de lave ou une cavité créée par une faille. Un petit chenal mène à cette fosse depuis la gauche (nord) et un chenal similaire s'étend à droite (sud) de la fosse. L'origine de ce chenal est inconnue, mais il ressemble aux chenaux volcaniques des coulées de lave sur Terre et sur Mars. La vallée large et peu profonde qui traverse la partie inférieure droite de cette image est probablement un graben indiquant une faille d'extension.

UNA FOSSA E UN CRATERE

Questa immagine proviene dal fianco sud-est di Alba Mons sul bordo di Ceraunius Fossae. Mostra diverse evidenti caratteristiche dei terreni vulcanici di Marte. In alto a destra c'è un cratere da impatto riempito che presenta un fondo piatto. Adiacente ad esso (in alto al centro) si osserva una fossa che è stata probabilmente prodotta dal collasso di un vuoto sotto la superficie, probabilmente un tubo di lava o uno creato da una faglia. Un piccolo canale conduce a questa fossa da sinistra (nord) e un canale simile si estende a destra (sud) della fossa. L'origine di questo canale è sconosciuta, ma è simile nell'aspetto ai canali vulcanici formati nelle colate di lava sulla Terra e su Marte. La valle ampia e poco profonda che taglia la parte inferiore destra di questa immagine è probabilmente un graben che indica una faglia normale (cioè estensionale).

ЯМА И КРАТЕР

Это снимок юго-восточного склона горы Альба на краю Борозд Керавна. На нем видны несколько структур, характерных для вулканических рельефов Марса. Справа вверху находится заполненный ударный кратер с плоским дном. Рядом с ним (вверху в центре) находится яма, которая, вероятно, образовалась в результате обрушения подповерхностных пустот — возможно, лавовой трубки, или пустот, созданных тектоническим разломом. В эту яму слева (с севера) ведет небольшой канал, и такой же канал тянется справа (с юга) от ямы. Происхождение этого канала неизвестно, но по внешнему виду он похож на вулканические каналы в лавовых потоках на Земле и Марсе. Широкая, неглубокая долина, пересекающая правую нижнюю часть снимка, вероятно, является грабеном, маркирующим разлом растяжения.

7.2   A PIT AND A CRATER

ACQUISITION TIME: 2021-03-13T23:24:17.756
LONGITUDE: 251.6 E   LATITUDE: 37.4 N
REGION: THARSIS

259

BERGRÜCKEN UND SCHICHTEN IN ARABIA TERRA

Dieses Bild zeigt Sedimentschichten im Zentrum eines Kraters in Arabia Terra. Das Bild zeigt auch ausgeprägte Felsgrate. Solche Grate werden oft mit Gesteinsgängen (vertikalen Platten aus intrudiertem Magma) in Verbindung gebracht, aber die Nähe zu den Sedimentablagerungen macht diese Idee zu einer weniger attraktiven Erklärung. Ihr Entstehungsmechanismus bleibt daher derzeit unklar, könnte aber mit der Verhärtung des Gesteins zusammenhängen, als Flüssigkeiten entlang von Sedimentrissen flossen.

RIDGES AND LAYERS IN ARABIA TERRA

This image shows sedimentary layers at the centre of a crater in Arabia Terra. The image also shows prominent ridges. Ridges of this type are often associated with dikes (vertical sheets of intruded magma) but the close proximity of sedimentary deposits make this idea here a less attractive explanation. Their formation mechanism therefore remains unclear at present but may be related to hardening rocks as fluids flowed along cracks in the sediments.

CRÊTES ET STRATES DANS LA RÉGION D'ARABIA TERRA

Cette image montre des couches sédimentaires au centre d'un cratère d'Arabia Terra. L'image montre également des crêtes proéminentes. Les crêtes de ce type sont souvent associées à des dikes (intrusions verticales de magma), mais la proximité des dépôts sédimentaires rend cette interprétation peu probable ici. Leur mécanisme de formation reste donc peu clair à l'heure actuelle mais pourrait être lié au durcissement des roches lors de l'écoulement de fluides le long de fissures dans les sédiments.

CRESTE E STRATI IN ARABIA TERRA

Questa immagine mostra strati sedimentari al centro di un cratere in Arabia Terra. L'immagine mostra anche creste prominenti. Le creste di questo tipo sono spesso associate a dicchi (strati verticali di magma intruso) ma la vicinanza dei depositi sedimentari rende questa ipotesi una spiegazione meno attraente. Il loro meccanismo di formazione rimane quindi poco chiaro al momento, ma potrebbe essere legato alla solidificazione di rocce avvenuta mentre dei liquidi scorrevano lungo le crepe nei sedimenti.

ХРЕБТЫ И СЛОИ В ЗЕМЛЕ АРАВИЯ

На этом снимке показаны осадочные слои в центре кратера на Земле Аравия. На снимке также видны внушительные хребты. Подобного типа возвышенности часто ассоциируются с дайками субвертикальными телами, сформированными внедрениями магмы, но близкое соседство осадочных отложений делает эту идею менее привлекательным объяснением. Поэтому механизм образования гребней в настоящее время остается неясным. Он может быть связан с затвердеванием пород при движении флюидов по трещинам в отложениях и последующей топографической инверсией.

7.3 RIDGES AND LAYERS IN ARABIA TERRA

ACQUISITION TIME: 2021-04-19T05:59:33.250
LONGITUDE: 27.9 E   LATITUDE: 28.2 N
REGION: ARABIA

261

GRATE UND MULDEN BEIM TERBY-KRATER

Dieses Bild stammt aus der Region Tyrrhena Terra nordöstlich von Hadriacus Palus. Es zeigt den Boden gegen das Zentrum eines ≈ 90 km grossen Kraters. Die Oberfläche weist zahlreiche Felsgrate und Schichten auf, die unterschiedlich widerstandsfähig gegen Erosion sind. Beim helleren, einfach erodierten Material handelt es sich wahrscheinlich um Sedimente, möglicherweise von einem See, der sich in diesem Krater gebildet hat. Das kleine dunkle Plateau in der Mitte des Bildes erinnert daran, wie Lavaströme Tafelberge bedecken, aber es ist auch möglich, dass es sich um widerstandsfähigere Sedimente handelt, die möglicherweise aus gröberem Material bestehen, das von dynamischeren Strömen eingebracht wurde.

RIDGES AND DEPRESSIONS NEAR TERBY CRATER

This image is from the Tyrrhena Terra region north-east of Hadriacus Palus. It shows the floor towards the centre of an ≈ 90 km crater. The surface has numerous ridges and layers that are variably resistant to erosion. The lighter toned, easily eroded material is likely to be sediments, possibly from a lake that formed in this crater. The small dark plateau in the center of this image is reminiscent of how lava flows cap mesas but it is also possible that these are more resistant sediments, possibly made up of coarser materials brought in by more energetic flows.

CRÊTES ET DÉPRESSIONS PRÈS DU CRATÈRE TERBY

Cette image provient de la région de Tyrrhena Terra au nord-est d'Hadriacus Palus. Elle montre le sol au centre d'un cratère d'environ 90 km. La surface présente de nombreuses crêtes et couches qui sont plus ou moins résistantes à l'érosion. Les matériaux plus clairs, facilement érodés, sont probablement des sédiments, peut-être issus d'un lac qui s'était formé dans ce cratère. Le petit plateau sombre au centre de l'image rappelle la façon dont les coulées de lave coiffent les mésas, mais il est également possible qu'il s'agisse de sédiments plus résistants, peut-être composés de matériaux plus grossiers apportés par des courants plus énergiques.

CRESTE E DEPRESSIONI VICINO AL CRATERE TERBY

Questa immagine proviene dalla regione di Tyrrhena Terra a nord-est di Hadriacus Palus. Mostra il pavimento vicino al centro di un cratere di ≈ 90 km. La superficie ha numerose creste e stratificazioni che sono variamente resistenti all'erosione. Il materiale più chiaro e facilmente erodibile è probabilmente costituito da sedimenti, forse provenienti da un lago che si formò in questo cratere. Il piccolo altopiano scuro al centro di questa immagine ricorda il modo in cui le colate di lava ricoprono la mesa, ma è anche possibile che si tratti di sedimenti più resistenti, forse costituiti da materiali più grossolani portati da correnti più intense.

ХРЕБТЫ И ВПАДИНЫ В РАЙОНЕ КРАТЕРА ТЕРБИ

Это снимок части Тирренской Земли к северо-востоку от Адриатического болота (классическая деталь альбедо Hadriacus Palus). На нём запечатлено дно вблизи центра 90-километрового кратера. Эта поверхность имеет многочисленные гребни и слои, в разной степени устойчивые к эрозии. Более светлый, легко поддающийся эрозии материал, вероятно, является отложениями озера, образованного в этом кратере. Небольшое темное плато в центре изображения напоминает покрытую застывшей лавой столовую гору, хотя так же вероятно, что это просто устойчивые отложения, состоящие, возможно, из грубых материалов, принесенных мощными потоками.

7.4 RIDGES AND DEPRESSIONS NEAR TERBY CRATER

ACQUISITION TIME: 2020-12-18T23:56:37.333
LONGITUDE: 80.8 E   LATITUDE: 24.1 S
REGION: TYRRHENA

MULDE IN NOACHIS

Dies ist ein Bild von Nirgal Vallis in Noachis Terra. Dieses Tal ist etwa 610 km lang und hätte sein Wasser in das Uzboi Vallis und von dort in den Holden-Krater abgeleitet. Dieses Bild zeigt Erosionsrinnen auf der Nordseite (linke Seite) des Tals und zeigt auch die Vielfalt der Zusammensetzung in der Ebene im Norden. Die Nebenflüsse des Nirgal Vallis sind sehr kurz und gedrungen. Dies wurde als Hinweis auf die Entstehung des Tals durch eine unterirdische Wasserquelle und nicht durch Niederschläge oder schmelzendes Eis an der Oberfläche gedeutet.

TROUGH IN NOACHIS

This is an image of Nirgal Vallis in Noachis Terra. This valley is around 610 km long and would have discharged its water into Uzboi Vallis and from there into Holden Crater. This image shows gullies on the north side (left side) of the valley and also shows compositional diversity on the plain to the north. The tributaries of Nirgal Vallis are very short and stubby. This has been interpreted as indicating valley formation through a sub-surface source of water rather than precipitation or melting ice on the surface.

FOSSE DANS NOACHIS

Ceci est une image de Nirgal Vallis dans Noachis Terra. Cette vallée fait environ 610 km de long et Image de Nirgal Vallis dans Noachis Terra. Cette vallée fait environ 610 km de long et aurait déversé ses eaux dans Uzboi Vallis, puis dans le cratère Holden. Cette image montre des ravines sur le côté nord (côté gauche) de la vallée et montre également la diversité de composition de la plaine au nord. Les affluents de Nirgal Vallis sont très courts. Cela a été interprété comme indiquant une formation de la vallée par une source d'eau souterraine plutôt que par des précipitations ou la fonte de la glace en surface.

VALLE IN NOACHIS

Questa è un'immagine di Nirgal Vallis in Noachis Terra. Questa valle è lunga circa 610 km e avrebbe scaricato le sue acque nella Uzboi Vallis e da lì nel Cratere Holden. Questa immagine mostra calanchi sul lato nord (lato sinistro) della valle e mostra anche una diversità compositiva sulla pianura a nord. Gli affluenti di Nirgal Vallis sono molto corti e tozzi. Questo è stato interpretato come una indicazione che la formazione della valle sia avvenuta attraverso una sorgente d'acqua sub-superficiale piuttosto che dovuta a precipitazioni o scioglimento del ghiaccio in superficie.

ВПАДИНА В ЗЕМЛЕ НОЯ

Это изображение долины Нергала на земле Ноя. Протяженность ее порядка 610 км, а ее воды должны были сбрасываться в долину Узбой и далее в кратер Холден. Это изображение показывает овраги на северной стороне долины (на её левом склоне), а также композиционное разнообразие равнины к северу. Притоки долины Нергала очень коротки и обрывисты, что интерпретируется как свидетельство формирования долины за счет подповерхностного источника воды, а не осадков или таяния льда на поверхности.

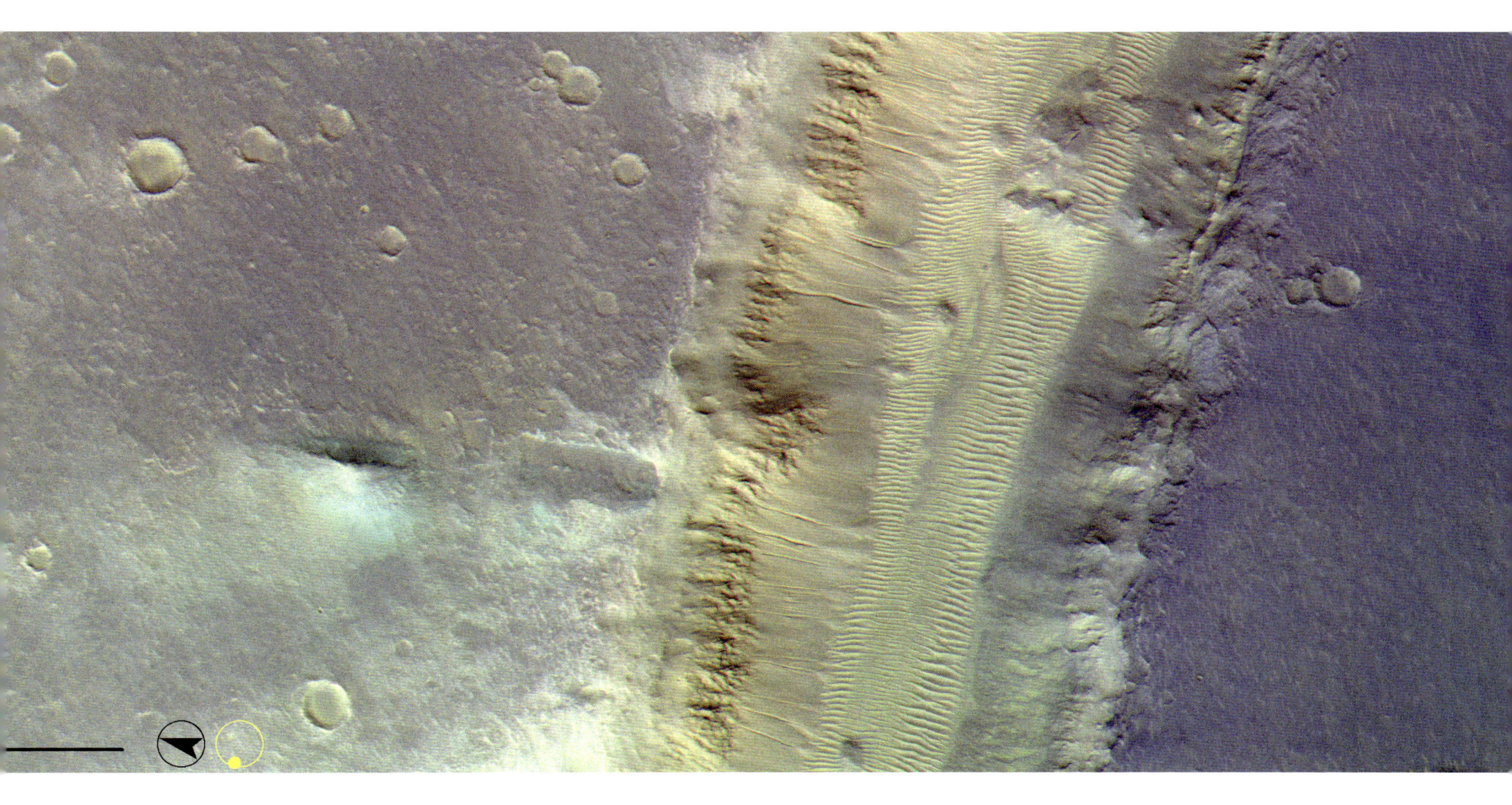

7.5   TROUGH IN NOACHIS

ACQUISITION TIME: 2020-11-22T21:06:28.980
LONGITUDE: 318.6 E   LATITUDE: 27.9 S
REGION: NOACHIS

ERODIERTER GRABEN MIT INNEREN SCHICHTEN

Dieser Ausschnitt eines langen Grabens in den Cerberus Fossae zeigt Schichten innerhalb des Grabens an der Nordwand (links). Der längste Bruch ist ≈ 1200 km lang. Dieser Teil der Cerberus Fossae befindet sich an der Spitze der Athabasca Valles und wird als Quelle von Lava- und/oder Wasserfluten interpretiert. Die Brüche und Ströme scheinen für den Mars extrem jung zu sein, möglicherweise nur wenige Millionen Jahre alt. Man beachte den einzelnen relativ frischen kleinen Krater unten rechts.

ERODED GRABEN WITH INTERIOR LAYERS

This image of part of a long graben in Cerberus Fossae shows layers within the graben in the north (left) wall. The longest fracture is ≈ 1200 km long. This part of the Cerberus Fossae is at the head of the Athabasca Valles and is interpreted to be the source of floods of lava and/or water. The fractures and flows appear extremely young for Mars, possibly as little as a few million years old. Note the lone relatively fresh small crater to the bottom right.

GRABEN ÉRODÉ AVEC STRATES INTÉRIEURES

Cette image d'une partie d'un long graben dans Cerberus Fossae montre les couches à l'intérieur du graben dans sa paroi nord (gauche). La plus longue fracture s'étend sur environ 1200 km. Cette partie de Cerberus Fossae se trouve à l'extrémité d'Athabasca Valles et est interprétée comme étant la source de l'eau ou de la lave qui a jadis inondé Athabasca Valles. Les fractures et les écoulements semblent extrêmement jeunes pour Mars, peut-être aussi récents que quelques millions d'années. Notez le petit cratère isolé relativement frais en bas à droite.

GRABEN EROSO CON STRATI INTERNI

Questa immagine di parte di un lungo graben a Cerberus Fossae mostra la stratificazione all'interno di un graben nella parete nord (sinistra). La frattura più lunga è lunga ≈ 1200 km. Questa parte del Cerberus Fossae si trova all'origine delle Athabasca Valles ed è interpretata come la sorgente di passate inondazioni di lava e/o acqua. Le fratture e le colate appaiono estremamente giovani per Marte, forse di pochi milioni di anni. Si noti l'unico piccolo cratere relativamente fresco in basso a destra.

ЭРОДИРОВАННЫЙ ГРАБЕН С ВНУТРЕННИМИ СЛОЯМИ

Этот снимок участка протяжённого грабена в бороздах Цербера демонстрирует наличие слоистой толщи в северной (левой) стене. Протяженность самой длинной трещины между слоями составляет около 1200 км. Данная часть борозд Цербера находится в верховьях долины Атабаска и интерпретируется как источник потоков лавы и/или воды. Разломы и застывшие потоки выглядят чрезвычайно молодыми для Марса — возможно, их возраст составляет всего несколько миллионов лет. Обратите внимание на небольшой относительно недавний кратер в правом нижнем углу.

7.6 ERODED GRABEN WITH INTERIOR LAYERS

ACQUISITION TIME: 2021-04-21T22:56:47.420
LONGITUDE: 157.6 E   LATITUDE: 10.4 N
REGION: ELYSIUM

GRABEN IN DEN SIRENUM FOSSAE

Die Mitte dieses Bildes wird von einem der Gräben der Sirenum Fossae in der Nähe von Gorgonum Chaos in Terra Sirenum durchschnitten. Es wird angenommen, dass die Gräben zur Zeit der Entwicklung von Tharsis entstanden sind. Talnetzwerke scheinen die Gräben zu verändern und dürften daher aus der Zeit nach ihnen stammen. In diesem Bild sichtbare hell getönte Ablagerungen könnten mit wässrigen Prozessen zusammenhängen, die die Täler gebildet haben könnten.

GRABEN IN SIRENUM FOSSAE

Cutting across the centre of this image is one of the graben of Sirenum Fossae near Gorgonum Chaos in Terra Sirenum. The graben are thought to have originated at the time of the development of Tharsis. Valley networks appear to modify the graben and thus should post-date them. Light-toned deposits outcropping in this image could be related to aqueous processes that might have formed the valleys.

GRABEN DANS SIRENUM FOSSAE

Au centre de cette image se trouve l'un des grabens de Sirenum Fossae près de Gorgonum Chaos dans Terra Sirenum. On pense que ce système extensif date de l'époque du développement de Tharsis. Les réseaux de vallées semblent modifier les grabens et devraient donc leur être postérieurs. Les dépôts de couleur claire qui affleurent sur cette image pourraient être liés à des processus aqueux qui ont pu former les vallées.

GRABEN IN SIRENUM FOSSAE

Quello che taglia il centro di questa immagine è uno dei graben di Sirenum Fossae vicino a Gorgonum Chaos in Terra Sirenum. Si pensa che il graben abbia avuto origine al tempo dello sviluppo della regione Tharsis. Le reti di valli sembrano modificare i graben e quindi questi ultimi sono probabilmente più antichi. I depositi di colore chiaro che affiorano in questa immagine potrebbero essere legati a processi acquosi che hanno potenzialmente formato queste valli.

ГРАБЕН В БОРОЗДАХ СИРЕН

Через центр этого изображения проходит один из грабенов борозд Сирен возле Хаоса Горгоны на земле Сирен. Считается, что эти грабены возникли во времена формирования Фарси, ды. Сети долин изменяют грабен следовательно, они должны были возникнуть позже него. Светлые отложения, выходящие на поверхность на этом снимке, могут быть связаны с водными процессами, которые возможно сформировали долины.

7.7 GRABEN IN SIRENUM FOSSAE

ACQUISITION TIME: 2020-12-31T22:30:29.994
LONGITUDE: 190.4 E   LATITUDE: 37.6 S
REGION: SIRENUM

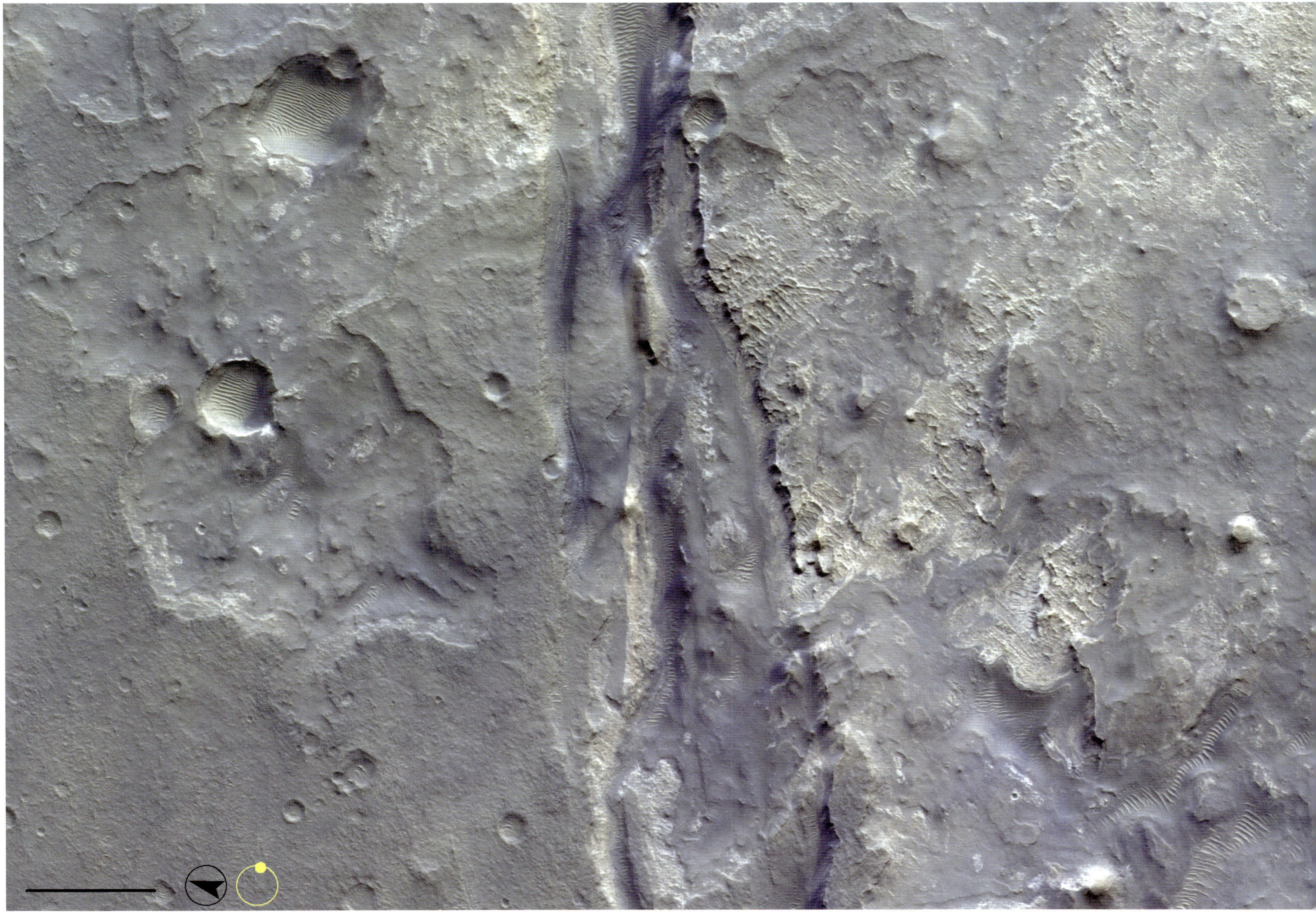

GRABEN UND EIN KRATER IN DEN LABEATIS FOSSAE

Mulden, manchmal auch Graben genannt, bilden sich, wenn sich die Kruste über die Bruchstelle hinaus ausdehnt. Es entstehen zwei steile Klippen, wobei der Bereich dazwischen nach unten sinkt. Diese besondere Gruppe von Gräben zeigt raues Gelände auf den Böden mit Hangstreifen (engl. slope streaks) und möglichen Erosionsrinnen an den steilen Wänden. Oben links ist ein Krater mit flachem Boden zu sehen, in dem sich möglicherweise eine Ablagerung befindet.

GRABEN AND A CRATER IN LABEATIS FOSSAE

Troughs, sometimes also called graben, form when the crust stretches past the breaking point. Two steep cliffs are produced with the area between sinking downwards. This particular set of graben show rough terrain on the floors with slope streaks and possible gullies on the steep walls. A flat-bottomed crater with a possible interior deposit is seen to the top left.

GRABEN ET CRATÈRE DANS LABEATIS FOSSAE

Les fossés, parfois aussi appelés grabens, se forment lorsque la croûte s'étire au-delà de son point de rupture. Deux falaises abruptes sont produites, la zone située entre les deux s'enfonçant vers le bas. Cet ensemble particulier de grabens présente un terrain rugueux au fond, avec des stries de pente et de possibles ravines sur les parois abruptes. Un cratère à fond plat avec un possible dépôt interne est visible en haut à gauche.

GRABEN E UN CRATERE IN LABEATIS FOSSAE

Le depressioni tettoniche, a volte chiamate anche graben, si formano quando la crosta si distende oltre il punto di rottura. Vengono così prodotte due ripide scarpate con un'area tra di esse che sprofonda verso il basso. Questa particolare serie di graben mostra un terreno irregolare sul pavimento, con strisce scure simili a frane sulle pareti e possibili canaloni sulle scarpate ripide. In alto a sinistra si può vedere un cratere a fondo piatto con un possibile deposito interno.

ГРАБЕН И КРАТЕР В БОРОЗДАХ ЛАБЕАТИС

Желоба, называемые грабенами, возникают, когда кора планеты растягивается до предела прочности. Образуются два протяжённых крутых уступа, область между которыми опускается вниз. Этот конкретный набор грабенов демонстрирует неровный рельеф дна со склоновыми полосами и возможными оврагами на крутых стенах. Слева вверху виден кратер с плоским дном и признаками осадочных пород.

7.8 GRABEN AND A CRATER IN LABEATIS FOSSAE

ACQUISITION TIME: 2019-09-25T09:35:09.905
LONGITUDE: 285.5 E   LATITUDE: 30.6 N
REGION: TEMPE

TAFELBERGÄHNLICHE QUASI-KREISFÖRMIGE STRUKTUR IN ELYSIUM

Dieses Bild zeigt ein rätselhaftes Gelände entlang dem südlichen Rand von Elysium Planitia, wo es auf die Medusae-Fossae-Formation trifft. Der Boden in diesem Gebiet ist chaotisch angehoben und in einem Muster zerrissen, das auf der Erde nicht vorkommt. Es gibt derzeit keine gute Erklärung dafür, wie diese Morphologie entsteht, aber eine Hypothese besagt, dass Lava unter eisreiche Sedimente eindringt. Die Hebung durch die Intrusion in Verbindung mit der Hitze, die das Eis verdrängt, könnte diese Art von Zerrüttung hervorrufen.

MESA-LIKE QUASI-CIRCULAR STRUCTURE IN ELYSIUM

This image shows enigmatic terrain along the southern margin of Elysium Planitia where it encounters the Medusae Fossae Formation. The ground in this area is chaotically uplifted and disrupted in a pattern that is not seen on Earth. There is currently no good explanation for how this morphology forms but one hypothesis is that lava intrudes underneath ice-rich sediment. The uplift from the intrusion coupled with the heat driving out the ice might plausibly create this type of disruption.

STRUCTURE QUASI-CIRCULAIRE EN FORME DE MÉSA À ELYSIUM

Cette image montre un terrain énigmatique le long de la marge sud d'Elysium Planitia où elle rencontre la formation de Medusae Fossae. Le sol dans cette zone possède une topographie chaotique et perturbée d'une façon non observée sur Terre. Il n'y a actuellement aucune explication définitive de la façon dont cette morphologie se forme, mais une hypothèse est que la lave fait intrusion sous des sédiments riches en glace. Le soulèvement dû à l'intrusion, associé à la fonte de la glace, pourrait vraisemblablement créer ce type de perturbation.

STRUTTURA QUASI CIRCOLARE SIMILE A UNA MESA IN ELYSIUM

Questa immagine mostra una regione enigmatica lungo il margine meridionale di Elysium Planitia dove incontra la Formazione Medusae Fossae. Il terreno in questa zona è sollevato caoticamente e fratturato in un disegno che non si osserva sulla Terra. Attualmente non esiste una buona spiegazione per come si forma questa morfologia, ma un'ipotesi è che la lava si intruda sotto sedimenti ricchi di ghiaccio. Il sollevamento causato dall'intrusione accoppiato al calore che spinge fuori il ghiaccio potrebbe plausibilmente creare questo tipo di fratturazione.

СТОЛОВАЯ ОКРУГЛАЯ СТРУКТУРА НА ПЛАТО ЭЛИЗИЙ

На этом снимке показан загадочный ландшафт вдоль южного края плато Элизий, где последний сближается с районом борозд Медузы. Земля в этой области хаотично поднята и нарушена с необычной закономерностью, которая не встречается на Земле. В настоящее время нет хорошего объяснения тому, как формируется такой морфологический тип поверхности, но одна из гипотез заключается в том, что лава проникает под богатые льдом отложения. Подъем, вызванный интрузией (внедрением), в сочетании с теплом, удаляющим лёд, может привести к образованию такого типа нарушений поверхности.

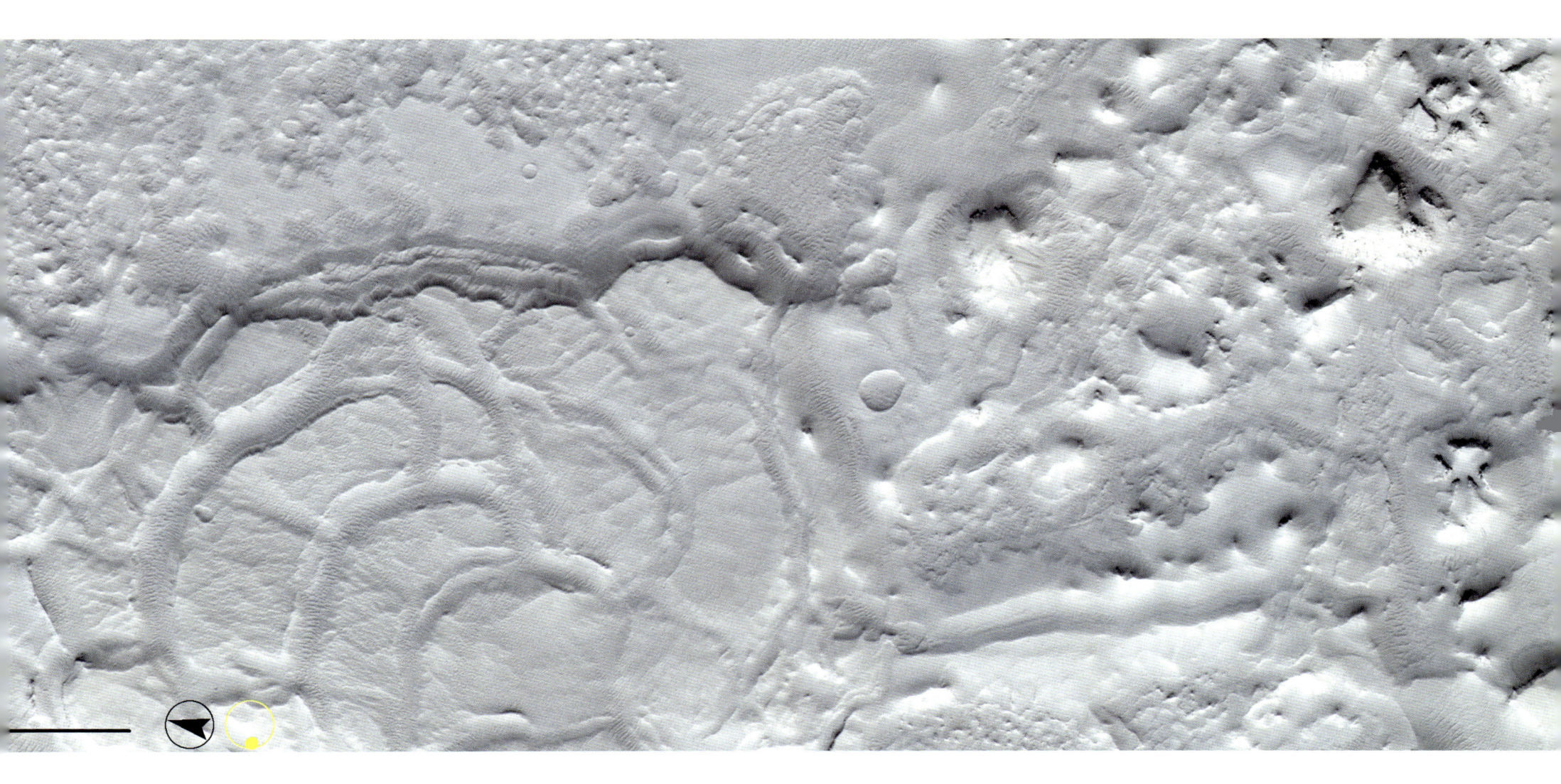

7.9   MESA-LIKE QUASI-CIRCULAR STRUCTURE IN ELYSIUM

ACQUISITION TIME: 2020-11-23T06:47:18.001
LONGITUDE: 168.2 E   LATITUDE: 2.4 S
REGION: ELYSIUM

GRABEN IN DEN CLARITAS FOSSAE

Dieses Bild zeigt den Rand eines Kraters mit einem Durchmesser von 50 km, der von einem Teil des Warrego-Valles-Systems durchquert wird. Der Krater mit einem Durchmesser von 3 km auf der linken Seite des Bildes ist das Ergebnis eines Einschlags in den Rand des grösseren Kraters. Das Bild befindet sich offiziell in den Claritas Fossae, aber an der Grenze zu den Thaumasia Fossae. Die Region Claritas Fossae ist ein dicht zergliedertes Gebiet, wobei die Brüche in der Kruste wahrscheinlich durch die Verformung durch das Gewicht der Tharsis-Ausbuchtung entstanden sind. Warrego Valles ist ein Netzwerk von verzweigten Tälern in den Thaumasia Fossae, die auf einen Wasserfluss in diesem Gebiet hindeuten, aber es gibt auch alte Täler, die möglicherweise vor der Entstehung der Gräben entstanden sind. Die Quelle(n) des Wassers, das die Täler gebildet hat, ist umstritten.

GRABEN IN CLARITAS FOSSAE

This image is of the rim of a 50-km-diameter crater which is crossed by part of the Warrego Valles system. The 3-km-diameter crater to the left of the image is the result of an impact into the rim of the larger crater. The image is formally in Claritas Fossae but on the border with Thaumasia Fossae. The Claritas Fossae region is densely dissected terrain with the breaks in the crust probably arising from deformation by the weight of the Tharsis bulge. Warrego Valles is a network of branching valleys in Thaumasia Fossae that suggest flow of water in this area but there are old valleys possibly created before the graben were created. The source(s) of water to produce the valleys is disputed.

GRABEN DANS CLARITAS FOSSAE

Cette image représente le bord d'un cratère de 50 km de diamètre qui est traversé par une partie du système de Warrego Valles. Le cratère de 3 km de diamètre à gauche de l'image est le résultat d'un impact sur le bord du plus grand cratère. L'image se trouve dans Claritas Fossae, à la frontière avec Thaumasia Fossae. La région de Claritas Fossae présente des terrains densément fracturés, les cassures de la croûte provenant probablement de la déformation due au poids du bombement de Tharsis. Warrego Valles est un réseau de vallées ramifiées dans Thaumasia Fossae qui suggère un écoulement d'eau dans cette région, mais il existe d'anciennes vallées qui peuvent prédater la formation du graben. La source de l'eau à l'origine de ces vallées est encore débattue.

GRABEN IN CLARITAS FOSSAE

Questa immagine rappresenta il bordo di un cratere di 50 km di diametro che è attraversato da una parte del sistema Warrego Valles. Il cratere di 3 km di diametro a sinistra dell'immagine è il risultato di un impatto sul bordo del cratere più grande. L'immagine è formalmente in Claritas Fossae ma al confine con Thaumasia Fossae. La regione di Claritas Fossae è un terreno densamente crepato, con rotture nella crosta che probabilmente derivano dalla sua deformazione dovuta al peso delle catene montuese di Tharsis. Warrego Valles è una rete di valli ramificate in Thaumasia Fossae che suggeriscono un flusso d'acqua in quest'area, ma ci sono vecchie valli probabilmente create prima che si formasse il graben. L'origine (o le origini) dell'acqua necessaria a creare queste valli è dibattuta.

ГРАБЕН В БОРОЗДАХ КЛАРИТАС

Это изображение вала кратера диаметром 50 км, который пересекает часть системы долин Уоррего. Кратер диаметром 3 км слева на снимке – результат удара в обод большего кратера. Изображение формально относится к бороздам Кларитас, но лежит на границе с бороздами Тавмасии. Регион борозд Кларитас представляет собой сильно расчленённую местность с разрывами в коре, вероятно, возникшими в результате деформаций под действием веса купола Фарсиды. Уоррего – это сеть разветвленных долин в бороздах Тавмасии, которые указывают на потоки воды в этой области, но тут также есть и старые долины, возможно, возникшие до формирования грабена. Источник воды для образования долин остаётся неясным.

7.10 GRABEN IN CLARITAS FOSSAE

ACQUISITION TIME: 2021-01-01T18:11:22.040
LONGITUDE: 262.8 E   LATITUDE: 42.8 S
REGION: AONIA

275

CYANE CATENA

Cyane Catena ist eine Schachtkraterkette in einem Gebiet, das als Cyane Fossae bekannt ist und sich südwestlich von Alba Patera befindet. Wie aus dem 1-km-Massstab (unten links) ersichtlich ist, erstrecken sich die Krater über mehr als 40 km. Diese Gruben sind wahrscheinlich das Ergebnis eines Absinkens oder Einsturzes der Oberfläche in darunter liegende Hohlräume. Solche Hohlräume können sich entlang vulkanischer Gräben öffnen, wenn Magma die Kruste auseinander drückt und dann abfliesst, obwohl dies nicht der einzige mögliche Mechanismus zur Erklärung dieser Gruben ist. Sie lassen sich von Einschlagkratern unterscheiden, da sie in der Regel weder einen erhöhten Rand noch Anzeichen von Auswurfmaterial aufweisen. Ketten von Schachtkratern sind entlang vulkanischer Grabenbrüche auf der Erde zu sehen, zum Beispiel auf Hawaii.

CYANE CATENA

Cyane Catena is a pit crater chain within an area known as Cyane Fossae which is southwest of Alba Patera. As can be seen from the 1-km scale bar (lower left), the pits extend over more than 40 km. These pits are probably the result of sinking or collapse of the surface into voids below. Such voids can open along volcanic rifts as magma pushes the crust apart and then drains away, although this is not the only possible mechanism to explain these pits. They can be distinguished from impact craters because they usually lack an elevated rim or evidence of ejecta. Chains of pit craters can be seen along volcanic rift zones on Earth in Hawaii, for example.

CYANE CATENA

Cyane Catena est une chaîne de cratères de fosse dans une zone connue sous le nom de Cyane Fossae qui se trouve au sud-ouest d'Alba Patera. Comme on peut le voir sur la barre d'échelle de 1 km (en bas à gauche), les fosses s'étendent sur plus de 40 km. Ces fosses sont probablement le résultat de l'enfoncement ou de l'effondrement de la surface dans des cavités situées en dessous. De telles cavités peuvent s'ouvrir le long de rifts volcaniques lorsque le magma écarte la croûte et s'écoule ensuite, bien que ce ne soit pas le seul mécanisme possible pour expliquer ces fosses. On peut les distinguer des cratères d'impact car elles n'ont généralement pas de bord surélevé ni de trace d'éjecta. On peut observer des chaînes de cratères de fosses le long des zones de rift volcanique sur Terre, à Hawaï par exemple.

CYANE CATENA

Cyane Catena è una catena di crateri a pozzo all'interno di un'area conosciuta come Cyane Fossae che si trova a sud-ovest di Alba Patera. Come si può vedere dalla scala di 1 km (in basso a sinistra), i crateri a pozzo si estendono per più di 40 km. Queste depressioni sono probabilmente il risultato dello sprofondamento o del collasso della superficie in vuoti sottostanti. Tali vuoti possono aprirsi lungo i rift vulcanici quando il magma spinge la crosta a formare camere magmatiche e poi drena via, anche se questo non è l'unico meccanismo possibile per spiegare questi crateri a pozzo. Possono essere distinti dai crateri da impatto perché di solito mancano di un bordo elevato o di prove di ejecta. Catene di crateri a pozzo possono essere osservate lungo le zone di spaccatura vulcanica sulla Terra, per esempio alle Hawaii.

ЦЕПОЧКА КИАНЫ

Цепочка Кианы – это ряд ям (кратеров обрушения) в области, известной как борозды Кианы, которая находится к юго-западу от патеры Альба. Как видно из масштабной линейки в 1 км (слева внизу), цепочка простирается более чем на 40 км. Данные ямы, вероятно, являются результатом опускания или обрушения поверхности в пустоты под ней. Такие пустоты могут формироваться вдоль вулканических разломов, когда магма раздвигает кору, а затем опускается, хотя это не единственный возможный вариант объяснения. Кратеры обрушения можно отличить от ударных кратеров, поскольку у них обычно отсутствует приподнятый обод или следы выбросов. На Земле подобные цепочки обычно тянутся вдоль вулканических рифтовых зон, например, на Гавайях.

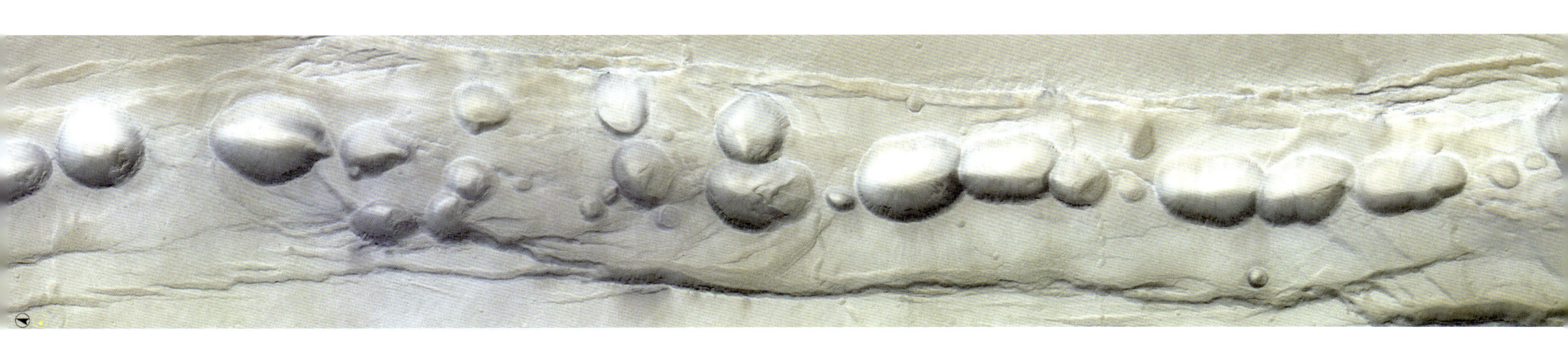

7.11 **CYANE CATENA**

ACQUISITION TIME: 2019-09-17T08:58:14.531
LONGITUDE: 241.6 E   LATITUDE: 36.4 N
REGION: CYANE

GRUBENKRATER IN DEN COPRATES CATENAE

Die Coprates Catenae bestehen aus einer Reihe von Schachtkraterketten, die ungefähr parallel zu Coprates Chasma auf der Südseite der Valles Marineris verlaufen. Dieses Bild zeigt einen tiefen Krater, der an der Südwand helles Material freilegt (rechts). Der Krater unmittelbar westlich davon (unterhalb des hier gezeigten Ausschnitts) zeigt Anzeichen von Sedimentfächern, und ein Kanal scheint in der Vergangenheit in diesen Krater gemündet zu haben. Die Grube im Osten ist grösser und weist ebenfalls Sedimentablagerungen auf. Dies ist hier nicht zu sehen, was darauf hindeutet, dass lokale Effekte die Oberfläche in dieser Region stark beeinflusst haben.

PIT CRATER IN COPRATES CATENAE

Coprates Catenae comprise lines of pit crater chains that run roughly parallel to Coprates Chasma on the southern side of Valles Marineris. This particular image is centered on a deep pit crater that exposes some light-toned material in the south wall (right). The pit crater immediately to the west (below the sub-frame shown here) shows evidence of sedimentary fan deposits and a channel appears to have discharged into that particular pit in the past. The pit to the east is larger and also has sedimentary deposits. This is not seen here suggesting local effects were important in influencing the surface in this region.

CRATÈRE D'EFFONDREMENT DANS COPRATES CATENAE

Coprates Catenae comprend des chaînes de cratères en forme de cuvette qui sont subparallèles à Coprates Chasma au sud de Valles Marineris. Cette image particulière est centrée sur un cratère profond qui expose des roches claires dans sa paroi sud (à droite). Le cratère de fosse immédiatement à l'ouest (en dessous du champ de vue montré ici) montre des signes de cônes de dépôts sédimentaires et un chenal semble s'être déversé dans cette fosse particulière par le passé. La fosse à l'est est plus grande et présente également des dépôts sédimentaires. Ici, d'autres processus locaux semblent avoir influencé la surface.

CRATERE A FOSSA CENTRALE IN COPRATES CATENAE

Coprates Catenae comprende catene di crateri a pozzo che corrono approssimativamente parallele a Coprates Chasma sul lato meridionale delle Valles Marineris. Questa particolare immagine è centrata su un profondo cratere a pozzo che esibisce del materiale chiaro nella parete sud (a destra). Il cratere a pozzo immediatamente a ovest (sotto la sottocornice mostrata qui) mostra prove di conoidi di deiezione e un canale sembra aver sversato in quel particolare cratere a pozzo in passato. Il cratere a pozzo a est è più grande e ha anche depositi sedimentari. Questi non si vedono invece in questa immagine, suggerendo che gli effetti locali sono stati importanti nell'influenzare la superficie in questa regione.

КРАТЕР ОБРУШЕНИЯ В ЦЕПОЧКЕ КОПРАТ

Цепочка Копрата состоит из ряда ям – кратеров обрушения, которые идут примерно параллельно каньону Копрата на южной стороне долин Маринера. На этом конкретном снимке в центре находится глубокий кратер, в южной стене которого обнажен материал светлых тонов (справа). В кратере обрушения, расположенном непосредственно на западе (ниже приведенного здесь субкадра), видны следы осадочных веерных отложений, а в прошлом в него, по-видимому, впадал канал. Кратерная яма на востоке (выше субкадра) больше и также имеет осадочные отложения. На данном снимке этих отложений не наблюдается, это говорит о том, что локальные эффекты оказывали существенное влияние на формирование поверхности в этом регионе.

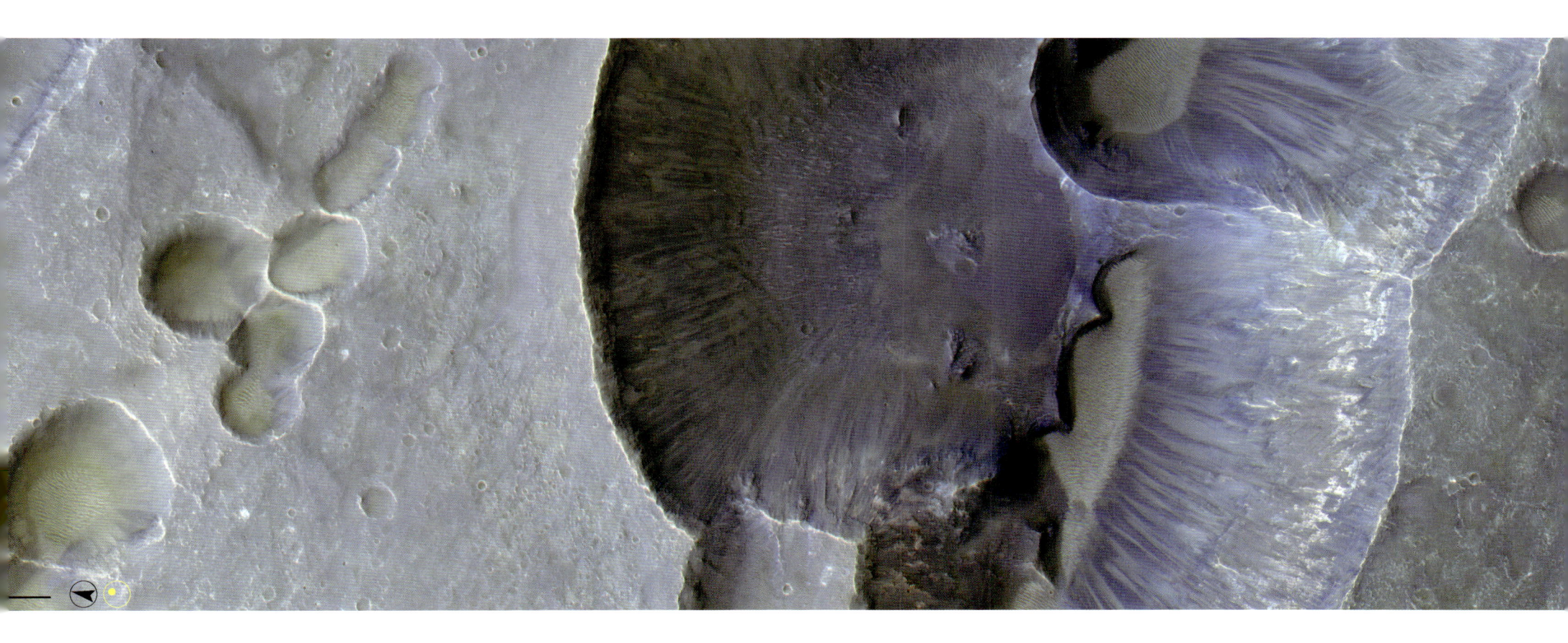

7.12 PIT CRATER IN COPRATES CATENAE

ACQUISITION TIME: 2021-04-02T05:29:40.272
LONGITUDE: 300.3 E LATITUDE: 14.6 S
REGION: MARINERIS

SICH KREUZENDE GRUBENKRATERKETTEN

Dieses Bild zeigt zwei sich kreuzende Schachtkraterketten an der Nordostflanke des Ascraeus Mons. Die grösseren Krater (circa in Nord-Süd-Richtung) scheinen älter und aufgefüllt zu sein. Die kleineren Krater (circa in Ost-West-Richtung) scheinen jünger zu sein und die ältere Kette zu durchschneiden. Auf dem gesamten Bild sind Anzeichen von Lavaströmen zu erkennen. Die Kraterketten können durch kollabierende Lavaröhren oder durch Hohlräume entstehen, die durch die Ausdehnung des Vulkangebäudes entstanden sind. In diesem Fall verläuft die ältere, von Norden nach Süden verlaufende Kette ungefähr parallel zur Richtung der Lavaströme, während die von Osten nach Westen verlaufende Kette senkrecht zur erwarteten Richtung der Lavaröhren verläuft.

CROSSING PIT CRATER CHAINS

This image shows two pit crater chains crossing on the northeastern flank of Ascraeus Mons. The larger pits (oriented roughly north-south) appear to be older and filled. The smaller pits (oriented roughly east-west) seem younger and cut across the older chain. There is evidence of lava flows throughout the image. The pit crater chains may be produced by collapsing lava tubes or voids opened by extension of the volcanic edifice. In this case, the older, north-south chain is roughly parallel to the direction of the lava flows but the east-west chain runs perpendicular to the expected direction of any lava tubes.

CROISEMENT DE CHAÎNES DE CRATÈRES D'EFFONDREMENT

Cette image montre deux chaînes de cratères d'effondrement se croisant sur le flanc nord-est d'Ascraeus Mons. Les plus grandes fosses (orientées grossièrement nord-sud) semblent être plus anciennes et remplies. Les plus petites fosses (orientées grossièrement est-ouest) semblent plus jeunes et coupent au travers des fosses anciennes. Il y a des traces de coulées de lave sur toute l'image. Les chaînes de cratères à fosse peuvent être produites par l'effondrement de tubes de lave ou de cavités ouvertes par l'extension de l'édifice volcanique. Dans ce cas, la chaîne plus ancienne, nord-sud, est à peu près parallèle à la direction des coulées de lave, mais la chaîne est-ouest est perpendiculaire à la direction attendue des tubes de lave.

CATENE DI CRATERI A POZZO

Questa immagine mostra due catene di crateri a pozzo che si incrociano sul fianco nord-orientale di Ascraeus Mons. I crateri a pozzo più grandi (orientati approssimativamente da nord a sud) sembrano essere più vecchi e riempiti. I crateri a pozzo più piccoli (orientati approssimativamente est-ovest) sembrano più giovani e tagliano la catena più vecchia. Ci sono prove di colate di lava in tutta l'immagine. Le catene di crateri a pozzo possono essere prodotte dal collasso di tubi di lava o da vuoti aperti dall'espansione dell'edificio vulcanico. In questo caso, la catena più vecchia, nord-sud, è approssimativamente parallela alla direzione delle colate di lava, ma la catena est-ovest corre perpendicolare alla direzione prevista dei tubi di lava.

ПЕРЕСЕКАЮЩИЕСЯ ЦЕПОЧКИ КРАТЕРОВ ОБРУШЕНИЯ

На этом снимке показаны две цепочки кратеров обрушения, пересекающихся на северо-восточной стороне горы Аскрийской. Крупные ямы (ориентированные примерно с севера на юг) кажутся более старыми и заполненными. Меньшие ямы (ориентированные примерно с востока на запад) кажутся более молодыми и пересекают старую цепь кратеров. По всему изображению видны следы лавовых потоков. Цепочки ям могут быть образованы обрушением лавовых трубок или пустот, образовавшихся в вулканической постройке в областях растяжения. В данном случае, старая цепочка, идущая с севера на юг, примерно параллельна направлению течению лавы, а цепочка, идущая с востока на запад, проходит перпендикулярно предполагаемому направлению любых лавовых трубок.

7.13 **CROSSING PIT CRATER CHAINS**

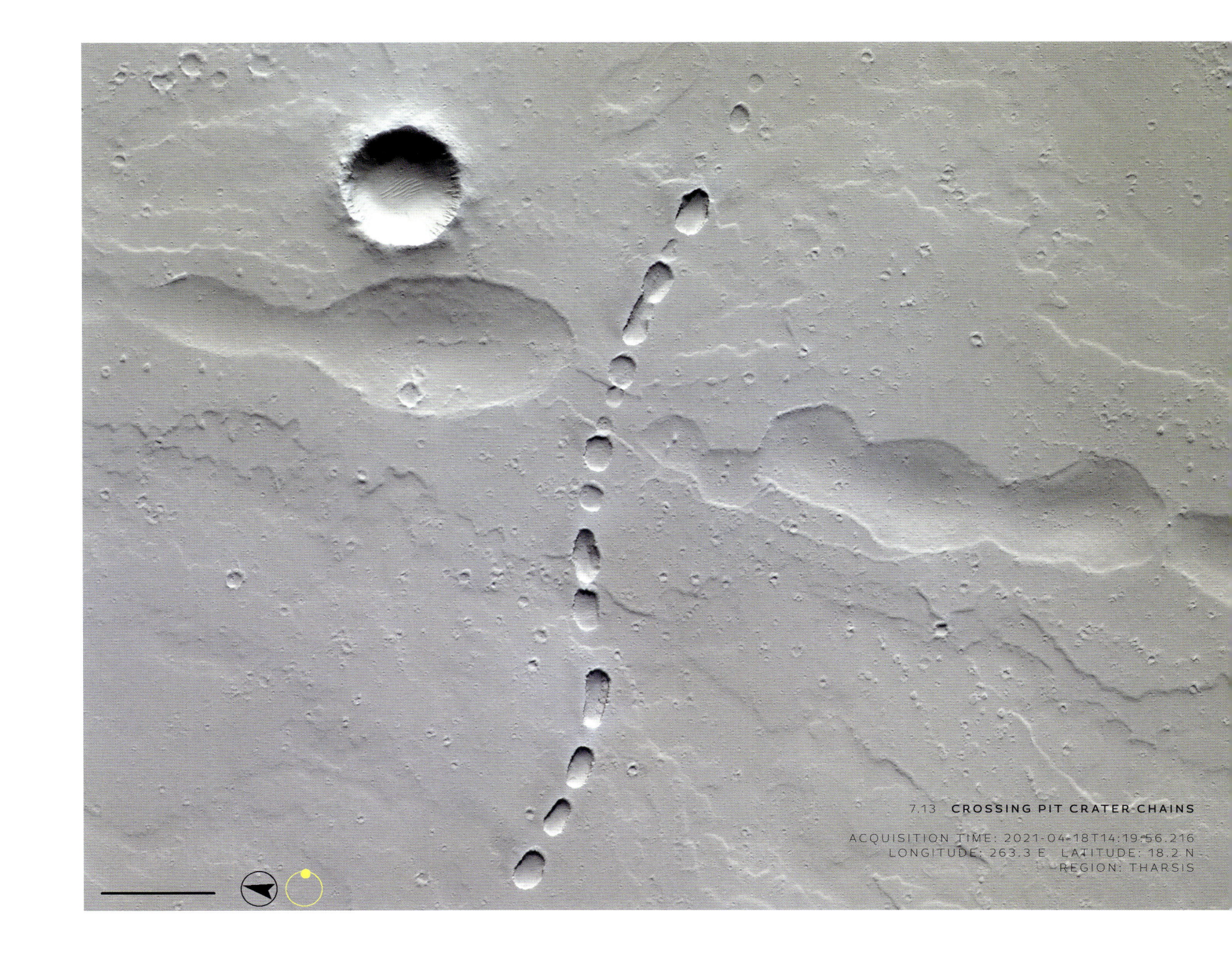

ACQUISITION TIME: 2021-04-18T14:19:56.216
LONGITUDE: 263.3 E   LATITUDE: 18.2 N
REGION: THARSIS

DIE OBERLICHTER DER MARSIANISCHEN LAVARÖHREN

Dieses Bild zeigt zwei dunkle Einsturzkrater auf der Oberfläche von Tharsis, südöstlich von Jovis Tholus. Man nimmt an, dass diese Krater durch den Einsturz von Lavaröhren entstanden sind. Ursprünglich wurden diese Röhren, die heute besser als Lavaröhren bekannt sind, als Pyrodukte bezeichnet. Sie haben als möglicher Lebensraum für künftige Astronauten grosses Interesse geweckt, da sie Schutz vor Strahlungsgefahren und dramatischen Temperaturschwankungen bieten.

MARTIAN LAVA TUBE SKYLIGHTS

This image shows two dark pits in the surface of Tharsis, southeast of Jovis Tholus. These pits are thought to have been produced by collapse of lava tubes and are sometimes known as pyroducts. They have provoked considerable interest as a possible habitat for future astronauts because of the protection offered from radiation hazards and dramatic swings in temperature.

LES TUBES DE LAVE MARTIENS

Cette image montre deux puits sombres à la surface de Tharsis, au sud-est de Jovis Tholus. On pense que ces puits ont été produits par l'effondrement de tubes de lave. Ils ont suscité un intérêt considérable en tant qu'habitat possible pour de futurs astronautes, en raison de la protection offerte contre les risques de radiation et les fortes variations de température à la surface.

IL COLLASSO DI TUBI DI LAVA MARZIANI

Questa immagine mostra due scuri crateri a pozzo sulla superficie di Tharsis, a sud-est di Jovis Tholus. Si pensa che queste strutture collassate siano state prodotte dal collasso di tubi di lava (una volta erano conosciuti come pirodotti, ma ora meglio definiti come tubi di lava). Hanno suscitato un notevole interesse come possibile habitat per futuri astronauti a causa della protezione che offrono dal rischio di radiazioni e dalle drammatiche oscillazioni di temperatura.

ПРОСВЕТЫ МАРСИАНСКИХ ЛАВОВЫХ ТРУБОК

На этом снимке показаны два темных провала на поверхности провинции Фарсида к юго-востоку от купола Юпитера. Считается, что эти колодцы образовались в результате обрушения лавовых трубок, ранее называвшихся лавовыми туннелями. Они вызывают значительный интерес, как возможное место обитания для будущих астронавтов, благодаря создаваемой внутри защите от радиационных угроз и резких колебаний температуры.

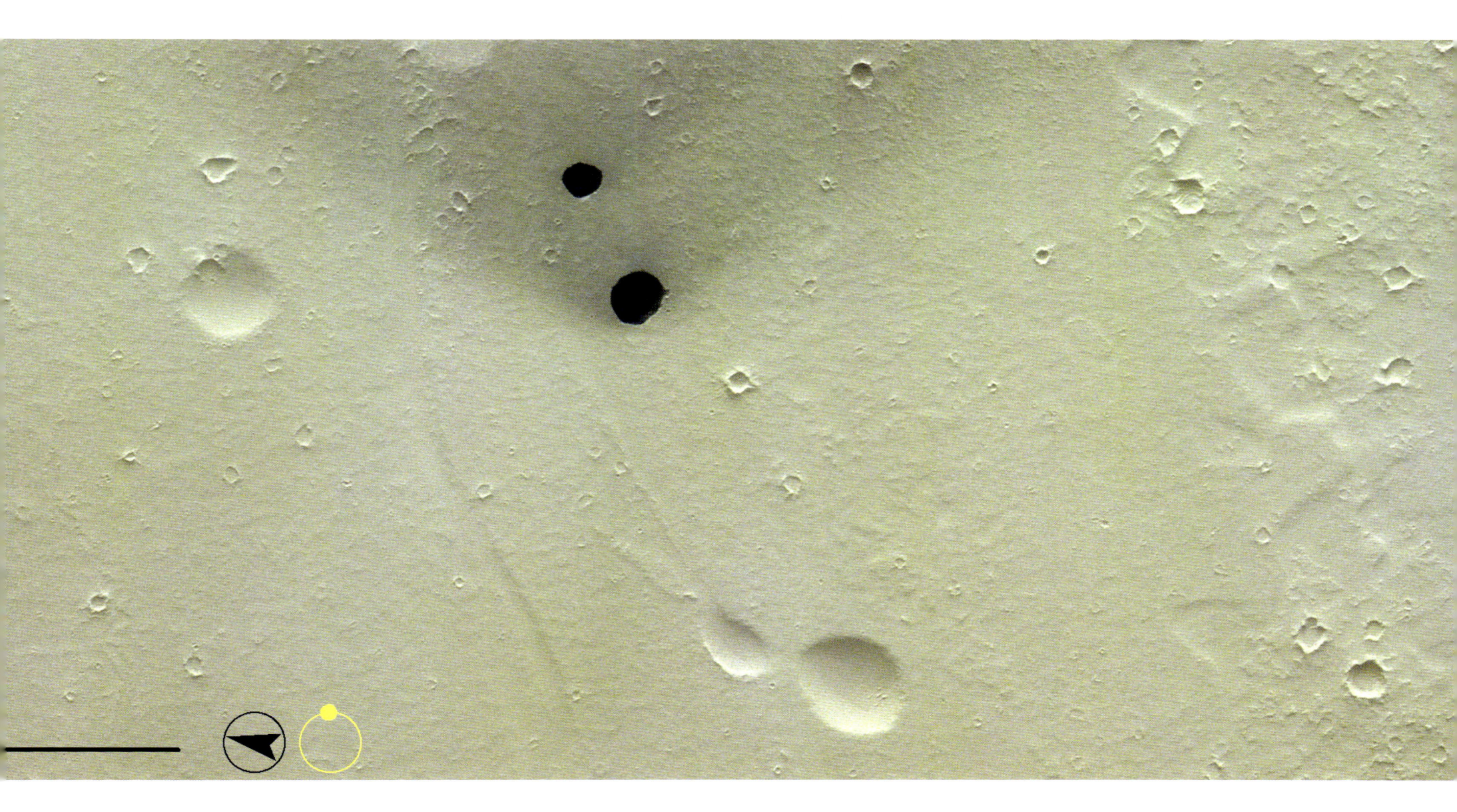

7.14 MARTIAN LAVA TUBE SKYLIGHTS

ACQUISITION TIME: 2021-04-26T18:50:50.678
LONGITUDE: 247.4 E   LATITUDE: 17.5 N
REGION: THARSIS

283

GRABEN MIT EINEM ZENTRALEN KANAL

Dies ist ein Bild eines Grabens, der Teil des Olympica-Fossae-Systems ist. Es ist besonders interessant wegen der Mulde, die in der Mitte des Grabens verläuft. Man beachte auch die Mulden, die etwa orthogonal zum Graben ganz rechts verlaufen, und die Albedo-Merkmale links, bei denen es sich fast sicher um Lavaströme handelt. Das wahrscheinliche Szenario ist, dass Lavaströme den Graben füllten, ihn überfluteten und dann abflossen, wobei ein dünner Panzer vulkanischer Merkmale über den grösseren und älteren tektonischen Merkmalen zurückblieb.

GRABEN WITH A CENTRAL CHANNEL

This is an image of a graben which is part of the Olympica Fossae system. It is particularly interesting because of the trough that runs along the centre of the graben. Note also the troughs that are roughly orthogonal to the graben on the extreme right and the albedo features visible to the left on what are almost certainly lava flows. The likely scenario here is that lava flows filled and overflowed the graben, and then drained out, leaving a thin carapace of volcanic features draped over the larger and older tectonic features.

GRABEN AVEC UN CHENAL CENTRAL

Image d'un graben qui fait partie du système d'Olympica Fossae, particulièrement intéressante en raison de la dépression qui court au centre du graben. Notez également les creux qui sont à peu près perpendiculaires au graben tout à droite de l'image et les variations d'albédo de la surface à gauche, sur de probables coulées de lave. Le scénario privilégié ici est que les coulées de lave ont rempli et débordé le graben, puis se sont écoulées, laissant une fine carapace de morphologies volcaniques drapées sur les morphologies tectoniques plus grandes et plus anciennes.

GRABEN CON UN CANALE CENTRALE

Questa è un'immagine di un graben che fa parte del sistema Olympica Fossae. È particolarmente interessante a causa del canale che corre lungo il centro del graben. Si notino anche i canali che sono approssimativamente ortogonali al graben all'estrema destra e le variazioni di albedo visibili a sinistra su quelle che sono quasi certamente colate di lava. Lo scenario più probabile qui è che le colate di lava abbiano riempito e allagato il graben, e poi si siano prosciugate, lasciando un sottile mantello di duro materiale vulcanico, coprendo le caratteristiche tettoniche più grandi e antiche.

ГРАБЕН С ЦЕНТРАЛЬНЫМ КАНАЛОМ

Это изображение грабена, который является частью системы борозд Олимпики (Olympica Fossae). Грабен особенно интересен благодаря желобу, проходящему по его центру. Обратите внимание также на желоба, которые примерно перпендикулярны грабену в крайней правой части, и на особенности альбедо, видимые слева, которые почти наверняка являются застывшими лавовыми потоками. Вероятный сценарий здесь заключается в том, что лавовые потоки заполняли и переполняли грабен, а затем утекали, оставляя тонкий панцирь вулканических материалов, покрывающий более крупные и старые тектонические формы.

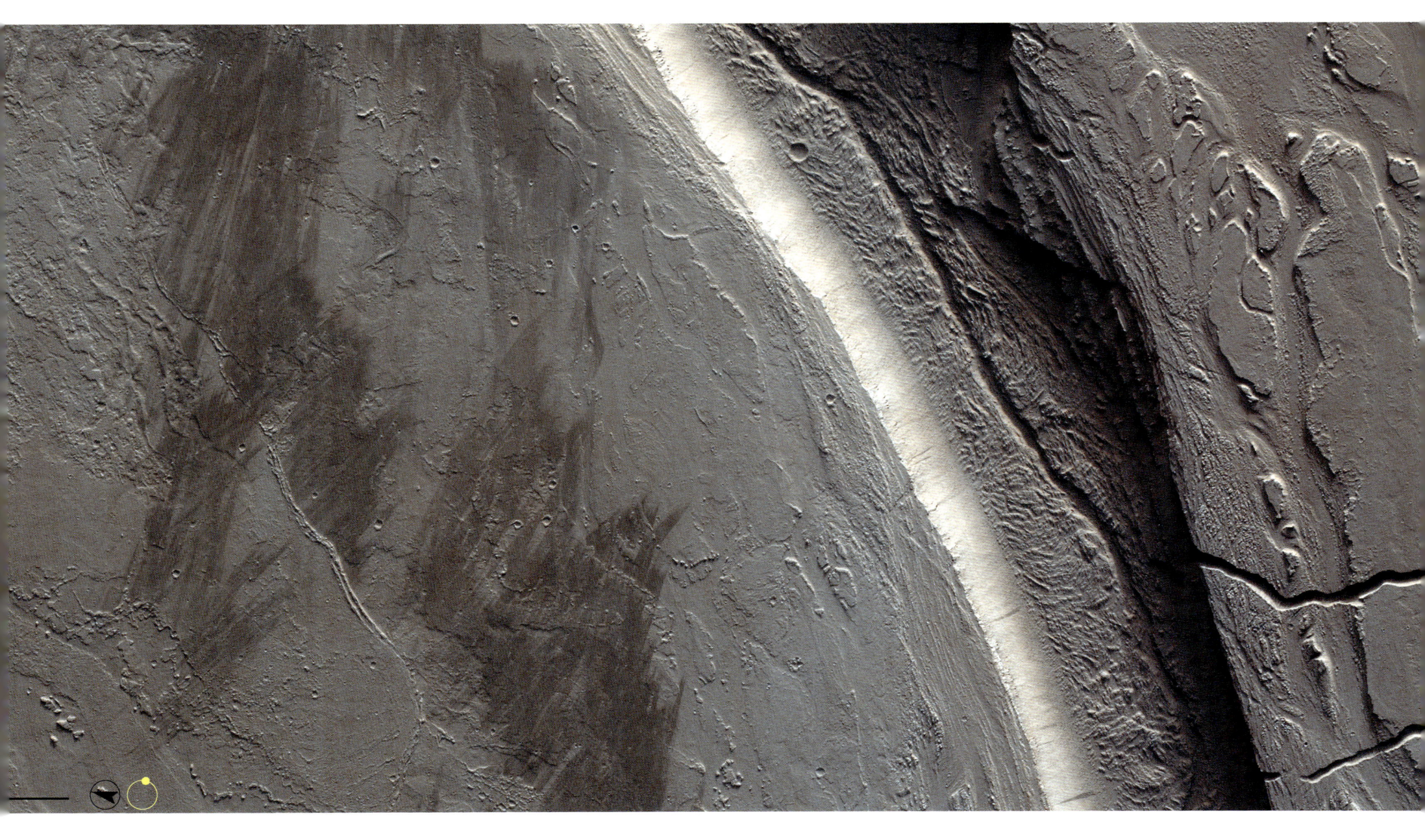

7.15 GRABEN WITH A CENTRAL CHANNEL

ACQUISITION TIME: 2018-05-09T19:12:26.000
LONGITUDE: 244.3 E   LATITUDE: 24.4 N
REGION: THARSIS

DIE CALDERA VON ALBA MONS

Dies ist ein Bild des westlichen Randes der Caldera von Alba Mons (früher bekannt als Alba Patera). Der Rand weist eine gewisse Farbvielfalt auf. Die Prozesse, die die inneren Felsgrate in der Caldera hervorbringen, sind nicht bekannt, aber es könnte sich um Horste handeln – die hoch aufragenden Erhebungen, die zwischen Gräben zu finden sind.

THE CALDERA OF ALBA MONS

This is an image of the western rim of the caldera of Alba Mons (formerly known as Alba Patera). There is some diversity in colour seen in the rim. The processes producing the internal ridges in the caldera are not known but they could be horsts – the high-standing ridges that are found between graben.

LA CALDEIRA D'ALBA MONS

Image de la bordure ouest de la caldeira d'Alba Mons (anciennement connue sous le nom d'Alba Patera). On observe une certaine diversité de couleurs sur le bord. Les processus à l'origine des crêtes internes de la caldeira ne sont pas connus, mais il pourrait s'agir de horsts – les crêtes surélevées que l'on trouve entre les grabens.

LA CALDERA DI ALBA MONS

Questa è un'immagine del bordo occidentale della caldera di Alba Mons (precedentemente conosciuta come Alba Patera). C'è una certa diversità di colore nel bordo. I processi che producono le creste interne alla caldera non sono noti, ma potrebbero essere horst, cioè pilastri tettonici che si trovano tra i graben.

КАЛЬДЕРА ГОРЫ АЛЬБА

Это изображение западного края кальдеры горы Альба (ранее известной как патера Альба). На ободе видно некоторое разнообразие цветов. Процессы, приводящие к образованию внутренних хребтов в кальдере, неизвестны, но последние могут быть горстами – высокими поднятиями, встречающимися между грабенами.

7.16 THE CALDERA OF ALBA MONS

ACQUISITION TIME: 2019-09-20T09:39:30.669
LONGITUDE: 248.8 E   LATITUDE: 40.5 N
REGION: THARSIS

RINNEN WESTLICH VON SULCI GORDII

Diese gewundenen Kanäle befinden sich in Lavaströmen, die sich von Süden nach Norden erstrecken (in diesem Bild von rechts nach links). Lavakanäle sind effiziente Wege für den Transport von Lava vom Schlot zu den distalen Teilen eines Lavastroms. Das erhöhte Plateau im oberen rechten Quadranten dieses Bildes ist ein Beispiel für den typischen distalen Teil eines Lavastroms mit Lappen mit gekerbten Rändern. Lavakanäle und Lappen dieser Grösse lassen auf Lavaströme schliessen, die grösser sind als die meisten Eruptionen auf der Erde, die von Menschen aufgezeichnet wurden.

CHANNELS WEST OF SULCI GORDII

These sinuous channels are found within lava flows extending from the south to the north (right to left in this image). Lava channels are efficient pathways to transport lava from the vent to the distal parts of a lava flow. The raised plateau in the upper right quadrant of this image is an example of the typical distal part of a lava flow with lobes with crenulated margins. Lava channels and lobes of this size suggest lava fluxes larger than most eruptions on Earth that humans have recorded.

CHENAUX À L'OUEST DE SULCI GORDII

Ces chenaux sinueux se trouvent dans les coulées de lave qui s'étendent du sud vers le nord (de droite à gauche sur cette image). Les chenaux de lave sont des voies efficaces pour transporter la lave de l'évent aux parties distales d'une coulée. Le plateau surélevé dans la portion supérieure droite de cette image est un exemple de la partie distale typique d'une coulée de lave avec des lobes aux marges crénelées. Des canaux de lave et des lobes de cette taille suggèrent des flux de lave plus importants que la plupart des éruptions enregistrées sur Terre.

CANALI A OVEST DI SULCI GORDII

Questi canali sinuosi si trovano all'interno di colate laviche che si estendono da sud a nord (da destra a sinistra in questa immagine). I canali di lava sono percorsi efficienti per trasportare la lava dalla sorgente al fronte di una colata. L'altopiano sopraelevato nel quadrante superiore destro di questa immagine è un esempio del tipico fronte di una colata lavica con lobi dai margini merlati. Canali e lobi di lava di queste dimensioni suggeriscono flussi di lava ben più grandi della maggior parte delle eruzioni sulla Terra.

КАНАЛЫ К ЗАПАДУ ОТ РЫТВИН ГОРДИЯ

Эти извилистые каналы найдены в застывших лавовых потоках, тянущихся с юга на север (справа налево на этом изображении). Лавовые каналы являются естественными путями транспорта лавы из жерла в дистальные (внешние) части лавового потока. Приподнятое плато в правом верхнем квадранте этого изображения является примером типичной дистальной части лавового потока с зубчатыми лопастями. Лавовые каналы и лопастевидные покровы таких размеров предполагают потоки лавы, превышающие большинство зафиксированных человеком извержений на Земле.

7.17 CHANNELS WEST OF SULCI GORDII

ACQUISITION TIME: 2021-02-10T00:20:23.081
LONGITUDE: 232.3 E LATITUDE: 19.3 N
REGION: THARSIS

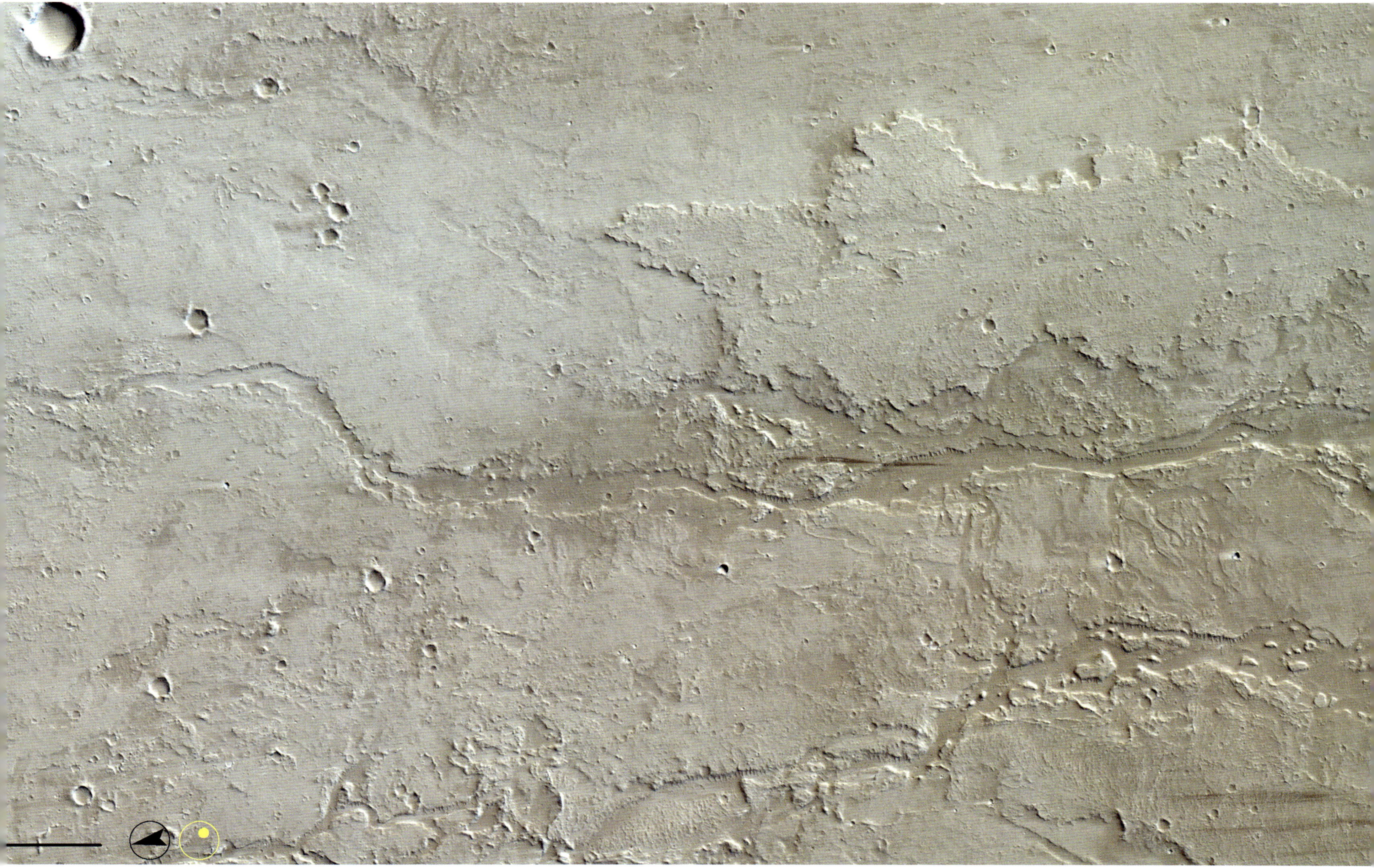

AUSGERICHTETE VULKANKEGEL

Dieses Bild zeigt Ketten von Kegeln in der Region Isidis Dorsa. Diese Art von grossen Kuppenbergen bedeckt einen Grossteil des Gebiets von Isidis. Ähnliche Merkmale sind auch im angrenzenden Utopia-Becken zu finden, unter anderem in der Nähe der Stelle, an welcher der chinesische Marsrover (Tianwen-1) im Jahr 2021 landete. Sie sind alle ungefähr gleich gross und häufig sowohl in einer Linie ausgerichtet als auch überlappend, was es äusserst unwahrscheinlich macht, dass sie mit einem Einschlag zusammenhängen. Sie ähneln in Grösse und Form den Schlackenkegeln, die in Gruppen vorkommen.

ALIGNED VOLCANIC CONES

This image shows chains of cones in the Isidis Dorsa region. These types of large knobs cover much of the area within Isidis and similar features can be found in the adjacent Utopia basin, including near where the Chinese Mars rover (Tianwen-1) landed in 2021. They are all roughly the same size and are frequently both aligned and overlapping, making it exceedingly improbable that they are impact related. They are similar in scale and shape to cinder cones which can be found in clusters.

CÔNES VOLCANIQUES ALIGNÉS

Cette image montre des chaînes de cônes dans la région d'Isidis Dorsa. Ces types de grands cônes couvrent une large fraction de la région d'Isidis et des morphologies similaires peuvent être trouvées dans le bassin adjacent d'Utopia, y compris près de l'endroit où le rover chinois Tianwen-1 a atterri en 2021. Elles ont toutes à peu près la même taille et sont fréquemment à la fois alignées et superposées, ce qui rend extrêmement improbable qu'elles soient liées à un impact. Leur échelle et leur forme sont similaires à celles des cônes de cendres, que l'on peut trouver en grappes.

PSEUDOCRATERI ALLINEATI

Questa immagine mostra catene di strutture coniche nella regione di Isidis Dorsa. Questi tipi di grandi tumuli coprono gran parte dell'area all'interno di Isidis e caratteristiche simili possono essere trovate nell'adiacente bacino di Utopia, anche vicino a dove il rover marziano cinese (Tianwen-1) è atterrato nel 2021. Sono tutti all'incirca della stessa dimensione e sono spesso allineati e sovrapposti, rendendo estremamente improbabile che siano correlati a eventi da impatto. Sono simili in dimensione e forma ai coni di scorie che possono essere talvolta trovati in gruppi.

СОГЛАСОВАННЫЕ ВУЛКАНИЧЕСКИЕ КОНУСЫ

На этом снимке показаны цепочки бугров в регионе гряд Исиды. Подобные крупные выступы покрывают большую часть поверхности гряд Исиды, и аналогичные объекты можно найти в соседней низменности Утопия, в том числе – недалеко от места посадки китайского марсохода (Тяньвэнь-1) в 2021 году. Все бугры примерно одинакового размера и часто располагаются и регулярно и с наложением друг на друга, что чрезвычайно уменьшает вероятность того, что они связаны с уда-ром метеорита.

7.18   ALIGNED VOLCANIC CONES

ACQUISITION TIME: 2021-03-07T08:21:03.408
LONGITUDE: 83.8 E   LATITUDE: 16.6 N
REGION: ISIDIS

8 DÜNEN, STAUB UND HANGSTREIFEN

DUNES, DUST, AND SLOPE STREAKS

DUNES, POUSSIÈRE ET TRAÎNÉES DE TALUS

DUNE, POLVERE E STRISCE DI PENDENZA

ДЮНЫ, ПЫЛЬ И ПОЛОСЫ НА СКЛОНАХ

STAUBTEUFELSPUREN IN PROMETHEI TERRA

Staubteufel entstehen, wenn heiße Luft in Oberflächennähe schnell durch kühlere Luft darüber aufsteigt. Dadurch entsteht ein Aufwind, der dann zu rotieren beginnen kann. Der Aufwind kann Staubpartikel von der Oberfläche in die Atmosphäre befördern. Die Kombination aus der Rotationsbewegung und dem Abtransport von Staub führt zu charakteristischen Spuren auf der Oberfläche. Auf dem Mars sind die Spuren in der Regel dunkler als die Umgebung. Hier sehen wir eine Stelle in Promethei Terra, an der die Oberfläche vollständig mit Staubteufelspuren bedeckt ist. Beachten Sie, dass die Spuren weder linear noch vorhersehbar verlaufen.

DUST DEVIL TRACKS IN PROMETHEI TERRA

Dust devils form when hot air near the surface rises quickly through cooler air above it. This produces an updraught that can then begin to rotate. The updraught can lift dust particles from the surface into the atmosphere. The combination of the rotational motion and the removal of dust leads to characteristic tracks on the surface. On Mars, the tracks are typically darker than the surroundings. Here we see a site in Promethei Terra where the surface is absolutely covered in dust devil tracks. Notice that they do not follow linear and predictable paths.

TRACES DE TORNADES DE POUSSIÈRE DANS PROMETHEI TERRA

Les tornades de poussière se forment lorsque de l'air chaud proche de la surface s'élève rapidement à travers l'air plus frais au-dessus d'elle. Cela produit un courant d'air ascendant qui peut alors commencer à tourbillonner. Ce courant ascendant peut soulever des particules de poussière de la surface vers l'atmosphère. La combinaison du mouvement de rotation et de l'élimination de la poussière entraîne des traces caractéristiques à la surface. Sur Mars, les traces sont généralement plus sombres que les alentours. Nous voyons ici un site de Promethei Terra où la surface est couverte de traces de tornades de poussière. Remarquez qu'elles ne suivent pas des chemins linéaires et prévisibles.

TRACCE DI DIAVOLI DI POLVERE IN PROMETHEI TERRA

I diavoli di polvere si formano quando l'aria calda vicino alla superficie sale rapidamente attraverso l'aria più fredda sopra di essa. Questo produce una corrente ascensionale che può poi iniziare a roteare e può sollevare particelle di polvere dalla superficie all'atmosfera. La combinazione del movimento di rotazione e la rimozione della polvere porta a tracce caratteristiche sulla superficie, e su Marte le tracce sono tipicamente più scure dei dintorni. Qui vediamo un sito in Promethei Terra dove la superficie è completamente coperta di tracce di diavoli di polvere. Si noti come le tracce non seguono percorsi lineari e prevedibili.

СЛЕДЫ ПЫЛЕВЫХ ДЬЯВОЛОВ НА ЗЕМЛЕ ПРОМЕТЕЯ

Пылевые дьяволы образуются, когда горячий воздух с поверхности быстро поднимается вверх через слои более холодного воздуха. В результате образуется восходящий поток, который начинает вращаться и может поднимать частицы пыли с поверхности в атмосферу. Совокупность вращения воздуха и сметания пыли приводит к появлению характерных следов. На Марсе такие треки, как правило, темнее, чем окружающая поверхность. Здесь мы видим участок земли Прометея, где поверхность буквально испещрена следами пылевых дьяволов. Обратите внимание, что их траектория совершенно непредсказуема и далека от линейной.

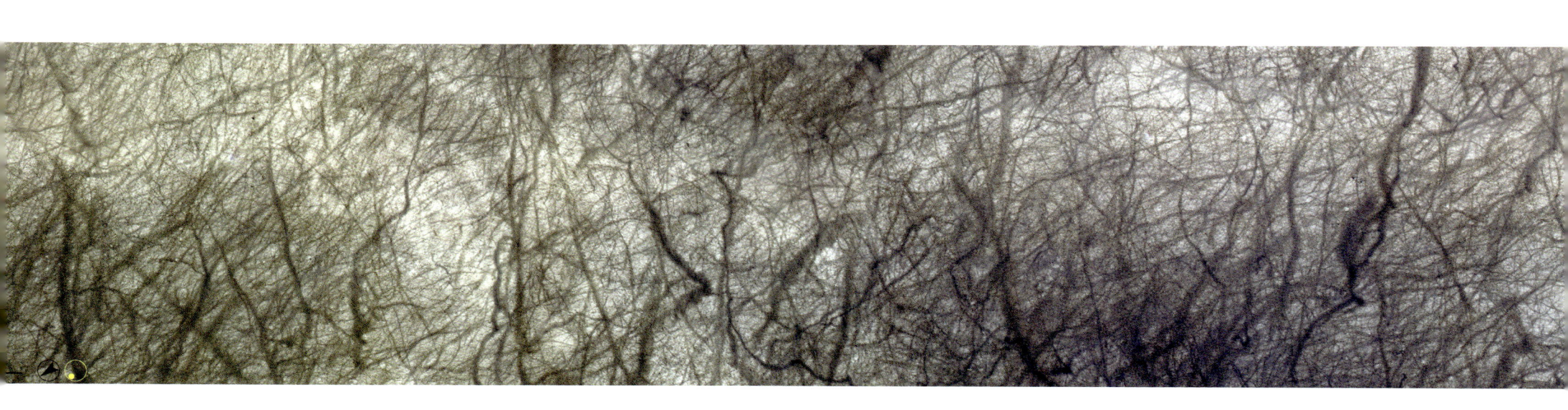

8.1   DUST DEVIL TRACKS IN PROMETHEI TERRA

ACQUISITION TIME: 2018-12-16T09:14:02.071
LONGITUDE: 117.7 E   LATITUDE: 59.9 S
REGION: PROMETHEI

RUSSELL-KRATER-DÜNEN IM FRÜHLING

Dieses bemerkenswerte Bild zeigt die Dünen im Russell-Krater in Noachis Terra. Das glänzende Aussehen ist das Ergebnis der jahreszeitlich bedingten Kohlendioxid-Eisbedeckung der Dünen, die im südlichen Winter die Oberfläche bedeckt und im Frühjahr sublimiert. Die dunkleren Flecken sind Fächerablagerungen, die durch einen Geysirmechanismus entstehen, der Druck unter der Eisdecke entwickelt und dann die Eisschicht durchbricht. Links im Bild sind rinnenartige Strukturen zu sehen, die wahrscheinlich auf die Sublimation von CO_2-Eis zurückzuführen sind, gefolgt von einem Schuttstrom, der sich von der Oberfläche löst.

RUSSELL CRATER DUNES IN SPRING

This remarkable image shows the dunes in Russell Crater in Noachis Terra. The shiny appearance is the result of the dunes being covered in seasonal carbon dioxide ice that covers the surface in southern winter and sublimes during spring. The darker spots are fan deposits resulting from a geyser mechanism that develops pressure under the ice sheet and then breaches the ice layer. To the left of the image we can see gully-like structures that are probably related to the sublimation of CO_2 ice followed by a flow of debris that breaks loose from the surface.

DUNES DU CRATÈRE RUSSELL AU PRINTEMPS

Cette image remarquable montre les dunes du cratère Russell dans Noachis Terra. L'aspect brillant est le résultat de la couverture des dunes par la glace saisonnière de dioxyde de carbone qui recouvre la surface pendant l'hiver austral et se sublime au printemps. Les taches plus sombres sont des dépôts en éventail résultant d'un mécanisme de geyser qui se forment suite à une surpression sous la couche de glace. Cette dernière finissant par se fracturer, le sable sombre présent à la surface est éjecté et se dépose en créant ces taches sombres. À gauche de l'image, on peut voir des structures ressemblant à des ravines qui sont probablement liées à la sublimation de la glace de CO_2.

DUNE DEL CRATERE RUSSELL IN PRIMAVERA

Questa notevole immagine mostra le dune nel Russell Crater in Noachis Terra. L'aspetto «lucido» è il risultato del fatto che le dune sono coperte dal ghiaccio stagionale di anidride carbonica che copre la superficie di Marte nell'inverno meridionale e sublima in primavera. Le macchie più scure sono depositi a ventaglio risultanti da un meccanismo simil geyser, derivante dalla pressione sviluppata sotto lo strato di ghiaccio e che alla fine lo frattura. A sinistra dell'immagine possiamo vedere strutture simili a calanchi che sono probabilmente legate alla sublimazione del ghiaccio di CO_2 seguita da un flusso di detriti che si stacca dalla superficie.

ДЮНЫ КРАТЕРА РАССЕЛ ВЕСНОЙ

Этот впечатляющий снимок показывает дюны кратера Рассел Земли Ноя. Их блеск – результат покрытия дюн сезонным углекислым льдом. Он образуется на поверхности во время южной зимы и сублимируется весной. Тёмные пятна – это аллювиальные отложения, образовавшиеся в результате гейзерной активности, вызванной прорывающим верхний слой льда сильным давлением газа (высвобожденного при сублимации подповерхностной углекислоты). Слева на снимке видны оврагоподобные структуры, которые, вероятно, связаны с возгонкой углекислого льда, спровоцировавшей обвалы высвобожденных из поверхности обломков.

8.2   RUSSELL CRATER DUNES IN SPRING

ACQUISITION TIME: 2018-05-22T02:36:24.000
LONGITUDE: 12.6 E   LATITUDE: 54.5 S
REGION: NOACHIS

RUSSELL-KRATER-DÜNEN – EISFREI

Dünen im Russell-Krater gegen Ende des südlichen Sommers. Zu diesem Zeitpunkt ist die Kohlendioxid-Eisschicht, die sich im südlichen Winter über den Dünen bildet, vollständig sublimiert. Erosionsrinnen sind oben links im Bild deutlich zu erkennen, aber bei näherer Betrachtung kann man feststellen, dass es auch an vielen anderen Stellen im abgebildeten Gebiet Erosionsrinnen gibt. Auch Staubteufelspuren sind überall in den Dünen zu sehen.

RUSSELL CRATER DUNES – ICE-FREE

Dunes in Russell Crater towards the end of southern summer. By this time, the carbon dioxide ice layer that forms over the dunes in southern winter has completely sublimed. Gullies are clearly visible to the top left of the image but close inspections show that there are gullies in many other place in the imaged area. Dust devil tracks can be also seen all over the dunes.

DUNES DU CRATÈRE RUSSELL – SANS GLACE

Dunes du cratère Russell vers la fin de l'été austral. À cette saison, la couche de glace de dioxyde de carbone qui se forme sur les dunes pendant l'hiver est complètement sublimée. Des ravines sont clairement visibles en haut à gauche de l'image, mais une analyse détaillée montre aussi des ravines à de nombreux autres endroits dans la zone imagée. Des traces de tornades de poussière sont également visibles sur l'ensemble des dunes.

DUNE PRIVE DI GHIACCIO NEL CRATERE RUSSELL

Dune nel Russell Crater verso la fine dell'estate meridionale. In questo periodo, lo strato di ghiaccio di anidride carbonica che si forma sopra le dune nell'inverno meridionale è ormai completamente sublimato. I calanchi sono chiaramente visibili in alto a sinistra dell'immagine, ma una ispezione ravvicinata mostra che ci sono calanchi in molti altri posti dell'immagine. Si possono anche vedere tracce di diavoli di polvere su tutto il campo di dune.

ДЮНЫ КРАТЕРА РАССЕЛ БЕЗ ЛЬДА

Дюны в кратере Рассел в конце южного лета. К этому времени слой льда из углекислого газа, который образуется на поверхности южной зимой, полностью сублимируется. В левой верхней части снимка хорошо различимы овраги, а при детальном рассмотрении можно обнаружить их присутствие и во многих других местах. Следы пылевых дьяволов также видны по всей площади дюн.

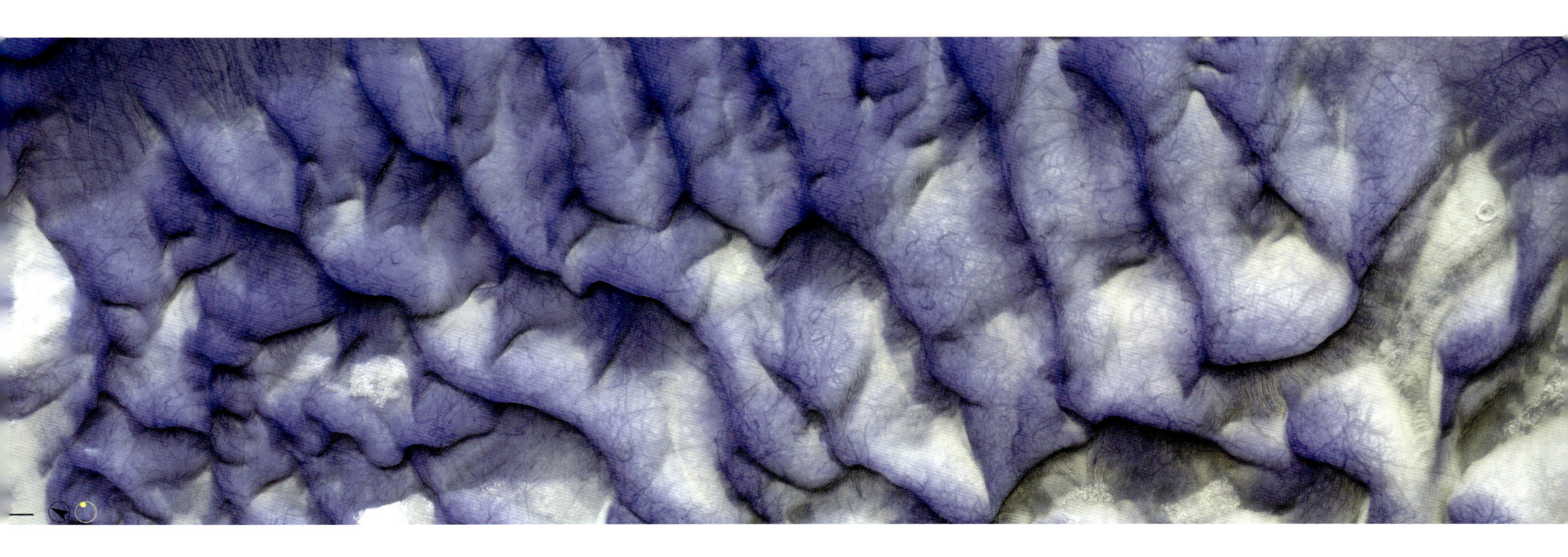

8.3 RUSSELL CRATER DUNES – ICE-FREE

ACQUISITION TIME: 2020-12-24T07:51:01.609
LONGITUDE: 12.5 E   LATITUDE: 54.0 S
REGION: NOACHIS

KLETTERNDE DÜNEN

Dieses Bild zeigt die steilen Wände des Coprates Chasma im Valles-Marineris-Canyon-System. In der Mitte des Bildes scheinen die dunklen Basaltsanddünen bergauf zu «klettern». Auf der Erde entstehen solche Dünen durch einseitig gerichtete Winde, die Sandpartikel auf einer erhöhten Topografie ablagern. Man beachte auch das bunte Material auf der linken Seite, das auf die mineralogische Vielfalt hinweist – etwas, was an vielen Stellen in Coprates Chasma zu sehen ist.

CLIMBING DUNES

This image features the steep walls of Coprates Chasma within Valles Marineris canyon system. In the center of the image, the dark basaltic sand dunes appear to "climb" upslope. On Earth, these types of dunes are formed by unidirectional winds that deposit sand particles on elevated topography. Notice also the colourful material to the left indicating mineralogical diversity – something that is seen in many places in Coprates Chasma.

DUNES REMONTANT LA PENTE

Cette image montre les parois abruptes de Coprates Chasma dans le système de canyons de Valles Marineris. Au centre de l'image, les dunes foncées de sable basaltique semblent remonter la pente. Sur Terre, ce type de dunes est formé par des vents unidirectionnels qui déposent des particules de sable sur une topographie élevée. Remarquez également les roches colorées à gauche de l'image, ce qui indique une diversité minéralogique que l'on observe dans de nombreux endroits de Coprates Chasma.

DUNE RAMPICANTI

Questa immagine mostra le ripide pareti del Coprates Chasma all'interno del sistema di canyon Valles Marineris. Al centro dell'immagine, le dune scure di sabbia basaltica sembrano «salire» verso l'alto. Sulla Terra, questi tipi di dune sono formati da venti unidirezionali che depositano particelle di sabbia su una topografia elevata. Si noti il materiale colorato a sinistra che indica la diversità mineralogica, qualcosa che si osserva in molti posti nel Coprates Chasma.

ВЗБИРАЮЩИЕСЯ ДЮНЫ

На этом снимке показаны крутые стены каньона Копрат в системе долин Маринер. В центре видны темные дюны базальтового песка, которые словно взбираются вверх по склону. На Земле дюны такого типа образуются под действием ветров, дующих в одном направлении и собирающих частицы песка на возвышенности. Обратите внимание также на разноцветный материал слева, указывающий на минералогическое разнообразие, которое наблюдается во многих местах каньона Копрат.

8.4 CLIMBING DUNES

ACQUISITION TIME: 2021-02-20T23:27:22.645
LONGITUDE: 304.9 E   LATITUDE: 15.5 S
REGION: MARINERIS

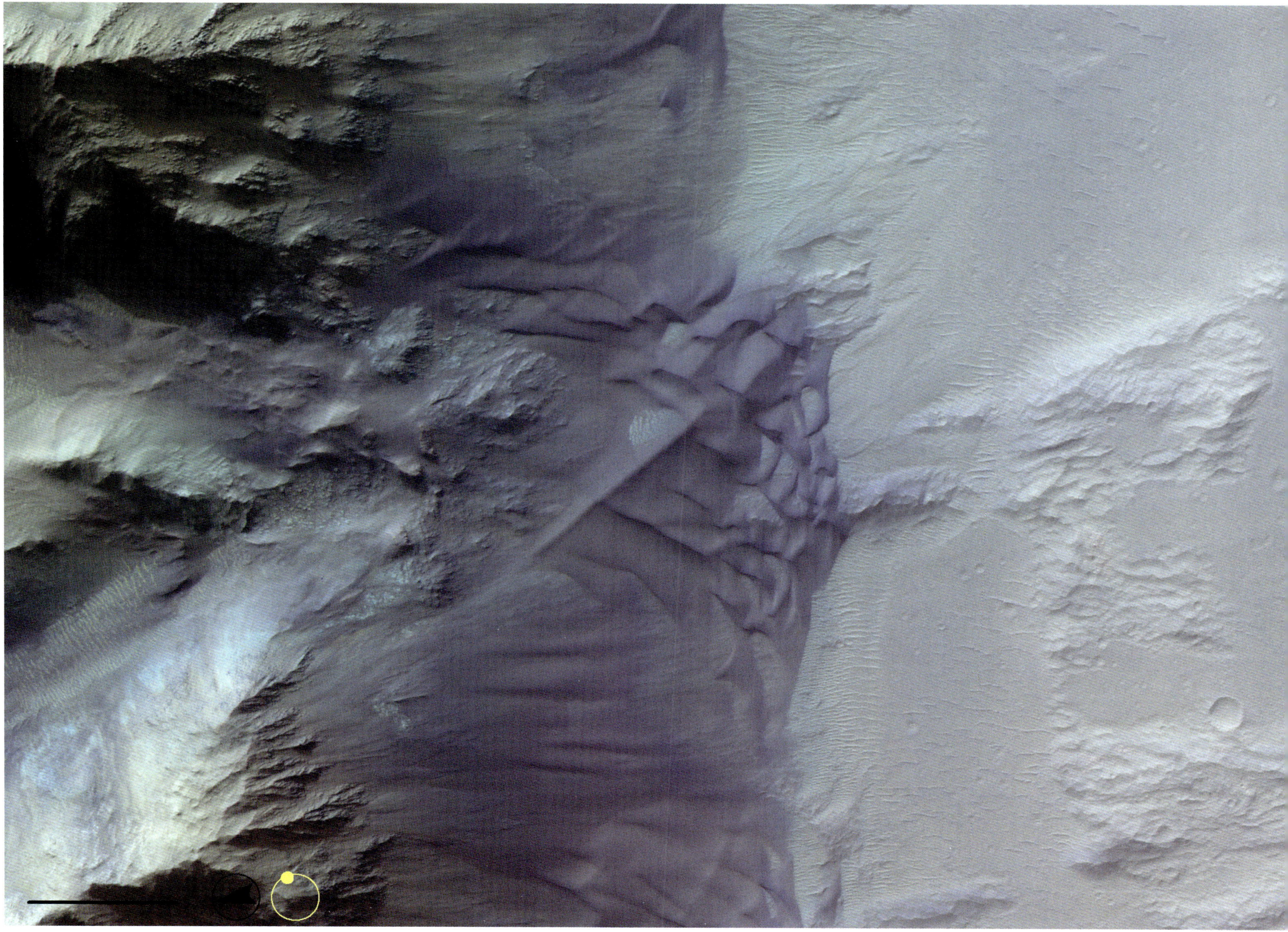

DÜNEN IM BRASHEAR-KRATER

Der Brashear-Krater beherbergt eine beeindruckende Reihe von Dünen, die in den CaSSIS-Farbprodukten wunderbar zur Geltung kommen. Der Krater hat einen Durchmesser von 77,5 km, und die Dünen befinden sich an der nordwestlichen Seite des Kraterbodens. Mit Hilfe des Windes und der Umlagerung von Sand können diese Dünen wandern, was durch die langen, diffusen Streifen angedeutet wird, die von den Dünenkämmen ausgehen.

DUNES IN BRASHEAR CRATER

Brashear Crater hosts an impressive set of dunes that stand out beautifully in CaSSIS colour products. The crater is 77.5 km in diameter and the dunes are on north-western side of the crater floor. With the help of wind and redeposition of sand these dunes can migrate, which is indicated by the long diffuse streaks emanating from the dune crests.

DUNES DU CRATÈRE BRASHEAR

Le cratère Brashear abrite un ensemble impressionnant de dunes qui ressortent magnifiquement dans les images couleur de la caméra CaSSIS. Le cratère a un diamètre de 77,5 km. Les dunes visibles ici se trouvent sur le côté nord-ouest du fond du cratère. Avec l'aide du vent et de la saltation des grains de sable, ces dunes peuvent migrer, ce qui est indiqué par les longues stries diffuses émanant de leurs crêtes.

DUNE NEL CRATERE BRASHEAR

Il cratere Brashear ospita un impressionante insieme di dune che risaltano meravigliosamente nei prodotti a colori di CaSSIS. Il cratere ha un diametro di 77,5 km e le dune si trovano sul lato nord-ovest del pavimento del cratere. Con l'aiuto del vento e la rideposizione della sabbia queste dune possono migrare, processo che è evidenziato dalle lunghe strisce diffuse che partono dalle creste delle dune.

ДЮНЫ В КРАТЕРЕ БРАШИР

В кратере Брашир находится группа дюн, которые красиво выделяются в цветовых фильтрах CaSSIS. Диаметр кратера составляет 77,5 км, а дюны находятся на северо-западной стороне дна. Под воздействием ветра и переноса песка эти дюны могут мигрировать. Об этом свидетельствуют длинные диффузные полосы, выходящие из гребней дюн.

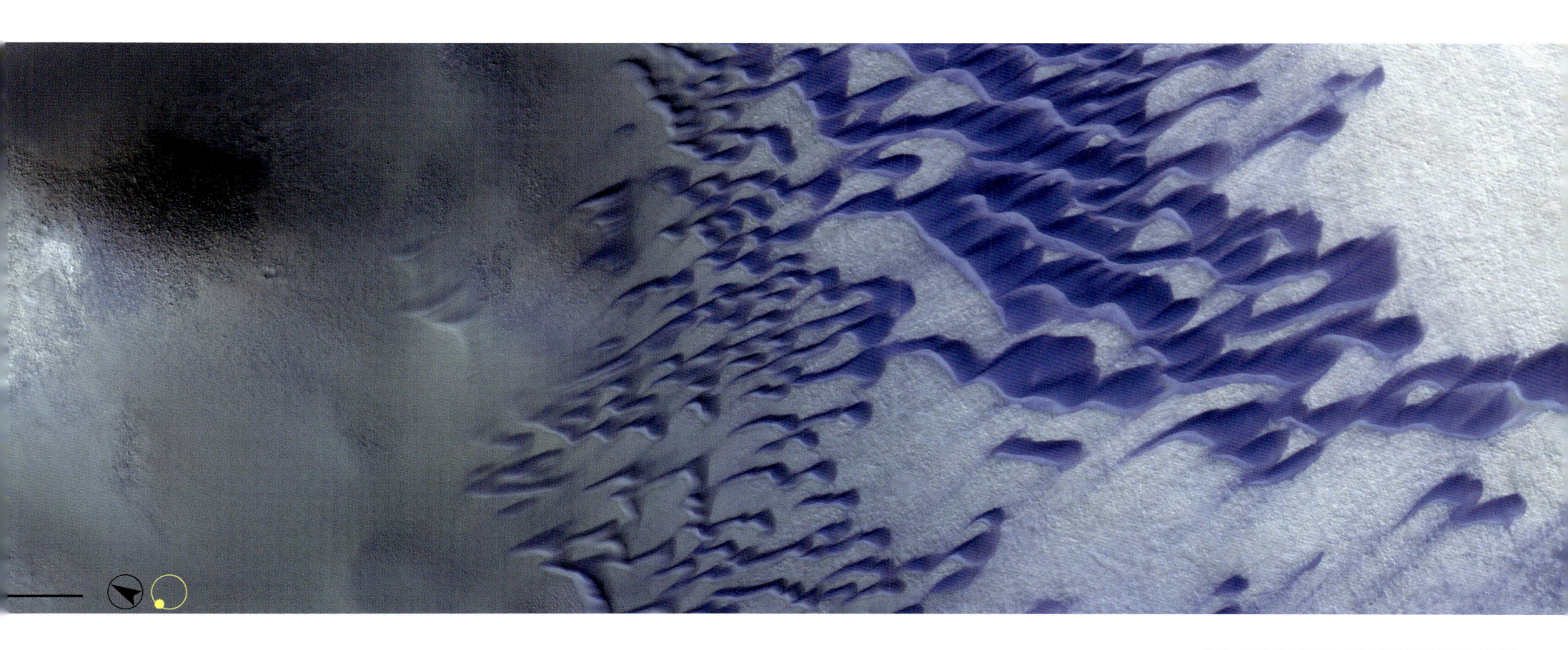

8.5 DUNES IN BRASHEAR CRATER

ACQUISITION TIME: 2020-11-25T04:16:35.099
LONGITUDE: 239.8 E   LATITUDE: 53.2 S
REGION: SIRENUM

GROSSE WINDSCHLIEREN AN DER FLANKE DES ARSIA MONS

Dieses Bild zeigt die Nordflanke des Vulkans Arsia Mons in Tharsis. Es zeigt die Auswirkungen sowohl der Staubablagerung als auch der Abtragung durch Oberflächenwinde. Beachten Sie die kleinen Krater auf der rechten Seite (Süden), die Quellen von hellerem Material zu sein scheinen. Alternativ könnten diese Krater auch Schutz vor den lokalen Winden geboten haben, sodass sich atmosphärischer Staub ansammeln konnte. Beachten Sie auch die hier vorherrschende Windrichtung.

LARGE WIND STREAKS ON THE FLANK OF ARSIA MONS

This image is of the northern flank of the Arsia Mons volcano in Tharsis. It shows the effects of both dust deposition and removal by surface winds. Notice the small craters to the right (south) which seem to be sources of brighter material. Alternatively, these craters could have provided protection from local winds and allowed atmospheric dust to accumulate. Note also the prevailing wind direction here.

GRANDES TRAÎNÉES DE VENT SUR LE FLANC D'ARSIA MONS

Flanc nord du volcan Arsia Mons dans la région de Tharsis. L'image montre les effets du dépôt de poussière et de son enlèvement par les vents de surface. Remarquez les petits cratères à droite (sud) qui semblent être des sources de matériaux plus clairs. Alternativement, ces cratères pourraient avoir fourni une protection contre les vents locaux et permis à la poussière atmosphérique de s'accumuler. Notez également la direction des vents dominants ici.

GRANDI STRISCE EOLICHE SUL FIANCO DI ARSIA MONS

Questa immagine è del fianco settentrionale del vulcano Arsia Mons a Tharsis. Mostra gli effetti sia della deposizione sia della rimozione della polvere da parte dei venti di superficie. Si notino i piccoli crateri a destra (sud) che sembrano essere fonti di materiale più chiaro. Oppure potrebbe essere che questi crateri hanno fornito protezione dai venti locali e aver permesso alla polvere atmosferica di accumularsi. Si noti anche la direzione dominante del vento in questa regione.

КРУПНЫЕ ВЕТРОВЫЕ ПОЛОСЫ НА СКЛОНЕ ГОРЫ АРСИЯ

Это изображение северной стороны вулкана Арсия в провинции Фарсида. На нем видны эффекты как осаждения пыли, так и ее удаления поверхностными ветрами. Обратите внимание на небольшие кратеры справа (на юге), которые, кажется, являются источниками более яркого материала. С другой стороны, эти кратеры, возможно, создают защиту от местных ветров и позволяют пыли накапливаться. Также стоит принять во внимание доминирующее направление ветра.

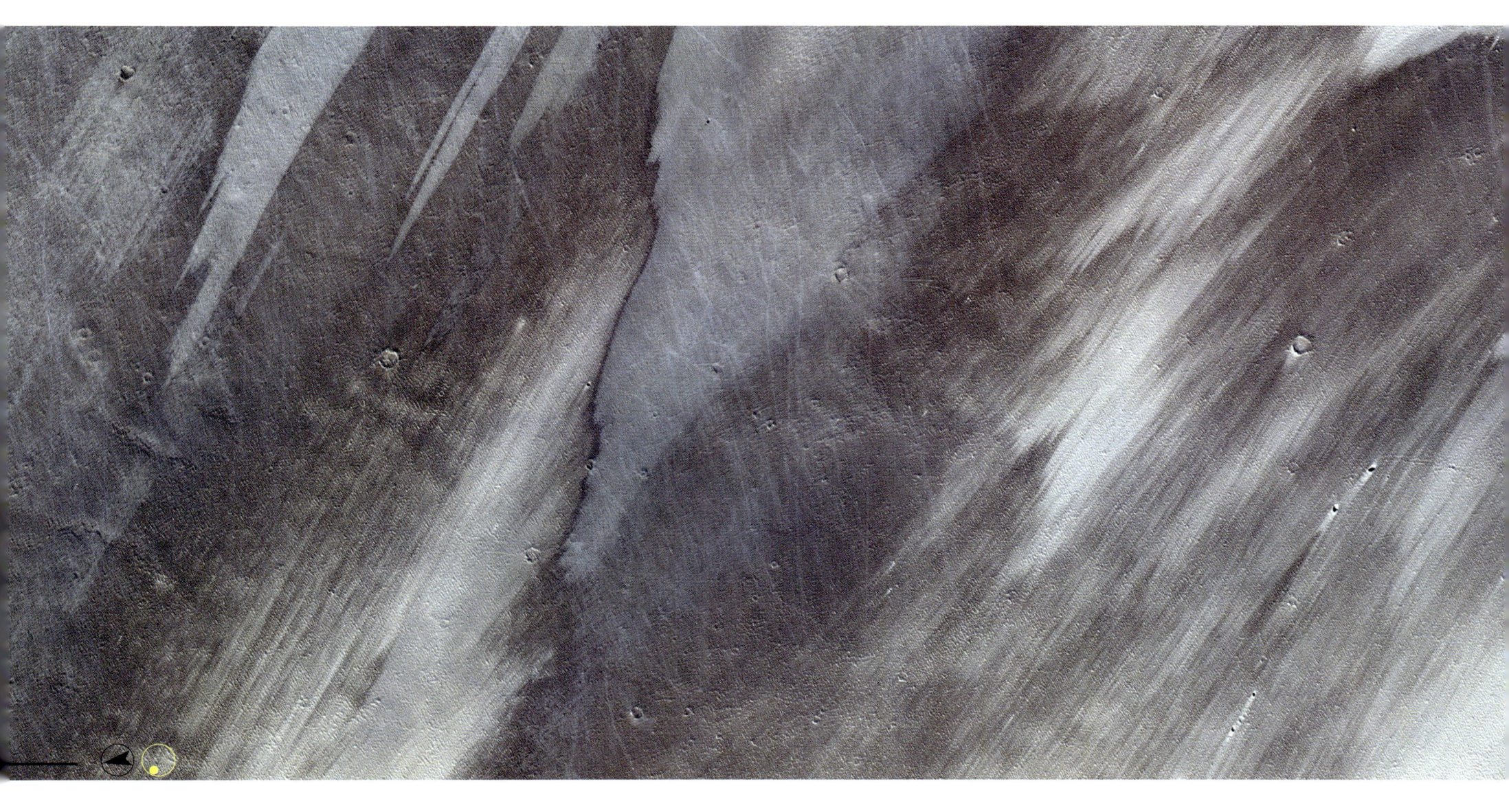

8.6   LARGE WIND STREAKS ON THE FLANK OF ARSIA MONS

ACQUISITION TIME: 2021-05-15T15:52:43.677
LONGITUDE: 236.4 E   LATITUDE: 3.6 S
REGION: THARSIS

305

MEHRFACHE DÜNENSTRUKTUREN

Dieses Bild zeigt Dünen in der Region Nereidum Montes südöstlich von Aonia Terra. Die Dünen hier weisen viele verschiedene Strukturen auf. Wenn man genau hinsieht, kann man an einigen Stellen das rauere (felsige) Substrat unter dem grossen (blauen) Dünenfeld erkennen.

MULTIPLE DUNE STRUCTURES

This image shows dunes in the Nereidum Montes region just to the south-east of Aonia Terra. The dunes here contain many different structures. Looking closely in some places, the rougher (rocky) substrate beneath the large (blue) dune field can be seen.

STRUCTURES DUNAIRES MULTIPLES

Cette image montre des dunes dans la région de Nereidum Montes, juste au sud-est d'Aonia Terra. Les dunes présentent ici de nombreuses structures différentes. En regardant de plus près à certains endroits, on peut voir le substrat plus rugueux (rocheux) sous le grand champ de dunes (bleu).

DIVERSE CONFORMAZIONI DI DUNE

Questa immagine mostra le dune nella regione di Nereidum Montes a sud-est di Aonia Terra. Le dune qui contengono molte strutture diverse. Guardando da vicino in alcuni punti, si può vedere il substrato più irregolare roccioso sotto il grande campo di dune (blu).

МНОГОЧИСЛЕННЫЕ ДЮННЫЕ СТРУКТУРЫ

На этом снимке изображены дюны в районе горного хребта Нереид к юго-востоку от земли Аонид. Дюны здесь содержат множество различных структур. При внимательном рассмотрении в некоторых местах можно увидеть более грубую каменистую основу под протяженным (синим) дюнным полем.

8.7　MULTIPLE DUNE STRUCTURES

ACQUISITION TIME: 2020-12-24T11:42:12.618
LONGITUDE: 306.8 E   LATITUDE: 40.7 S
REGION: AONIA

PARALLELE DÜNENFORMEN UND WECHSELWIRKUNGEN MIT EINEM KRATER IN AONIA TERRA

Dünen und ihre Bestandteile bewegen sich auf dem Mars. Die Geschwindigkeit der Dünen kann bis zu 5 m pro Marsjahr betragen. Auf der linken Seite des Bildes befinden sich viele Querdünen. Querdünen stehen normalerweise senkrecht zur vorherrschenden Windrichtung. Die Fortbewegung führt auch zur Interaktion mit anderen geologischen Merkmalen wie Kratern. Hier sehen wir eine lange Düne am Kraterrand. Bei näherer Betrachtung der nahen gelegenen Kraterwand sind viele eingeschnittene Rinnen zu erkennen, die vom Kraterrand weggeblasenen Sand enthalten. Unten rechts im Bild sind auch Dünen innerhalb des Kraters zu sehen, die die zentrale Erhebung umgeben.

PARALLEL DUNE FORMS AND INTERACTIONS WITH A CRATER IN AONIA TERRA

Dunes and their constituents move on Mars. Dune speeds can be up to about 5 m per Martian year. To the left of the image there are many transverse dunes. Transverse dunes are normally perpendicular to the prevailing wind direction. Motion leads also to interaction with other geologic features such as craters. Here we can see a long dune on the crater edge. Close inspection of the nearby crater wall shows that there are many incised gullies, which contain sand blown off from the crater edge. Towards the bottom right of the image dunes can be also seen inside the crater surrounding the central uplift.

FORMES DE DUNES PARALLÈLES ET INTERACTIONS AVEC UN CRATÈRE DANS AONIA TERRA

Les dunes et leurs constituants se déplacent sur Mars. La vitesse des dunes peut atteindre environ 5 mètres par année martienne. À gauche de l'image, on trouve de nombreuses dunes transversales, normalement perpendiculaires à la direction des vents dominants. Le mouvement entraîne également une interaction avec d'autres structures géologiques telles que les cratères. Ici, nous pouvons voir une longue dune sur le bord du cratère. Une inspection minutieuse de la paroi du cratère à proximité montre qu'il y a de nombreuses ravines incisées, qui contiennent du sable emporté par le vent depuis le bord du cratère. Vers le bas à droite de l'image, on peut également voir des dunes à l'intérieur du cratère entourant le soulèvement central.

CONFORMAZIONI DUNALI PARALLELE E INTERAZIONE CON UN CRATERE IN AONIA TERRA

Le dune e i loro costituenti si muovono sulla superficie di Marte. La velocità delle dune può arrivare a circa 5 m per anno marziano. A sinistra dell'immagine ci sono molte dune trasversali che sono normalmente perpendicolari alla direzione dominante del vento. Il movimento delle dune le porta anche a interagire con altre conformazioni geologiche come i crateri. Qui si può vedere una lunga duna sul bordo del cratere e un'ispezione ravvicinata della parete del cratere mostra che ci sono molti canaloni incisi, che contengono sabbia soffiata via dal bordo del cratere. Nella parte inferiore destra dell'immagine, si vedono anche dune che circondano l'altura centrale all'interno del crate.

ПАРАЛЛЕЛЬНЫЕ ДЮНЫ И ИХ ВЗАИМОДЕЙСТВИЕ С КРАТЕРОМ НА ЗЕМЛЕ АОНИД

Дюны и их компоненты (пески) на Марсе движутся. Скорость движения может доходить почти до 5 метров в марсианский год. На снимке слева находится множество поперечных дюн – как правило перпендикулярных преобладающему направлению ветра. Движение также приводит к их взаимодействию с другими геологическими объектами, например кратерами. Здесь мы видим протяженную дюну на краю кратера. Детальное рассмотрение его близлежащей стены показывает, что там находится много врезанных оврагов, в которых содержится песок, сдутый с края кратера. В правом нижнем углу снимка внутри кратера видны дюны, окружающие центральное поднятие.

8.8   PARALLEL DUNE FORMS AND INTERACTIONS WITH A CRATER IN AONIA TERRA

ACQUISITION TIME: 2020-12-25T13:17:58.673
LONGITUDE: 294.2 E   LATITUDE: 49.5 S
REGION: AONIA

DÜNEN UND RIFFELUNGEN

Dies ist ein Bild des Bodens und der Nordwand eines Kraters in Terra Sirenum, etwa östlich des Millman-Kraters. Die dunklen Rinnen auf der linken Seite (Norden) scheinen die Kraterwand stark erodiert und das Material hangabwärts transportiert zu haben. Beachten Sie auch die eingeschnittenen Wellenrippel auf den blau-grün gefärbten Dünen.

DUNES AND RIPPLES

This is an image of the floor and northern wall of a crater in Terra Sirenum, roughly east of Millman Crater. The dark gullies to the left (north) appear to have heavily eroded the crater wall and transported the material downslope. Notice also, the incised ripples on the blue-green coloured dunes.

DUNES ET ONDULATIONS

Image du fond et de la paroi nord d'un cratère de Terra Sirenum, à l'est du cratère Millman. Les ravines sombres à gauche (nord) semblent avoir fortement érodé la paroi du cratère et transporté les matériaux vers le bas de la pente. Remarquez également les ondulations incisées sur les dunes de couleur bleue-verte.

DUNE E INCRESPATURE

Questa è un'immagine del fondo e della parete settentrionale di un cratere in Terra Sirenum, approssimativamente ad est del cratere Millman. I calanchi scuri a sinistra (nord) sembrano aver eroso pesantemente la parete del cratere e trasportato del materiale verso il basso. Si notino anche le increspature scavate sulle dune di colore blu-verde.

ДЮНЫ И РЯБЬ

Это изображение дна и северной стены кратера земли Сирен, примерно к востоку от кратера Миллман. Тёмные овраги слева (на севере), похоже, сильно разрушили стену кратера и перенесли материал вниз по склону. Обратите внимание на дюнную рябь сине-зелёного цвета.

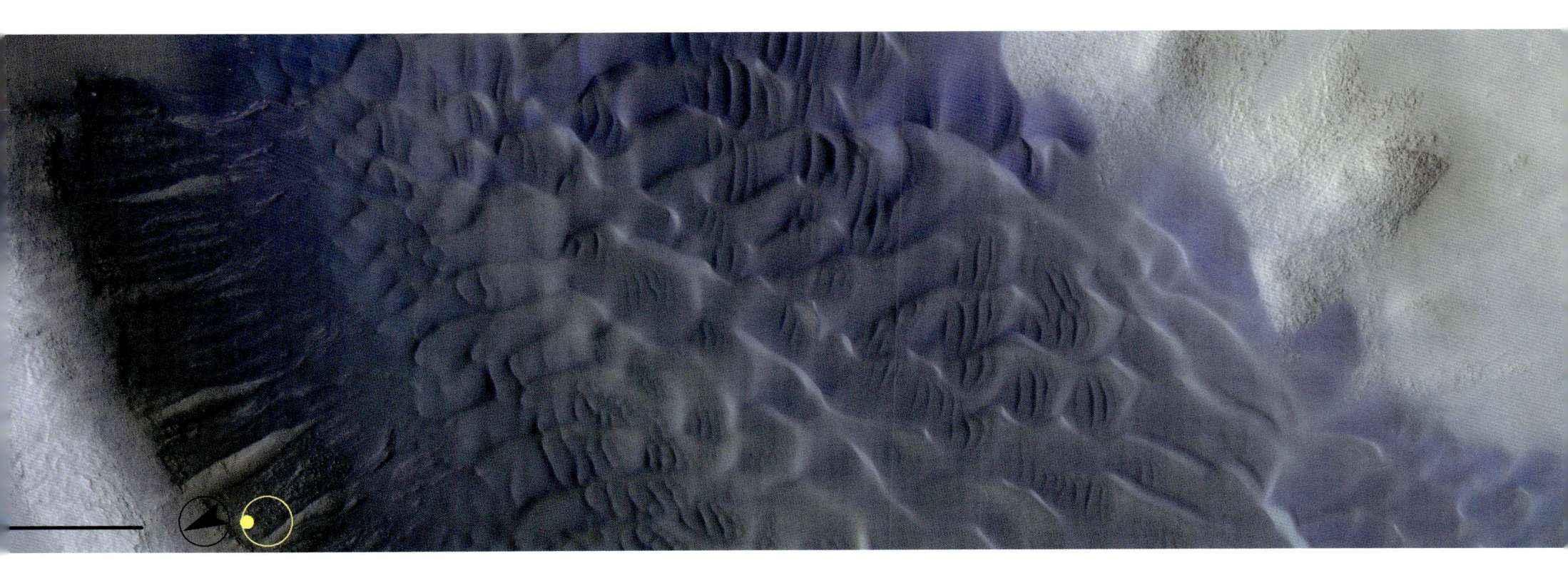

8.9   DUNES AND RIPPLES

ACQUISITION TIME: 2021-01-31T19:49:19.606
LONGITUDE: 215.5 E   LATITUDE: 53.0 S
REGION: SIRENUM

POLYGONALE DÜNEN

Dies ist ein Bild der zentralen Grube des Mazamba-Kraters in Thaumasia Planum. Diese Grube befindet sich in der Mitte der zentralen Erhebung des Kraters mit einem Durchmesser von 53,5 km. Auf dem Boden des Kraters befindet sich ein System von polygonalen Dünen. Sie haben sich wahrscheinlich aus Sand gebildet, der von Winden aus verschiedenen Richtungen herbeigeweht wurde. Dies ist in Kratern üblich, in denen sich Sand ansammeln kann.

POLYGONAL DUNES

This is an image of the central pit of Mazamba Crater in Thaumasia Planum. This pit is at the centre of the central uplift of the 53.5 km diameter crater. On the floor of the pit is a system of polygonal dunes. They likely formed from sand blown by winds coming from multiple directions. This is common in craters where sand can accumulate.

DUNES POLYGONALES

Fosse centrale du cratère Mazamba dans la région de Thaumasia Planum. Cette fosse se situe au milieu du soulèvement central du cratère de 53,5 km de diamètre. Sur le sol de la fosse se trouve un système de dunes polygonales. Elles se sont probablement formées à partir de sable soufflé par des vents venant de plusieurs directions. Ce phénomène est courant dans les cratères où le sable peut s'accumuler.

DUNE POLIGONALI

Questa è un'immagine della fossa centrale del cratere Mazamba nel Thaumasia Planum. Questa fossa è al centro dell'altura centrale del cratere di 53,5 km di diametro. Sul pavimento della fossa si osserva un sistema di dune poligonali. Probabilmente si sono formate dalla sabbia soffiata da venti provenienti da più direzioni. Questo processo è comune nei crateri dove la sabbia trova le condizioni ottimali per accumularsi.

ПОЛИГОНАЛЬНЫЕ ДЮНЫ

Это изображение центральной депрессии кратера Мазамба на плато Тавмасия. Она находится в центральном поднятии кратера диаметром 53,5 км. На дне депрессии видна система дюн с полигональным рисунком. Вероятно, они образованы песком, приносимым ветрами разных направлений. Для кратеров, в которых может накапливаться песок, это обычное явление.

8.10 **POLYGONAL DUNES**

ACQUISITION TIME: 2018-09-18T10:25:35.000
LONGITUDE: 290.3 E   LATITUDE: 27.9 S
REGION: THAUMASIA

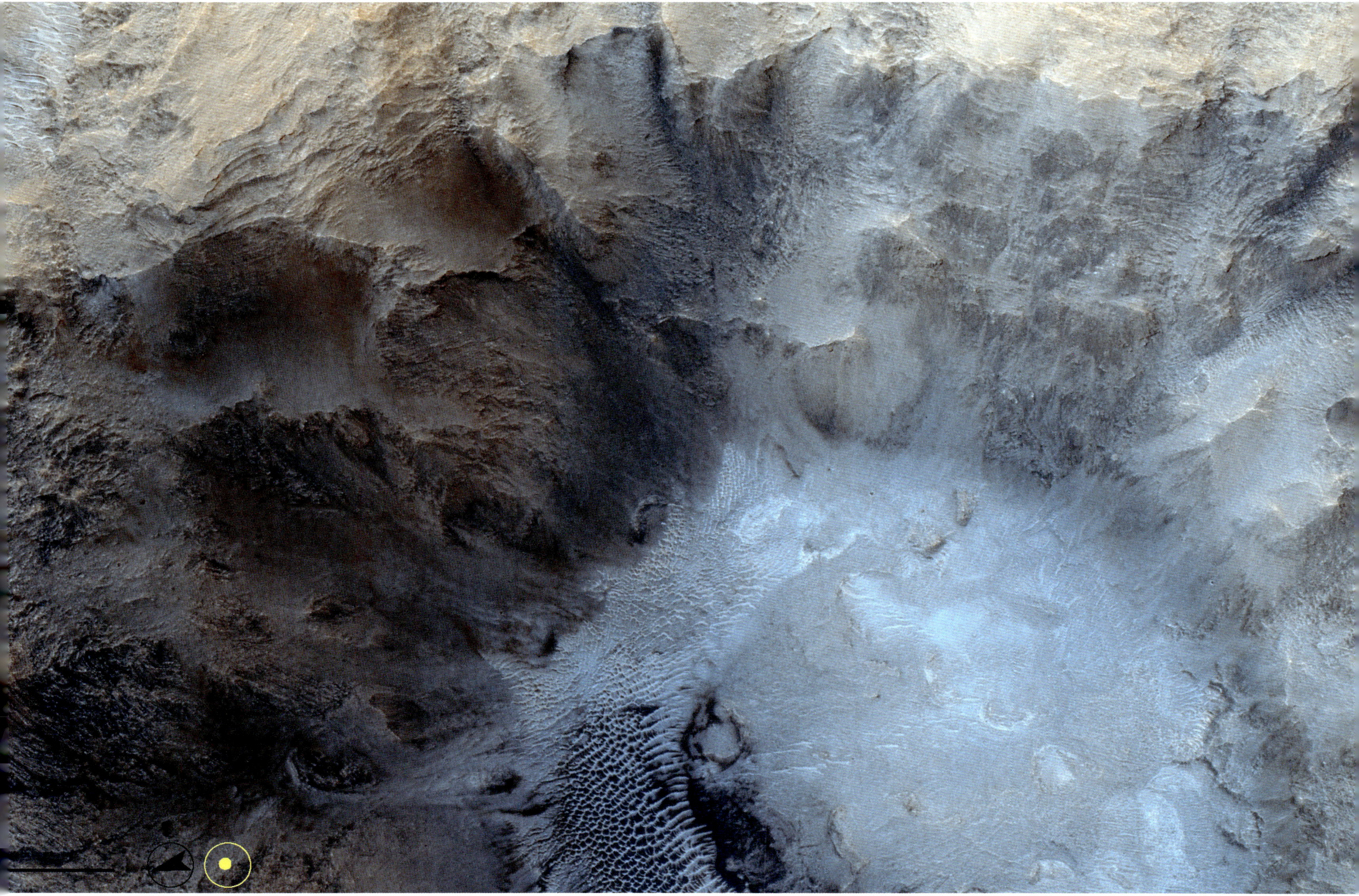

EIN STAUBTEUFEL IN AKTION

Dies ist ein Bild einer Stelle in Tyrrhena Terra nordwestlich des Savich-Kraters und zeigt einen Kanal (Mitte), der möglicherweise mit dem Vichada-Valles-System zusammenhängt. Auf diesem Bild ist jedoch auch ein Staubteufel zu sehen (Mitte oben). Der Schatten des Staubes in der Säule des Staubteufels ist gut zu erkennen und weist leichte Farbsäume auf (grüne/violette Ränder). Dies ist wahrscheinlich auf die Bewegung des Staubes während der Bildaufnahme durch CaSSIS zurückzuführen.

A DUST DEVIL IN ACTION

This is an image of a site in Tyrrhena Terra northwest of Savich Crater and shows a channel (centre), which may be related to the Vichada Valles system. However, this image also includes a dust devil (centre top). The shadow of the dust in the dust devil column can be seen well and shows some slight colour fringing (green/violet edges). This is probably a result of the movement of the dust while CaSSIS was imaging.

UNE TORNADE DE POUSSIÈRE EN ACTION

Cette image d'un site de Tyrrhéna Terra au nord-ouest du cratère Savich montre un chenal (au centre), qui pourrait être lié au système des vallées de Vichada. Cette image comprend également un tourbillon de poussière (au centre en haut). L'ombre de la poussière dans la colonne du tourbillon est bien visible et présente de légères franges de couleur (bords verts/violets). Ceci est probablement le résultat du mouvement de la poussière pendant que CaSSIS prenait les images.

UN DIAVOLO DI SABBIA IN AZIONE

Questa è un'immagine di un sito in Tyrrhena Terra a nord-ovest del Savich Crater e mostra un canale (centro), che potrebbe essere collegato al sistema Vichada Valles. Tuttavia, questa immagine include anche un diavolo di polvere (centro in alto), la cui ombra può essere identificata bene e mostra alcune leggere frange di colore (bordi verdi/viola). Questo è probabilmente il risultato del movimento della polvere mentre CaSSIS stava acquisendo le immagini.

ПЫЛЕВОЙ ДЬЯВОЛ В ДЕЙСТВИИ

Это изображение участка Тирренской земли к северо-западу от кратера Савич. В центре виден канал, который может относиться к системе долин Вичада. Но самое примечательное, что этот снимок запечатлел пылевого дьявола (в центре вверху). Тень от пыли в столбе этого вихря хорошо видна и имеет небольшое цветовое окаймление (зеленые/фиолетовые края). Вероятно, это результат движения пыли во время съемки CaSSIS.

8.11   A DUST DEVIL IN ACTION

ACQUISITION TIME: 2020-05-05T22:40:37.761
LONGITUDE: 91.5 E   LATITUDE: 23.2 S
REGION: TYRRHENA

STAUBTEUFEL UND EIN ELEFANT

Dieses Bild wurde in der Region Phlegra Dorsa aufgenommen. Das Gelände in diesem Gebiet ist geprägt von Spuren vergangenen Vulkanismus wie Lavaströmen und Vulkankegel. Diese besondere Stelle ist unter Marsfotografen wegen der merkwürdigen Struktur in der Mitte des Bildes, die ein wenig wie ein Elefantenkopf aussieht, gut bekannt. Der helle Fleck unterhalb und links des Elefanten ist ein Staubteufel mit einem langen Schatten, der sich nach Nordosten erstreckt.

DUST DEVIL AND AN ELEPHANT

This image was obtained in the Phlegra Dorsa region. The terrain in the area is dominated by evidence of past volcanism including lava flows and volcanic cones. This particular site is known well in the Mars imaging community because of the curious structure in the centre of the image here, which looks a little like the head of an elephant! The bright patch below and to the left of the elephant is a dust devil with a long shadow extending towards the north-east.

UNE TORNADE DE POUSSIÈRE ET UN ÉLÉPHANT

Cette image a été obtenue dans la région de Phlegra Dorsa. Le terrain de cette région est dominé par des traces de volcanisme passé, notamment des coulées de lave et des cônes volcaniques. Ce site particulier est bien connu des spécialistes de l'imagerie martienne en raison de la curieuse structure au centre de l'image, qui ressemble un peu à la tête d'un éléphant ! La tache brillante en dessous et à gauche de l'éléphant est un tourbillon de poussière avec une ombre s'étendant vers le nord-est.

UN DIAVOLO DI SABBIA E UN ELEFANTE

Questa immagine è stata ottenuta nella regione di Phlegra Dorsa. Il terreno nella zona è dominato da prove di vulcanismo passato, tra cui colate di lava e coni vulcanici. Questo particolare sito è ben noto nella comunità di osservazione di Marte a causa della curiosa struttura al centro dell'immagine, che sembra un po' la testa di un elefante! La macchia luminosa sotto e a sinistra dell'elefante è un diavolo di polvere con una lunga ombra che si estende verso nord-est.

ПЫЛЕВОЙ ДЬЯВОЛ И СЛОН

Это снимок района гряды Флегра. На местности в этом районе преобладают следы древнего вулканизма, включая лавовые потоки и вулканические конусы. Данный конкретный участок хорошо известен в сообществе специалистов по съемке Марса благодаря любопытной структуре в центре снимка, которая напоминает голову слона! Яркое пятно слева внизу от слона – это пылевой дьявол с длинной тенью, простирающейся на северо-восток.

8.12 DUST DEVIL AND AN ELEPHANT

ACQUISITION TIME: 2018-06-03T23:27:25.000
LONGITUDE: 173.0 E LATITUDE: 27.4 N
REGION: ARCADIA

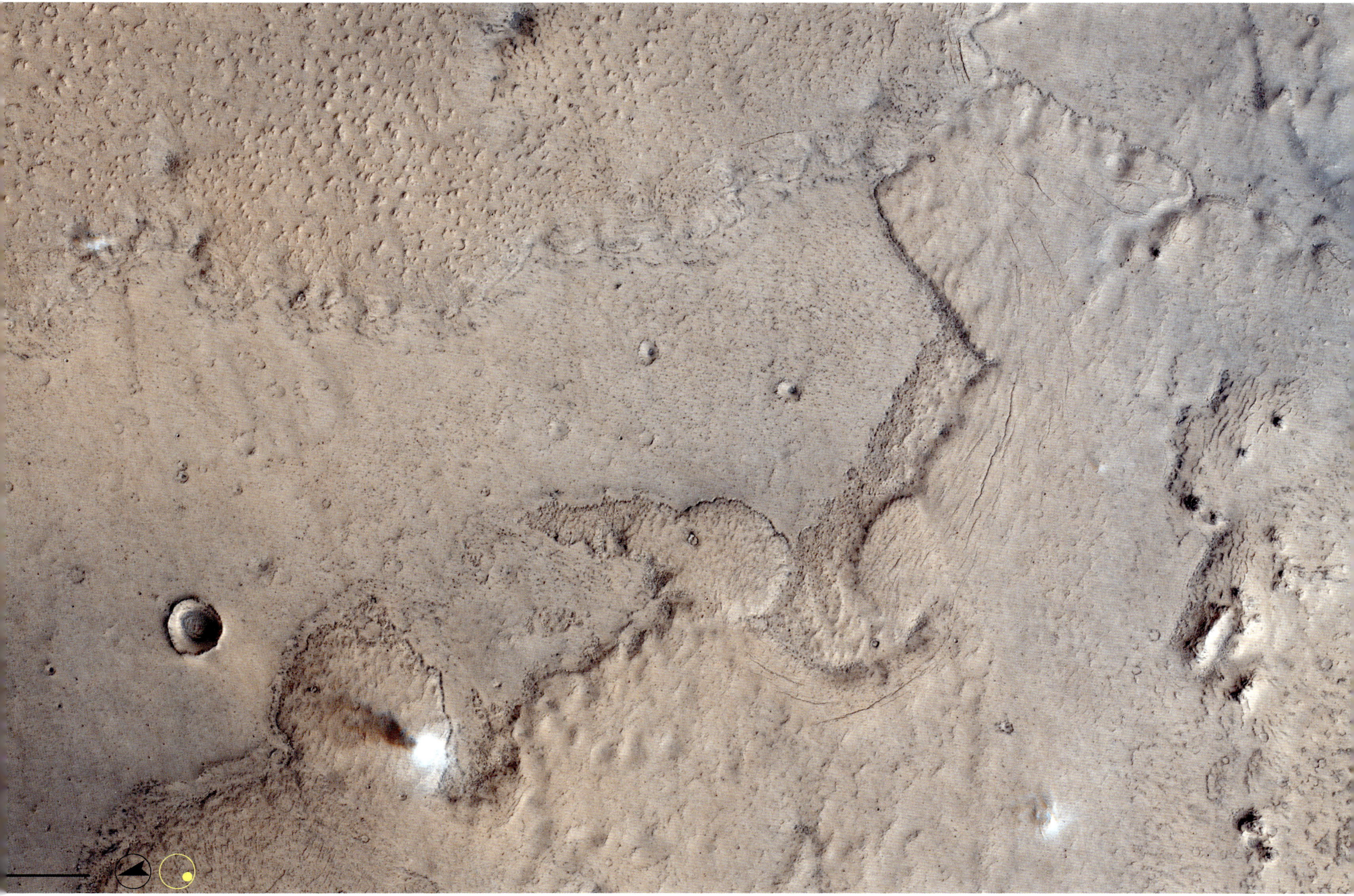

STAUBTEUFELSPUREN UND WOLKEN

Dies ist ein Bild von einem Standort in Argyre Planitia, der für die Suche nach Staubteufeln genutzt wurde. Hier sind einige lange Spuren zu sehen, die an den Nordhängen der Hügelkämme entstehen. Interessant an diesem Bild ist auch das Vorhandensein von hellen Wolken links oben im Bild, die die Spuren teilweise verdecken.

DUST DEVIL TRACKS AND CLOUDS

This is an image from a site in Argyre Planitia that was used to search for dust devils. Here we can see some long tracks originating on the north slopes of hill crests. What is also attractive about this image is the presence of bright clouds to the upper left of the frame that partially obscure the tracks.

TRACES DE TORNADES DE POUSSIÈRE ET NUAGES

Cette image provient d'un site d'Argyre Planitia qui a été utilisé pour rechercher des tourbillons de poussière. Ici, nous pouvons voir de longues traces qui prennent naissance sur le versant nord des crêtes des collines. Cette image est magnifiée par la présence de nuages en haut et à gauche du cadre qui masquent partiellement les traces à la surface.

TRACCE DI DIAVOLO DI SABBIA E NUVOLE

Questa è un'immagine da un sito in Argyre Planitia che è stata utilizzata per cercare i diavoli di polvere. Qui si possono vedere alcune lunghe tracce che hanno origine sulle pendici settentrionali delle creste delle colline. Ciò che è interessante è la presenza di nuvole luminose in alto a sinistra del fotogramma che oscurano parzialmente le tracce.

СЛЕДЫ ПЫЛЕВЫХ ДЬЯВОЛОВ И ОБЛАКА

Это снимок участка равнины Аргир, на котором часто шел поиск пылевых дьяволов. Здесь мы видим несколько длинных следов, берущих начало на северных склонах холмов. Привлекательным на этом снимке является наличие ярких облаков в верхней левой части кадра, которые частично заслоняют следы.

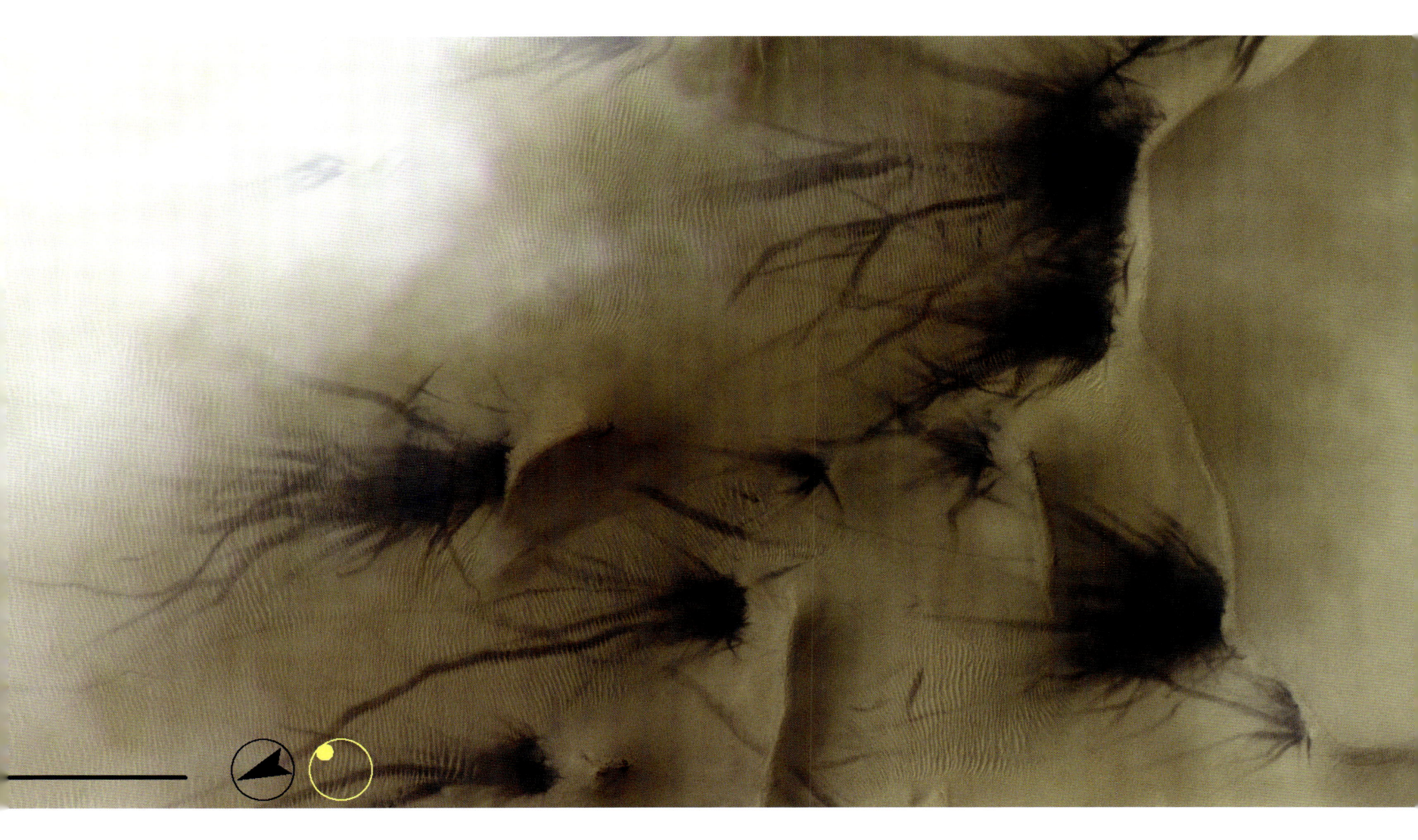

8.13 **DUST DEVIL TRACKS AND CLOUDS**

ACQUISITION TIME: 2019-01-07T05:43:59.039
LONGITUDE: 318.4 E LATITUDE: 46.3 S
REGION: ARGYRE

HELLE UND DUNKLE HANGSTREIFEN

Dieses Bild zeigt ein Gebiet in Terra Sabaea südlich des Locras-Valles-Systems und zeigt ein Phänomen, das als Hangstreifen bekannt ist. Man geht davon aus, dass diese durch Staublawinen aus feinkörnigem Material verursacht werden. Dieses Bild zeigt jedoch, dass Hangstreifen sowohl dunkel als auch hell sein können und dass dies wahrscheinlich nicht auf Beleuchtungseffekte zurückzuführen ist, da helle und dunkle Streifen in unmittelbarer Nähe zueinander in der gleichen Ausrichtung zur Sonne zu sehen sind. Eine Hypothese besagt, dass sich die Streifen im Laufe der Zeit durch einen Mechanismus, der nachfolgend eine Ablagerung und Verdichtung von Staub beinhaltet, von dunkel zu hell entwickeln.

BRIGHT AND DARK SLOPE STREAKS

This image is of an area in Terra Sabaea just south of the Locras Valles system and shows a phenomenon known as slope streaks. These are thought to be caused by dust avalanches of finely grained material. However, this image shows that slope streaks can be both dark and bright and that this is unlikely to be a result of illumination effects because bright and dark streaks are seen in close proximity to each other in the same orientation with respect to the Sun. One hypothesis is that the streaks evolve with time from dark to bright through a mechanism involving later deposition and compaction of dust.

TRAÎNÉES DE TALUS CLAIRES ET SOMBRES

Cette image juste au sud du système de Locras Valles (zone de Terra Sabaea) montre un phénomène connu sous le nom de traînées de talus (« slope streaks » en anglais). On pense que ces marques sont laissées par des avalanches de poussières. Cette image montre que les traînées de talus peuvent être sombres ou claires et qu'il est peu probable que cela soit le résultat d'effets d'illumination, car traces claires et sombres sont observées à proximité les unes des autres avec une orientation similaire par rapport au Soleil. Une hypothèse est que les traînées évoluent avec le temps en devenant plus claires par un mécanisme impliquant un dépôt et une compaction ultérieure de la poussière.

STRISCE DI PENDENZA CHIARE E SCURE

Questa immagine proviene da un'area in Terra Sabaea appena a sud del sistema Locras Valles e mostra un fenomeno noto come «slope streaks» (strisce di pendenza). Si pensa che queste siano causate da valanghe di materiale a grana fine. Tuttavia, questa immagine mostra che le strisce di pendenza possono essere sia scure che chiare e che è improbabile che questo sia il risultato di effetti di illuminazione perché le strisce chiare e scure sono osservate in stretta vicinanza l'una all'altra con lo stesso orientamento rispetto al Sole. Un'ipotesi è che le striature evolvano con il tempo da scure a chiare attraverso un meccanismo che coinvolge la successiva deposizione e compattamento della polvere.

СВЕТЛЫЕ И ТЕМНЫЕ СКЛОНОВЫЕ ПОЛОСЫ

На этом снимке Сабейской земли к югу от системы долин Локра, показано явление, известное под названием «склоновые полосы». Считается, что они вызваны пылевыми лавинами мелкозернистого материала. Однако это изображение демонстрирует, что полосы на склонах могут быть как темными, так и светлыми. Такое различие цвета вряд ли является следствием эффектов освещения, поскольку яркие и темные полосы видны на этом снимке в непосредственной близости друг от друга и при одинаковой ориентации по отношению к Солнцу. По одной из гипотез, полосы эволюционируют со временем от темных к светлым посредством механизма, включающего повторное отложение и уплотнение пыли.

8.14   BRIGHT AND DARK SLOPE STREAKS

ACQUISITION TIME: 2021-02-07T11:20:48.949
LONGTUDE: 47.5 E   LATITUDE: 6.2 N
REGION: SABAEA

HANGSTREIFEN ÜBERALL

Dieses Bild zeigt eine Vielzahl von Hangstreifen in einem Gebiet im Süden von Amazonis Planitia. Man geht davon aus, dass die Hangstreifen durch Staublawinen aus feinkörnigem Material verursacht werden. Man sieht, wie die Streifen von der Topografie beeinflusst werden. Insbesondere der dunkelste Streifen in der Mitte des Bildes gleitet um einen Hügel herum. Eine Hypothese besagt, dass dieser Streifen wahrscheinlich der jüngste ist, da er dunkler ist als die benachbarten Streifen und vermutlich die geringste Menge an ausfallendem atmosphärischem Staub aufweist.

SLOPE STREAKS EVERYWHERE

This image shows a multitude of slope streaks in an area in south Amazonis Planitia. Slope streaks are thought to be caused by dust avalanches of finely grained material. Note how the streaks are influenced by topography. In particular, the darkest streak in the centre of the frame slides around a mound. One hypothesis states that this streak is likely the youngest because it is darker than nearby streaks and presumably has the least amount of precipitating atmospheric dust.

DES TRAINÉES DE TALUS PARTOUT

Cette image montre une multitude de traînées de talus dans une zone au sud d'Amazonis Planitia. On pense que les traînées de talus sont causées par des avalanches de poussière fine. Notez comme ces structures sont influencées par la topographie, en particulier la traînée la plus sombre au centre de l'image qui glisse autour d'un monticule. Selon l'hypothèse actuelle, cette traînée est probablement la plus jeune parce qu'elle est plus sombre que les traînées voisines et qu'elle contient vraisemblablement la plus petite quantité de poussière atmosphérique précipitée.

STRISCE DI PENDENZA OVUNQUE

Questa immagine mostra una moltitudine di strisce di pendenza in un'area nel sud di Amazonis Planitia. Si pensa che le striature di pendenza siano causate da valanghe di polvere di materiale a grana fine. Si noti come le striature sono influenzate dalla topografia. In particolare, la striscia più scura al centro del fotogramma scorre intorno a un cumulo. Un'ipotesi afferma che questa striscia è probabilmente la più giovane perché è più scura delle strisce vicine e presumibilmente è coperta da una minore quantità di polvere atmosferica depositata.

СКЛОНОВЫЕ ПОЛОСЫ ПОВСЮДУ

На этом снимке показано множество склоновых полос на юге равнины Амазония. Считается, что полосы на склонах вызваны пылевыми лавинами мелкозернистого материала. Обратите внимание, как полосы зависят от топографии. В частности, видно, как самая тёмная полоса в центре кадра повторяет рельеф кургана. Согласно одной из гипотез, эта полоса, скорее всего, самая молодая, поскольку, она темнее соседних полос и, предположительно, покрыта наименьшим количеством оседающей атмосферной пыли.

8.15 SLOPE STREAKS EVERYWHERE

ACQUISITION TIME: 2021-03-29T10:58:59.075
LONGITUDE: 188.2 E   LATITUDE: 7.8 N
REGION: AMAZONIS

9 UNTERSCHIEDLICHE MINERALOGIEN UND SCHICHTEN

DIVERSE MINERALOGIES AND LAYERS

MINÉRALOGIES ET STRATES DIVERSES

DIVERSE MINERALOGIE E STRATIFICAZIONI

МНОГООБРАЗИЕ МИНЕРАЛОВ И СЛОЁВ

SPEKTRALE VIELFALT IM ÖSTLICHEN THAUMASIA PLANUM

Dieses CaSSIS-Bild war lang und dünn. Es ist so interessant und farblich so vielfältig (mineralische Zusammensetzung), dass wir es in zwei Teile aufgeteilt haben, um die gesamte Abbildung in hoher Auflösung zeigen zu können. Der nördlichste Abschnitt befindet sich zuoberst. Das Bild zeigt ein Gebiet, das reich an Schichtsilikaten (Tonen) und kalziumarmem Pyroxen sein könnte. Es befindet sich in der Nähe der Hochlandvulkane nordwestlich des Ibragimov-Kraters. Thaumasia Planum liegt südlich der Melas und Coprates Chasmata in den Valles Marineris, und seine Oberfläche wurde stark von basaltischen Lavaströmen beeinflusst.

SPECTRAL DIVERSITY IN EAST THAUMASIA PLANUM

This CaSSIS image was long and thin. It is so interesting and diverse in colour (mineral composition) that we have split it into two parts to be able to show the whole scene at high resolution. The northernmost section is at the top. It shows an area that may be rich in phyllosilicates (clays) and low calcium pyroxene. It is near the highland volcanoes to the northwest of Ibragimov Crater. Thaumasia Planum lies south of the Melas and Coprates Chasmata in Valles Marineris and its surface was strongly influenced by basaltic lava flows.

DIVERSITÉ SPECTRALE DANS L'EST DE THAUMASIA PLANUM

Cette image CaSSIS est longue et mince. Elle est si intéressante et diversifiée en termes de couleurs (et donc de composition minéralogique) que nous l'avons divisée en deux parties pour pouvoir montrer l'ensemble de la scène à haute résolution. La partie la plus au nord est en haut. Elle montre une zone qui pourrait être riche en phyllosilicates (argiles) et en pyroxènes à faible teneur en calcium. Elle est proche des hauts volcans au nord-ouest du cratère Ibragimov. Thaumasia Planum se trouve au sud de Melas et Coprates Chasmata dans Valles Marineris et sa surface a été fortement influencée par des coulées de lave basaltique.

DIVERSITÀ SPETTRALE NELLA THAUMASIA PLANUM ORIENTALE

Questa immagine CaSSIS è molto lunga e sottile. È così interessante e varia nei colori (composizione minerale) che è stata divisa in due parti per poter mostrare l'intera scena ad alta risoluzione. La sezione più a nord è in alto e mostra un'area che potrebbe essere ricca di fillosilicati (argille) e pirosseno povero di calcio. Si trova vicino ai vulcani dell'altopiano a nord-ovest del cratere Ibragimov. Thaumasia Planum si trova a sud del Chasmata Melas e Coprates nelle Valles Marineris e la sua superficie è stata fortemente influenzata da colate di lava basaltica.

СПЕКТРАЛЬНОЕ РАЗНООБРАЗИЕ В ВОСТОЧНОЙ ЧАСТИ ПЛАТО ТАВМАСИЯ

Данный протяженный и узкий снимок CaSSIS настолько интересен и разнообразен по цвету (фиксирующему минеральный состав), что мы разделили его на две части, с целью показать всю картину в высоком разрешении. Северная часть, находящаяся вверху, показывает область, которая может быть богата филлосиликатами (глинами) и низкокальциевым пироксеном. Она расположена рядом с высокогорными вулканами к северо-западу от кратера Ибрагимов. Плато Тавмасия, к которому принадлежит видимая область, лежит к югу от каньонов Мелас и Копрат, что в долинах Маринер. На его поверхность сильно повлияли базальтовые потоки лавы.

9.1 SPECTRAL DIVERSITY IN EAST THAUMASIA PLANUM

ACQUISITION TIME: 2020-12-17T08:37:51.248
LONGITUDE: 298.7 E   LATITUDE: 22.0 S
REGION: THAUMASIA

AUFGESCHLOSSENES HELLES MATERIAL IM SÜDLICHEN THAUMASIA PLANUM

Dieses Bild wurde südöstlich des Kraters Aniak in Thaumasia Planum aufgenommen. Im Norden (links) sind zahlreiche Talstrukturen zu sehen. Hier sehen wir unregelmässige Vertiefungen mit hellen Böden. Es gibt mehrere Merkmale, bei denen es sich wahrscheinlich um Faltenkämme (Dorsa) handelt, d.h. um niedrige, gewundene Kämme. Dies sind tektonische Merkmale, die entstehen, wenn basaltische Lava abkühlt und sich zusammenzieht. Das helle Material könnte das Ergebnis einer längeren Interaktion von flüssigem Wasser mit abkühlenden Lavaströmen sein.

EXPOSED BRIGHT MATERIAL IN SOUTH THAUMASIA PLANUM

This image was taken to the south-east of Aniak Crater in Thaumasia Planum. There are numerous valley structures to the north (left). Here we see irregular depressions with light-toned floors. There are several features that are probably wrinkle ridges which are low, sinuous ridges. They are tectonic features created when basaltic lava cools and contracts. The bright material may be the result of prolonged interaction of liquid water with cooling lava flows.

MATÉRIAUX CLAIRS EXPOSÉS DANS LE SUD DE THAUMASIA PLANUM

Cette image a été prise au sud-est du cratère Aniak dans Thaumasia Planum. De nombreuses structures de vallées sont visibles au nord (à gauche). Ici, nous observons des dépressions irrégulières avec des roches de couleur claire. Des structures allongées, basses et sinueuses, sont aussi visibles ici. Ce sont des structures tectoniques créées lorsque la lave basaltique refroidit et se contracte. Le matériau clair peut être le résultat d'une interaction prolongée de l'eau liquide avec des coulées de lave en cours de refroidissement.

MATERIALE CHIARO AFFIORANTE NEL THAUMASIA PLANUM MERIDIONALE

Questa immagine è stata presa a sud-est del cratere Aniak nel Thaumasia Planum. Ci sono numerose strutture vallive a nord (a sinistra). Qui si vedono depressioni irregolari con pavimenti chiari. Ci sono diverse caratteristiche che sono probabilmente dorsum, cioè creste basse e sinuose. Sono caratteristiche tettoniche create quando la lava basaltica si raffredda e si contrae. Il materiale chiaro può essere il risultato dell'interazione prolungata dell'acqua liquida con le colate di lava in raffreddamento.

ОБНАЖЕНИЯ ЯРКОГО МАТЕРИАЛА НА ЮГЕ ПЛАТО ТАВМАСИЯ

Это снимок области к юго-востоку от кратера Аниак на плато Тавмасия. К северу (слева) находятся многочисленные долинные структуры. Здесь мы видим впадины неправильной формы со светлым дном. Есть несколько характерных деталей рельефа, которые, вероятно, являются «морщинными хребтами» – низкими извилистыми гребнями. Они являются тектоническими структурами, возникающими при неравномерном остывании и сжатии потоков базальтовой лавы. Длительное взаимодействие с ними жидкой воды могло создать видимый яркий материал.

9.2 EXPOSED BRIGHT MATERIAL IN SOUTH THAUMASIA PLANUM

ACQUISITION TIME: 2020-11-22T23:06:14.984
LONGITUDE: 291.4 E   LATITUDE: 33.4 S
REGION: THAUMASIA

UNGEWÖHNLICH GEFÄRBTES UNTERGRUNDMATERIAL IN TERRA CIMMERIA

Dieses Bild zeigt ein Gebiet südwestlich des Molesworth-Kraters in Terra Cimmeria. Es gibt recht feine Farbunterschiede. Der hier gezeigte Teil des Bildes legt nahe, dass es sich hier um eine ältere Materialkappe (dunkelblau) über einem helleren Material (hellviolett) handelt. Es gibt relativ wenige Aufnahmen von Marskameras in diesem speziellen Gebiet, aber die meisten Krater in der Umgebung scheinen mit einer Ablagerung gefüllt worden zu sein und haben daher einen flachen Boden.

UNUSUALLY COLOURED UNDERLYING MATERIAL IN TERRA CIMMERIA

This image was taken of an area to the south-west of Molesworth Crater in Terra Cimmeria. There are rather subtle colour differences. The part of the image shown here illustrates that there may be an older cap of material (darkish blue) over a lighter-toned material (light purple). There are relatively few observations by Mars cameras of this specific area but most craters in the vicinity appear to have been filled with a deposit and are consequently flat-bottomed.

MATÉRIAU SOUS-JACENT DE COULEUR INHABITUELLE DANS TERRA CIMMERIA

Image d'une zone située au sud-ouest du cratère Molesworth dans la région de Terra Cimmeria. Les différences de couleur sont assez subtiles. L'extrait d'image présenté ici semble montrer une formation plus ancienne (bleu foncé) sur une formation plus claire (violet clair). Il y a relativement peu d'observations de cette zone spécifique mais la plupart des cratères des environs semblent avoir été remplis de dépôts et ont donc un fond plat.

SUBSTRATO INSOLITAMENTE COLORATO IN TERRA CIMMERIA

Questa immagine è stata presa da un'area a sud-ovest del cratere Molesworth in Terra Cimmeria. Ci sono alcune deboli differenze di colore. La parte dell'immagine mostrata qui illustra che ci può essere una copertura di materiale più vecchio (blu scuro) sopra un materiale più chiaro (viola chiaro). Ci sono relativamente poche osservazioni da parte delle fotocamere di Marte in questa specifica zona, ma qualche caratteristica si può evincere già da questa immagine: la maggior parte dei crateri nelle vicinanze sembra essere stata riempita con un deposito e sono di conseguenza a fondo piatto.

НЕОБЫЧНО ОКРАШЕННЫЙ ПОДСТИЛАЮЩИЙ МАТЕРИАЛ В КИММЕРИЙСКОЙ ЗЕМЛЕ

Это снимок района к юго-западу от кратера Моулсворт на Киммерийской земле. Здесь видны довольно тонкие цветовые различия. Часть приведенного снимка иллюстрирует, что здесь может существовать покров более старых пород (темно-синего цвета) поверх более светлого материала (светло-фиолетового цвета). Было проведено относительно немного наблюдений в этой конкретной области на Марсе, но похоже, что большинство кратеров в окрестностях были заполнены отложениями и, следовательно, имеют плоское дно.

9.3 UNUSUALLY COLOURED UNDERLYING MATERIAL IN TERRA CIMMERIA

ACQUISITION TIME: 2020-11-23T08:54:12.005
LONGITUDE: 145.3 E LATITUDE: 28.8 S
REGION: CIMMERIA

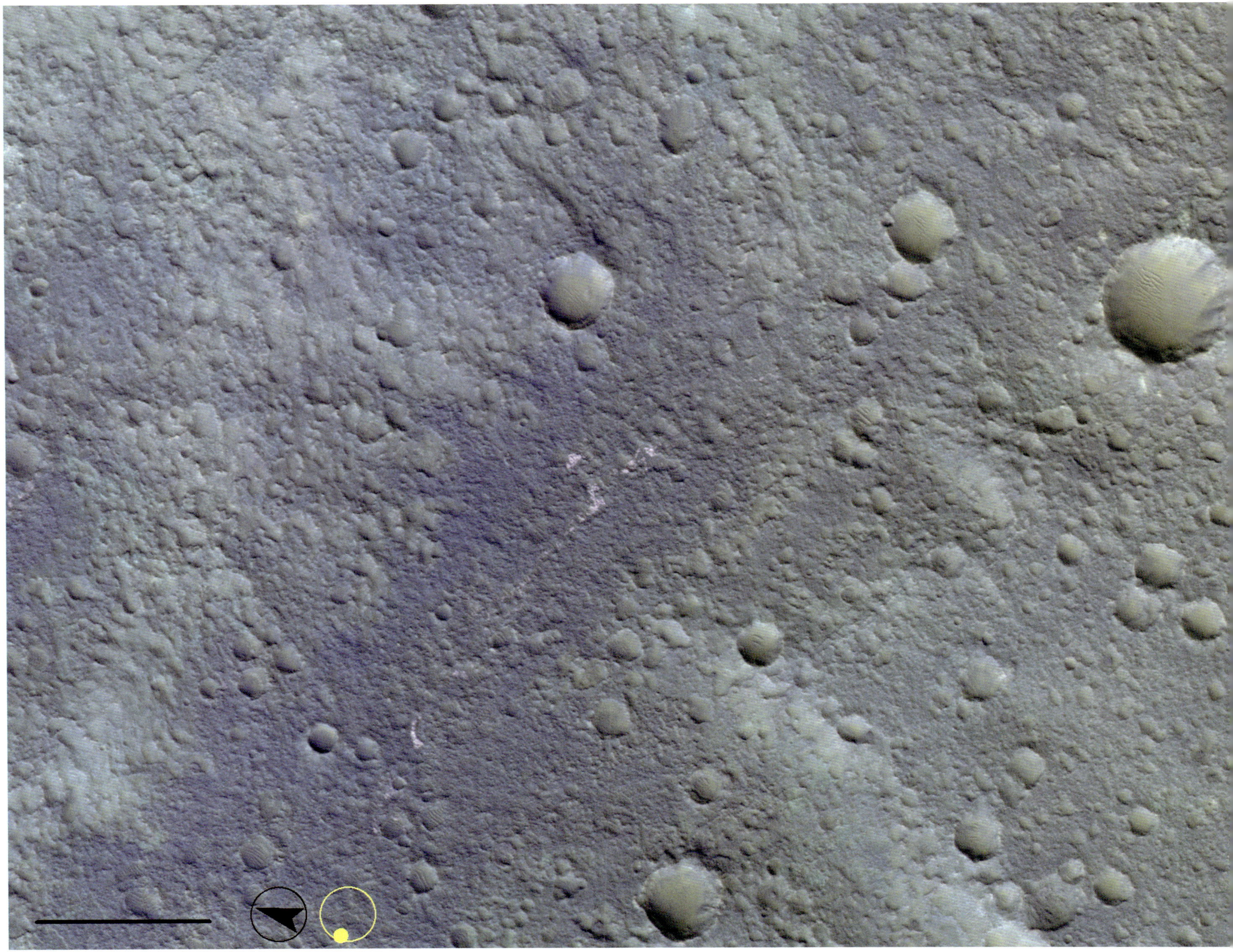

SCHICHTEN IN TYRRHENA TERRA

Dies ist ein Bild von geschichteten, hellen Ablagerungen nordöstlich des Hellas-Beckens in einer Region namens Hadriacus Cavi. Ein möglicher zukünftiger Roverlandeplatz (Hadriacus Palus) liegt direkt nördlich dieses Gebiets. Es sind zahlreiche Schichten sichtbar. Die Entstehung und Entwicklung dieses Gebiets ist nicht genau bekannt, aber die Schichten hier könnten vulkanischen Ursprungs sein und dann durch fluviale Aktivitäten erodiert und freigelegt worden sein. In der Nähe gibt es mehrere alte Talnetzwerke (z. B. Huallaga Vallis).

LAYERS IN TYRRHENA TERRA

This is an image of layered, light-toned deposits to the northeast of Hellas Basin in a region called Hadriacus Cavi. A possible future rover landing site (Hadriacus Palus) is just to the north of this area. Numerous layers are visible. The formation and evolution of this area is not well established but the layers here could be volcanic in origin and then eroded and exposed by fluvial activity. There are several ancient valley networks (e.g. Huallaga Vallis) nearby.

STRATES DANS TYRRHENA TERRA

Cette image montre des dépôts stratifiés de couleur claire au nord-est du bassin Hellas, dans une région appelée Hadriacus Cavi. Un possible futur site d'atterrissage de rovers (Hadriacus Palus) se trouve juste au nord de cette zone. De nombreuses couches sont visibles ici. La formation et l'évolution de cette zone n'est pas bien établie mais ces couches pourraient être d'origine volcanique et ensuite érodées et exposées par l'activité fluviale. Il existe plusieurs anciens réseaux de vallées (par exemple, Huallaga Vallis) à proximité.

STRATI IN TYRRHENA TERRA

Questa è un'immagine di depositi stratificati di colore chiaro a nord-est del Bacino di Hellas in una regione chiamata Hadriacus Cavi. Un possibile futuro sito di atterraggio del rover (Hadriacus Palus) è proprio a nord di questa zona. Si possono apprezzare numerosi strati e, sebbene la formazione e l'evoluzione di quest'area non siano ben accertate, gli strati qui potrebbero essere di origine vulcanica e conseguentemente erosi ed esposti dall'attività fluviale. Ci sono diverse reti di antiche valli (per esempio Huallaga Vallis) nelle vicinanze.

СЛОИСТЫЕ ОТЛОЖЕНИЯ В ТИРРЕНСКОЙ ЗЕМЛЕ

Это изображение слоистых, светлых отложений к северо-востоку от впадины Эллада в регионе Адриатических котловин. Возможное место будущей посадки марсохода (Адриатическое болото) находится к северу от этой области. Видны обнажения многочисленных слоёв породы. Формирование и эволюция этой области не слишком хорошо изучены, но данные слои могли иметь вулканическое происхождение, подвергнуться затем эрозии и обнажиться в результате флювиальной активности. Поблизости находится сеть нескольких древних долин (например, долина Уальяга).

9.4   LAYERS IN TYRRHENA TERRA

ACQUISITION TIME: 2020-12-21T01:05:14.439
LONGITUDE: 78.2 E   LATITUDE: 26.7 S
REGION: TYRRHENA

SCHICHTEN IM SPALLANZANI-KRATER

Spallanzani ist ein Krater mit 71,7 km Durchmesser im Südwesten von Promethei Terra. Malea Planum liegt im Westen. Dieser Krater hat die Aufmerksamkeit auf sich gezogen, weil der Kraterboden, wie auf diesem Bild zu sehen ist, stark geschichtet ist. Es gibt Hinweise auf fliessfähiges Material, das eindeutig irgendwann in seiner Geschichte erodiert wurde, aber der Ursprung der Schichtung ist nicht bekannt.

LAYERS IN SPALLANZANI CRATER

Spallanzani is a 71.7 km diameter crater in south-west Promethei Terra. Malea Planum is to the west. This crater has attracted attention because of the extensive layering on the crater floor as illustrated in this image. There is evidence of flowing material that has clearly been eroded at some stage in its history but the origin of the layering is not known.

STRATES DANS LE CRATÈRE SPALLANZANI

Spallanzani est un cratère de 71,7 km de diamètre situé au sud-ouest de Promethei Terra. Malea Planum se trouve à l'ouest. Ce cratère a attiré l'attention en raison de l'étendue de la stratification sur le plancher du cratère, comme illustré sur cette image. Il y a des preuves d'écoulement de matériaux érodés à un moment donné de son histoire, mais l'origine de la stratification n'est pas connue.

STRATI NEL CRATERE SPALLANZANI

Spallanzani è un cratere di 71,7 km di diametro nel sud-ovest di Promethei Terra. Malea Planum si trova ad ovest. Questo cratere ha attirato l'attenzione dei ricercatori a causa dell'estesa stratificazione sul pavimento del cratere, come illustrato in questa immagine. C'è evidenza di flussi di materiale che sono stati chiaramente erosi in qualche fase della loro storia, ma l'origine della stratificazione non è nota.

СЛОИ В КРАТЕРЕ СПАЛЛАНЦАНИ

Спалланцани – кратер диаметром 71,7 км на юго-западе земли Прометея. На западе от него находится плато Малея. Этот кратер привлек к себе внимание своим многослойным дном, видимым на этом снимке. Здесь есть следы текучего материала, который явно подвергался эрозии на каком-то этапе своей истории, но происхождение слоистости неизвестно.

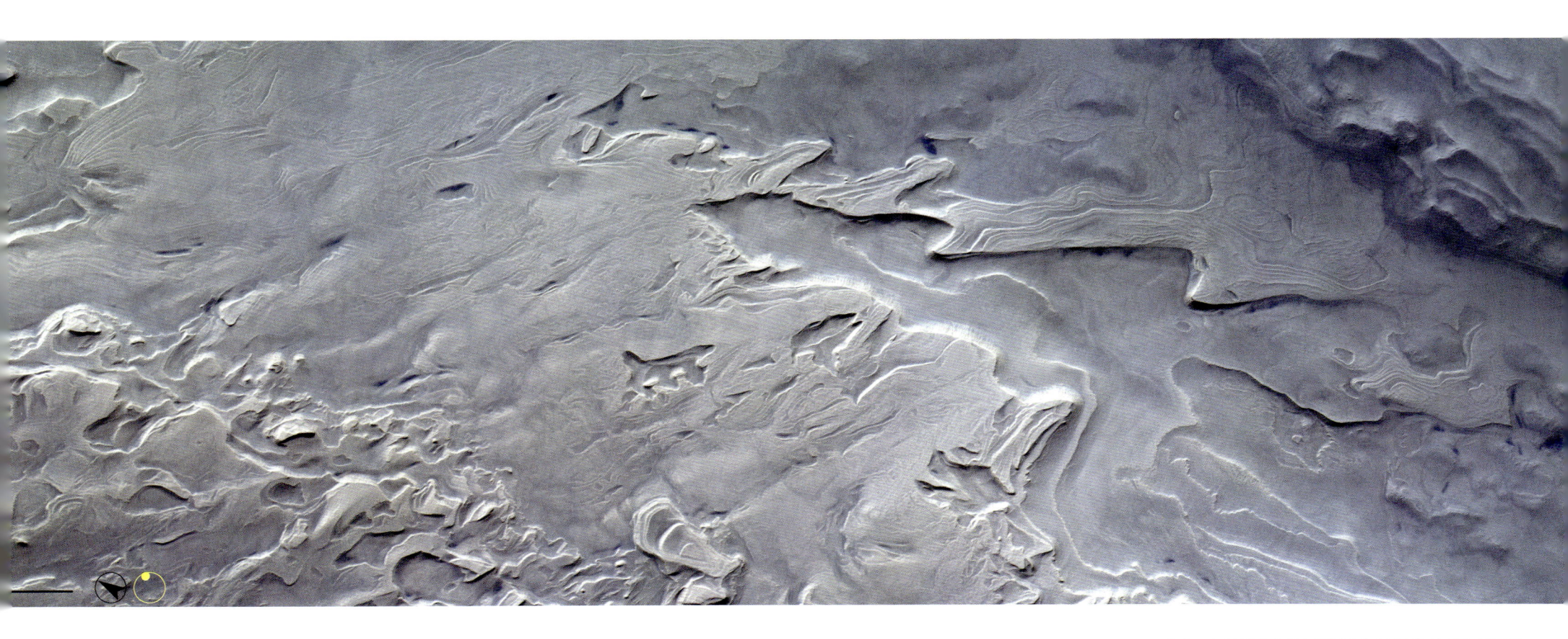

9.5 LAYERS IN SPALLANZANI CRATER

ACQUISITION TIME: 2020-12-25T03:31:23.652
LONGITUDE: 86.3 E   LATITUDE: 57.5 S
REGION: PROMETHEI

SCHICHTEN IN SISYPHI CAVI

Dieses Bild stammt aus Sisyphi Cavi in Sisyphi Planum. Dieses Gebiet ist für seine Tafelberge und Erosionsrinnen bekannt. In der Tat sind hier Erdrutsche und Fächerablagerungen deutlich sichtbar (z. B. oben links). Hier sehen wir jedoch auch geschichtetes Material zwischen zwei Tafelbergen (Mitte oben). Der Entstehungsmechanismus ist unklar, könnte aber mit historischen Erdrutschen und Erosion zusammenhängen.

LAYERS IN SISYPHI CAVI

This image is from Sisyphi Cavi within Sisyphi Planum. This area is known for its mesas and gullies. Indeed, there are landslides and fan deposits clearly visible here (top left, for example). However, here we also see layered material between two mesas (centre to top). The formation mechanism is unclear but could be related to ancient landslides and erosion.

STRATES DANS SISYPHI CAVI

Cette image provient de Sisyphi Cavi dans Sisyphi Planum. Cette zone est connue pour ses mésas et ses ravines. En effet, des glissements de terrain et des cônes de dépôt sont clairement visibles ici (en haut à gauche, par exemple). Cependant, nous observons également ici des matériaux stratifiés entre deux mésas (au centre vers le haut). Le mécanisme de formation n'est pas éclairci mais pourrait être lié à d'anciens glissements de terrain et à l'érosion.

STRATI IN SISYPHI CAVI

Questa immagine proviene da Sisyphi Cavi all'interno di Sisyphi Planum. Questa zona è nota per le sue mesa e i suoi calanchi. Infatti, ci sono frane e conoidi di deiezione chiaramente visibili in questa immagine (in alto a sinistra, per esempio). Tuttavia, vediamo anche del materiale stratificato tra le due mesa (dal centro in alto). Il meccanismo di formazione non è chiaro, ma potrebbe essere legato ad antiche frane ed erosioni.

СЛОИ В КОТЛОВИНАХ СИЗИФА

Это изображение относится к котловинам Сизифа на плато Сизифа. Эта область известна своими столовыми горами и оврагами. Действительно, здесь хорошо видны оползни и веерные отложения (например, вверху слева). Однако, мы также видим слоистый материал между двумя столовыми горами (в центре вверху). Механизм его образования неясен, но может быть связан с древними оползнями и эрозией.

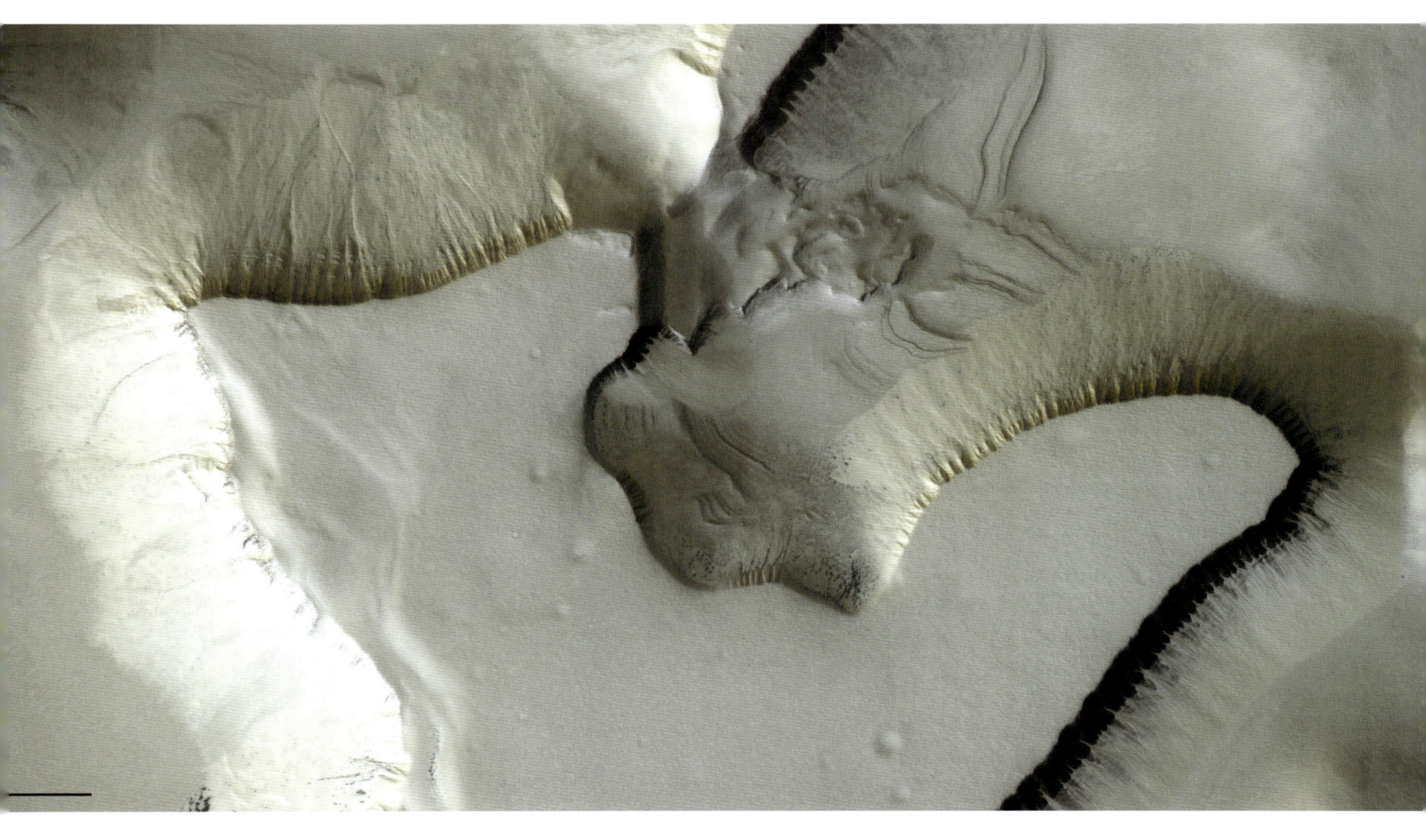

9.6 LAYERS IN SISYPHI CAVI

ACQUISITION TIME: 2020-06-21T03:17:17.082
LONGITUDE: 1.2 E   LATITUDE: 70.5 S
REGION: SISYPHI

NIEDRIGE PARALLELE GRATE IN HELLAS

Dieses Bild zeigt niedrige parallele Erhebungen am westlichen Rand des Hellas-Beckens. Die Entstehung dieser Strukturen ist noch wenig erforscht, aber es ist denkbar, dass ein zähes, fliessendes Material für die Schichtung verantwortlich ist. Möglicherweise besteht auch ein Zusammenhang mit dem «gebänderten Gelände», das an den tiefsten Stellen des Hellas-Beckens zu sehen ist.

LOW PARALLEL RIDGES IN HELLAS

This image shows low parallel ridges on the western edge of Hellas Basin. There has been little research on the formation of these structures but flow of a viscous material with layering is conceivable. There may also be some relationship to the "banded terrain" that is seen at the lowest points of Hellas Basin.

CRÊTES PARALLÈLES BASSES DANS HELLAS

Cette image montre des crêtes parallèles basses sur la bordure ouest du bassin Hellas. Peu de recherches ont été menées sur la formation de ces structures, mais l'écoulement d'un matériau visqueux avec formation de couches est concevable. Il peut également y avoir une relation avec le «terrain en bandes» que l'on observe aux points les plus bas du bassin d'Hellas.

BASSE CRESTE PARALLELE IN HELLAS

Questa immagine mostra basse creste parallele sul bordo occidentale del bacino di Hellas. Ci sono state poche ricerche sulla formazione di queste strutture, ma potrebbero essere dovute ad un flusso di un materiale viscoso con conseguente stratificazione. Potrebbe anche esserci qualche relazione con il «terreno a bande» che si vede nei punti più bassi del bacino di Hellas.

НИЗКИЕ ПАРАЛЛЕЛЬНЫЕ ХРЕБТЫ В ЭЛЛАДЕ

На этом снимке показаны низкие параллельные хребты на западном краю бассейна Эллада. Исследований об образовании этих структур было мало, но они похожи на следы потока вязкого расслоенного материала. Также может существовать определенная связь с «полосчатым рельефом», который наблюдается в самых низких точках бассейна Эллада.

9.7　LOW PARALLEL RIDGES IN HELLAS

ACQUISITION TIME: 2021-01-16T00:36:55.789
LONGITUDE: 49.0 E   LATITUDE: 42.3 S
REGION: HELLAS

STARK STRUKTURIERTE OBERFLÄCHE IM RODDY-KRATER

Der Roddy-Krater in Noachis Terra hat einen Durchmesser von 86 km und ist das Ergebnis eines Einschlags in der frühen Hesperian-Epoche. Es gibt Anzeichen für Schwemmfächer innerhalb des Kraters, und es scheinen invertierte Kanäle vorhanden zu sein. Das Bild zeigt einen Tafelberg (oben links) auf dem Kraterboden und zeigt erodierte Schichten. Beachten Sie auch die scharfe Kontaktfläche zwischen diesem geschichteten Terrain und dem hellen polygonalen Material im Süden (rechts).

HIGHLY STRUCTURED SURFACE IN RODDY CRATER

Roddy Crater is an 86 km diameter crater in Noachis Terra and is the result of an impact in the early Hesperian epoch. There is evidence of alluvial fans within the crater and inverted channels appear to be present. The image shows a mesa (top left) on the crater floor and shows eroded layers. Note also the sharp contact between this layered terrain and bright polygonal material to the south (right).

SURFACE TRÈS STRUCTURÉE DANS LE CRATÈRE RODDY

Le cratère Roddy est un cratère de 86 km de diamètre dans la région de Noachis Terra et est le résultat d'un impact au début de l'ère Hespérienne. Des cônes alluviaux sont observés dans le cratère et des chenaux inversés semblent également être présents. L'image montre une mésa (en haut à gauche) sur le fond du cratère ainsi que des couches érodées. Notez également le contact net entre ce terrain stratifié et le matériau clair à texture polygonale au sud (à droite).

SUPERFICIE ALTAMENTE DIVERSIFICATA NEL CRATERE RODDY

Roddy Crater è un cratere di 86 km di diametro in Noachis Terra ed è il risultato di un impatto nella prima periodo dell'epoca Esperiana. Ci sono prove di conoidi alluvionali all'interno del cratere e sembrano essere presenti canali invertiti. L'immagine mostra una mesa (in alto a sinistra) sul pavimento del cratere e strutture stratificate erose. Si noti anche il contrasto netto tra questo terreno stratificato e la superfice poligonale chiara a sud (a destra).

ВЫСОКОСТРУКТУРИРОВАННАЯ ПОВЕРХНОСТЬ В КРАТЕРЕ РОДДИ

Кратер Родди – это кратер диаметром 86 км на земле Ноя, возникший в результате удара метеорита в начале Гесперийского периода. Существуют свидетельства наличия в кратере аллювиальных вееров и, по-видимому, перевёрнутых каналов (прошедших через топографическую инверсию). На снимке показана столовая гора на дне кратера (вверху слева) и видны эродированные слои породы. Обратите внимание на резкий контакт между этим многослойным рельефом и поверхностью с ярким материалом в полигональных трещинах на юге (справа).

9.8 HIGHLY STRUCTURED SURFACE IN RODDY CRATER

ACQUISITION TIME: 2021-01-20T08:52:57.012
LONGITUDE: 320.5 E LATITUDE: 22.3 S
REGION: NOACHIS

GESCHICHTETE NOPPEN IN AMAZONIS PLANITIA

Dieses Bild stammt von Amazonis Planitia westlich von Olympus Rupes und zeigt einen Teil von Olympus Maculae. Die Maculae sind anomale, wahrscheinlich sehr junge Albedo-Merkmale. Dieses CaSSIS-Bild zeigt eine hellblaue Farbe, die für äolische Sandablagerungen charakteristisch ist, die dunkleres Material innerhalb von Vertiefungen überlagern. Die Umgebung besteht aus geschichtetem, gefestigtem Material, das wahrscheinlich mit Staub bedeckt ist. Der Ursprung der Maculae ist derzeit Gegenstand der Forschung.

LAYERED KNOBS IN AMAZONIS PLANITIA

This image is from Amazonis Planitia due west of Olympus Rupes and shows a part of Olympus Maculae. The maculae are anomalous, probably very young, albedo features. This CaSSIS image shows a bright blue colour, characteristic of aeolian sand deposits, superposed on darker material within depressions. The surroundings comprise layered consolidated material, likely covered with dust. The origin of the maculae is a current subject of research.

MONTICULES STRATIFIÉS DANS AMAZONIS PLANITIA

Cette image provient d'Amazonis Planitia, à l'ouest d'Olympus Rupes, et montre une partie de Olympus Maculae. Les « maculae » sont des anomalies d'albédo à la surface, probablement très jeunes. Cette image CaSSIS montre une couleur bleue vive, caractéristique des dépôts de sable éolien, superposée à un matériau plus sombre à l'intérieur de dépressions. La région comprend aussi des matériaux consolidés en couches, probablement recouverts de poussière. L'origine des maculae est un sujet de recherche actuel.

COLLINETTE STRATIFICATE IN AMAZONIS PLANITIA

Questa immagine proviene da Amazonis Planitia a ovest di Olympus Rupes e mostra una parte di Olympus Maculae. Le macule sono caratteristiche anomale di albedo (intensità luminosa riflessa), probabilmente molto giovani. Questa immagine di CaSSIS mostra un colore blu brillante, caratteristico dei depositi di sabbia eolica, sovrapposto a materiale più scuro all'interno di depressioni. I dintorni presentano materiale consolidato a strati, probabilmente coperto di polvere. L'origine delle macule è attualmente oggetto di ricerca.

СЛОИСТЫЕ БУГРЫ НА РАВНИНЕ АМАЗОНИЯ

Данное изображение области равнины Амазония к западу от уступов Олимпа показывает часть макулы (пятна) Олимпа. Макулы — это аномальные, вероятно, очень молодые особенности альбедо. На этом снимке CaSSIS ярко-синий цвет, характерный для эоловых песчаных отложений, наложен на более темный материал внутри впадин. Окружающая местность сложена плотным слоистым материалом, вероятно, покрытым пылью. Происхождение макул — предмет современных исследований.

9.9 LAYERED KNOBS IN AMAZONIS PLANITIA

ACQUISITION TIME: 2021-04-05T11:54:48.442
LONGITUDE: 216.3 E   LATITUDE: 17.9 N
REGION: AMAZONIS

GESCHICHTETER HÜGEL IN JUVENTAE CHASMA

Rechts auf diesem Bild ist der westliche Rand eines geschichteten Hügels in Juventae Chasma zu sehen – einer der interessantesten Canyons des Valles-Marineris-Systems. An den ungeschichteten Rändern befinden sich hell getönte Materialien. Auf der linken Seite sind Ausläufer zu sehen. Diese gehören zur nordwestlichen Wand des Canyons und weisen eine erhebliche lithologische Vielfalt auf.

STRATIFIED MOUND IN JUVENTAE CHASMA

To the right of this image can be seen the western edge of a stratified mound in Juventae Chasma – one of the most interesting canyons of the Valles Marineris system. There are light-toned materials around the unstratified edges. To the left, spurs can be seen. These belong to the northwestern wall of the canyon and show significant lithological diversity.

MONTICULE STRATIFIÉ DANS JUVENTAE CHASMA

À droite de cette image, on peut voir le bord ouest d'un monticule stratifié dans Juventae Chasma – l'un des canyons les plus intéressants du système de Valles Marineris. Des matériaux clairs sont visibles autour des bords non stratifiés. À gauche de l'image, on peut voir des éperons. Ils appartiennent à la paroi nord-ouest du canyon et montrent une diversité lithologique importante.

TUMULO STRATIFICATO IN JUVENTAE CHASMA

A destra di questa immagine si può vedere il bordo occidentale di un cumulo stratificato nel Juventae Chasma – uno dei canyon più interessanti del sistema delle Valles Marineris. Ci sono materiali chiari intorno ai margini non stratificati. A sinistra si possono vedere dei crinali affilati. Questi appartengono alla parete nord-occidentale del canyon e mostrano una significativa diversità litologica.

РАССЛОЕННЫЙ КУРГАН В КАНЬОНЕ ЮВЕНТА

Справа на этом снимке виден западный край многослойного кургана в каньоне Ювенты – одном из самых интересных каньонов системы долин Маринер. Вокруг нестратифицированных краёв имеются материалы светлых тонов. Слева видны отроги (ответвления). Они относятся к северо-западному склону каньона и демонстрируют значительное литологическое разнообразие.

9.10   STRATIFIED MOUND IN JUVENTAE CHASMA

ACQUISITION TIME: 2019-12-04T03:02:11.208
LONGITUDE: 297.5 E   LATITUDE: 4.6 S
REGION: MARINERIS

SCHICHTUNG IN MAWRTH VALLIS

Mawrth Vallis durchschneidet Gelände aus dem noachischen Zeitalter in Arabia Terra. Es ist normalerweise mit dicken, tonhaltigen Aufschlüssen und zahlreichen wässrigen Mineralien verbunden. Dieses Bild zeigt eine Fülle von Details. In der Mitte des Bildes ist ein invertierter Kanal zu sehen. Der Krater mit einem Durchmesser von 1 km (unten in der Mitte) weist mehrere hell getönte Schichten in seinen Wänden auf. Ganz links ist ebenfalls geschichtetes Material mit einer auffälligen hellen Schicht zu sehen. Die Vielfalt, die in Mawrth Vallis zu sehen ist, priorisiert dieses Gebiet für künftige In-situ-Erkundungen.

LAYERING IN MAWRTH VALLIS

Mawrth Vallis cuts through Noachian-aged terrain in Arabia Terra. It is usually associated with thick clay-bearing outcrops and numerous aqueous minerals. This image shows a wealth of detail. In the centre of the frame is an inverted channel. The 1 km diameter crater (bottom centre) shows multiple light-toned layers in its walls. Layered material with a prominent light-toned stratum is also seen to the far left. The diversity seen in Mawrth Vallis makes it a high priority for future in situ exploration.

STRATIFICATION DANS MAWRTH VALLIS

Mawrth Vallis traverse des terrains d'âge noachien dans la région d'Arabia Terra. Elle est généralement associée à d'épais affleurements argileux et à de nombreux minéraux aqueux. Cette image montre une grande richesse de détails. Au centre du cadre se trouve un chenal inversé. Le cratère de 1 km de diamètre (en bas au centre) montre de multiples couches de couleur claire dans ses parois. Un matériau stratifié avec une strate proéminente de couleur claire est également visible tout à gauche. La diversité minéralogique observée dans Mawrth Vallis en fait une priorité pour les futures missions d'exploration à la surface.

STRATIFICAZIONE DELLA MAWRTH VALLIS

La Mawrth Vallis taglia il terreno di età noachiana in Arabia Terra. È solitamente associata a spessi affioramenti argillosi e a numerosi minerali acquosi. Questa immagine mostra una ricchezza di dettagli: al centro del fotogramma c'è un canale invertito e un cratere di 1 km di diametro (in basso al centro) mostra molteplici strati di colore chiaro sulle sue pareti. All'estrema sinistra si vede una simile stratificazione di materiale con un prominente strato dai toni. La diversità osservata nella Mawrth Vallis la rende una priorità per la futura esplorazione in situ.

СЛОИСТОСТЬ В ДОЛИНЕ МАВРТ

Долина Маврт прорезает местность нойского возраста на земле Аравия. Обычно она ассоциируется с мощными глинистыми обнажениями и многочисленными водными минералами. На этом снимке видно множество деталей. В центре кадра – канал с перевёрнутым рельефом. В кратере диаметром 1 км (внизу в центре), в стенах видны многочисленные светлые слои. Стратифицированный материал с заметными светлыми пластами также виден слева. Разнообразие, наблюдаемое в долине Маврт, делает её приоритетной для будущих полевых исследований.

9.11 LAYERING IN MAWRTH VALLIS

ACQUISITION TIME: 2018-05-29T09:51:50.000
LONGITUDE: 340.9 E   LATITUDE: 23.8 N
REGION: ARABIA

HELL GETÖNTE SCHICHTEN BEI DEN LOUROS VALLES

Die Region Louros Valles trennt Sinai Planum von Ius Chasma in den Valles Marineris. Das Bild zeigt die Schichtung in den Wänden der Täler (z. B. unten links und Mitte rechts). In den Ebenen zwischen den Tälern befindet sich ebenfalls geschichtetes, helles Material, wie im mittleren Teil des Bildes zu sehen ist. Diese geschichteten Ablagerungen bedecken eine grosse Fläche und sind vermutlich etwa 100 m dick. Man nimmt an, dass es sich bei den Louros Valles um Sickerrinnen handelt, die man als «Mega-Erosionsrinnen» bezeichnen könnte, wobei Wasser die wahrscheinliche Erosionsflüssigkeit ist.

LIGHT-TONED LAYERS NEAR LOUROS VALLES

The Louros Valles region separates Sinai Planum from Ius Chasma in Valles Marineris. The image shows the layering in the walls of the valleys (e.g. bottom left and centre right). In the plains between the valleys there is also layered light-toned material as shown in the central part of this image. These layered deposits cover a large area and are thought to be ≈ 100 m thick. The Louros Valles are thought to be sapping channels and might be described as "mega-gullies" with water being the likely eroding fluid.

STRATES DE COULEUR CLAIRE PRÈS DE LOUROS VALLES

La région de Louros Valles sépare Sinai Planum de Ius Chasma dans Valles Marineris. L'image montre la stratification dans les parois des vallées (par exemple, en bas à gauche et au centre à droite). Dans les plaines entre les vallées, on trouve également des couches de roches claires, comme le montre la partie centrale de cette image. Ces dépôts en couches couvrent une grande surface et on pense qu'ils ont une épaisseur d'environ 100 m. Les vallées de Louros sont considérées comme des chenaux de sapement et pourraient être décrites comme des «méga-ravines», l'eau étant le fluide érosif le plus probable.

STRATIFICAZIONI CHIARE VICINO A LOUROS VALLES

La regione delle Louros Valles separa il Sinai Planum dallo Ius Chasma nelle Valles Marineris. L'immagine mostra la stratificazione delle pareti vallive (ad esempio in basso a sinistra e al centro a destra). Nelle pianure tra le valli si può vedere anche del materiale stratificato chiaro, come mostrato nella parte centrale di questa immagine. Questi depositi stratificati coprono una vasta area e si pensa che abbiano uno spessore di ≈ 100 m. Si pensa che le valli di Louros siano dei canali di erosione dovuta a infiltrazioni sotterranee di liquidi e potrebbero essere descritte come dei «mega-calanchi» con acqua come probabile liquido erodente.

СЛОИ СВЕТЛЫХ ТОНОВ В РАЙОНЕ ДОЛИН ЛУРОС

Регион долин Лурос отделяет плато Синай от каньона Ио в долинах Маринер. На снимке видны обнажения слоёв пород, слагающих склоны долины (например, внизу слева и в центре справа). На равнинах между долинами также имеется слоистый материал светлых тонов, как показано в центральной части этого изображения. Эти стратифицированные отложения покрывают большую площадь и, предположительно, имеют толщину ≈ 100 м. Считается, что долины Лурос представляют собой «каналы прорыва» и могут быть описаны как «мега-овраги» (образованные временным стоком), а вода являлась вероятной эрозионной жидкостью.

9.12   LIGHT-TONED LAYERS NEAR LOUROS VALLES

ACQUISITION TIME: 2021-05-09T10:30:06.353
LONGITUDE: 275.3 E   LATITUDE: 8.5 S
REGION: SINAI

HELL GETÖNTE BRUCHFLÄCHEN IN DEN CLARITAS FOSSAE

Dieses Bild zeigt den Boden eines unbenannten Kraters in den Claritas Fossae nahe der Grenze zu Solis Planum. Die Ablagerungen sind überfüllt von hell getöntem Material, durchgängiger Bruchbildung und mutmasslichen Gängen. Die regionalen Brüche, die Gänge (unten links), die nahe gelegenen Talnetzwerke und die Nähe zu Tharsis lassen vermuten, dass in dieser Region in der Vergangenheit hydrothermale Aktivitäten stattgefunden haben könnten.

LIGHT-TONED FRACTURED SURFACES IN CLARITAS FOSSAE

This image shows the floor of an unnamed crater in Claritas Fossae close to the boundary with Solis Planum. The deposits are chock-full of light-toned material, pervasive fracturing and putative dikes. The regional fracturing, the dikes (bottom left), nearby valley networks and the proximity to Tharsis support the idea that this region may have been subject to hydrothermal activity in the past.

SURFACES FRACTURÉES DE COULEUR CLAIRE DANS CLARITAS FOSSAE

Cette image montre le sol d'un cratère sans nom dans Claritas Fossae, près de la limite avec Solis Planum. Les dépôts présentent des matériaux de couleur claire, des fractures omniprésentes et des dykes présumés (en bas à gauche). La fracturation régionale, les dykes, les réseaux de vallées proches et la proximité de Tharsis soutiennent l'idée que cette région a pu être soumise à une activité hydrothermale dans le passé.

SUPERFICI FRATTURATE DAI TONI CHIARI NELLA CLARITAS FOSSAE

Questa immagine mostra il pavimento di un cratere senza nome in Claritas Fossae vicino al confine con Solis Planum. I depositi sono dominati dalla presenza di materiale di colore chiaro, fratturazioni pervasive e probabili dicchi. La fratturazione locale, i dicchi (in basso a sinistra), le reti di valli vicine e la vicinanza a Tharsis supportano l'idea che questa regione possa essere stata soggetta ad attività idrotermale in passato.

РАСТРЕСКАВШИЕСЯ СВЕТЛЫЕ ПОВЕРХНОСТИ В БОРОЗДАХ КЛАРИТАС

На этом снимке показано дно безымянного кратера в бороздах Кларитас, недалеко от границы с плато Солнца. Отложения заполнены материалом светлых тонов, повсеместными трещинами и предполагаемыми дайками. Региональная трещиноватость, дайки (внизу слева), близлежащие сети долин и соседство с провинцией Фарсида подсказывают идею о том, что в прошлом этот регион мог быть подвержен гидротермальной активности.

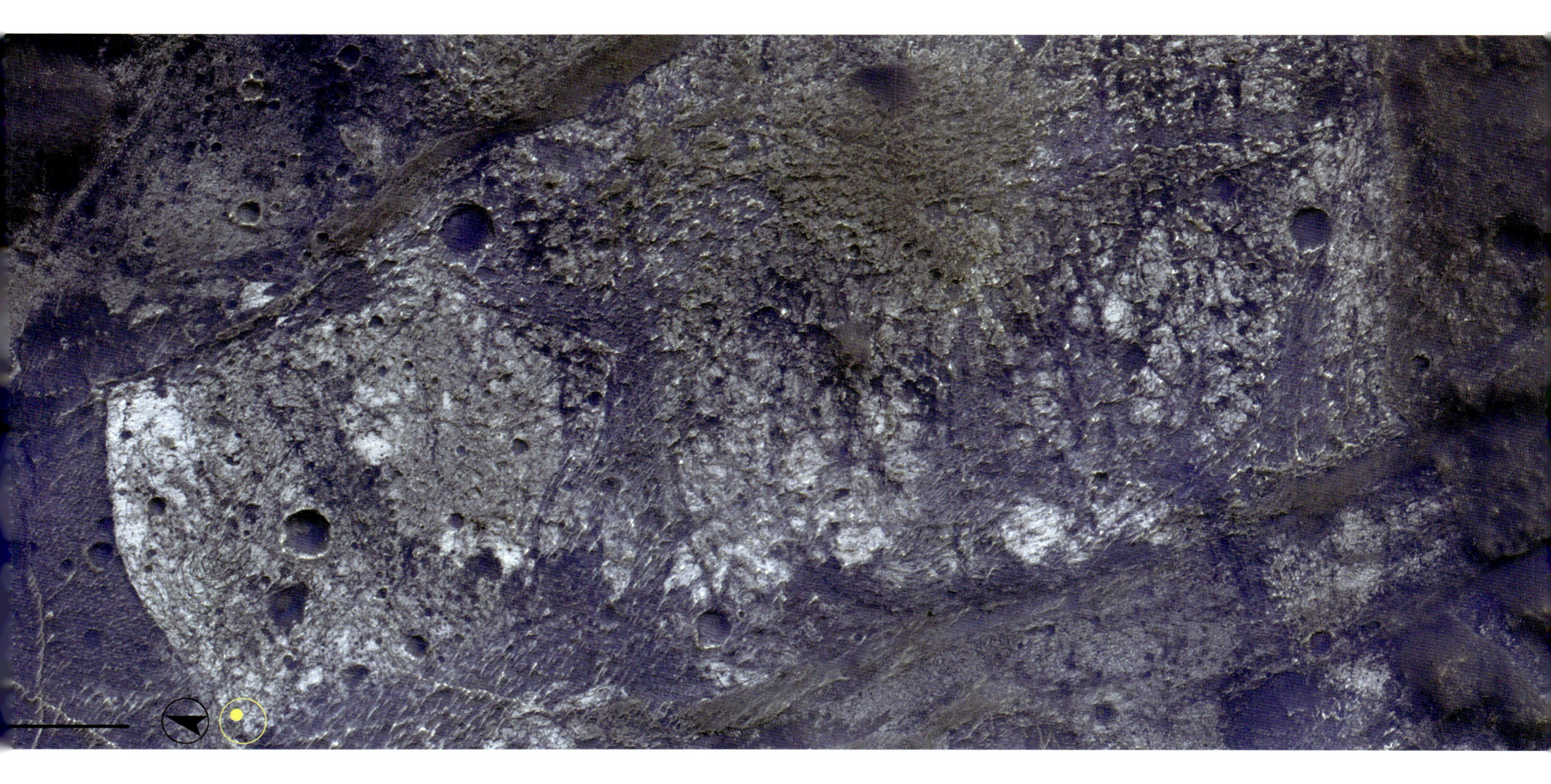

9.13 LIGHT-TONED FRACTURED SURFACES IN CLARITAS FOSSAE

ACQUISITION TIME: 2020-12-14T09:57:29.096
LONGITUDE: 266.2 E   LATITUDE: 34.6 S
REGION: CLARITAS

DER RAND UND DER SÜDWESTLICHE BODEN DES CROSS-KRATERS

Dies ist ein optimiertes synthetisches RGB-Bild des Cross-Kraters in Terra Sirenum. Der Kraterrand ist unten rechts zu sehen. Der Kraterboden in der Nähe des Kraterrands weist eine Schichtung auf. Der Krater hat einen Durchmesser von etwa 65 km und enthält vermutlich ein Mineral namens Alunit, das auf saure, schwefelhaltige Bedingungen zum Zeitpunkt seiner Entstehung hindeutet.

THE RIM AND SOUTH-WEST FLOOR OF CROSS CRATER

This is an enhanced synthetic RGB image of Cross Crater in Terra Sirenum. The crater rim can be seen to the lower right. The crater floor (to the left) close to the rim shows layering. The crater is around 65 km in diameter and is thought to contain a mineral called alunite suggesting acidic conditions rich in sulphur at the time of its formation.

LA BORDURE ET LE SUD-OUEST DU FOND DU CRATÈRE CROSS

Image du cratère Cross dans la région de Terra Sirenum. Le bord du cratère est visible en bas à droite. Le fond du cratère près du bord montre des couches. Le cratère a un diamètre d'environ 65 km et on pense qu'il contient un minéral appelé alunite, ce qui suggère des conditions acides et en environnement riche en soufre au moment de sa formation.

IL BORDO E IL PAVIMENTO SUD-OVEST DEL CROSS CRATER

Questa è un'immagine composta nei filtri RGB del Cross Crater in Terra Sirenum. Si può vedere il bordo del cratere in basso a destra. Il pavimento del cratere vicino al bordo mostra una certa stratificazione. Il cratere ha un diametro di circa 65 km e si pensa che contenga un minerale chiamato alunite la cui presenza suggerisce che si sia formato in condizioni acide ricche di zolfo.

ВАЛ И ЮГО-ЗАПАДНОЕ ДНО КРАТЕРА КРОСС

Это RGB-изображение с усиленными цветами кратера Кросс на земле Сирен. Вал кратера виден в правом нижнем углу. Вблизи него на дне кратера обнажаются слои пород. Кратер имеет около 65 км в диаметре и, предположительно, содержит минерал под названием алунит, что указывает на сернокислую среду, существовавшую во время его образования.

9.14 THE RIM AND SOUTH-WEST FLOOR OF CROSS CRATER

ACQUISITION TIME: 2018-10-30T23:08:18.943
LONGITUDE: 201.9 E   LATITUDE: 30.3 S
REGION: SIRENUM

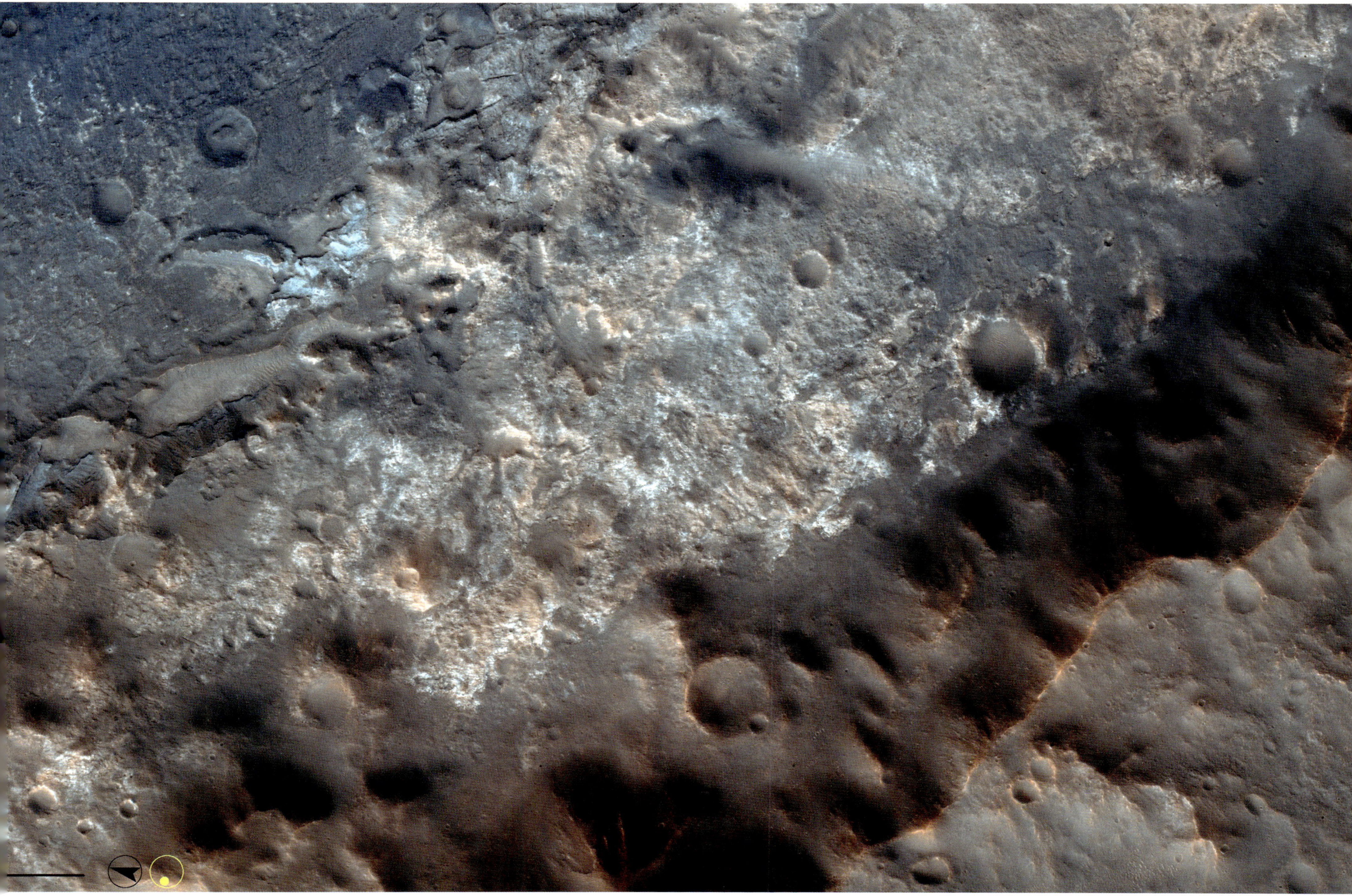

HELL GETÖNTE AUFSCHLÜSSE IM ROEMER-KRATER

Der Roemer-Krater ist ein unregelmässig geformter Krater von 120 km Durchmesser in Noachis Terra. Der Boden innerhalb des südlichen Rands scheint stark erodiert zu sein, und es gibt Überreste von Sedimentmaterial mit hellen Freilegungen.

LIGHT-TONED EXPOSURES IN ROEMER CRATER

Roemer Crater is an irregularly-shaped 120 km diameter crater in Noachis Terra. The floor inside the southern rim appears to be heavily eroded and there are remnants of sedimentary material with light-toned exposures.

AFFLEUREMENTS CLAIRS DANS LE CRATÈRE ROEMER

Le cratère Roemer est un cratère de forme irrégulière de 120 km de diamètre dans la région de Noachis Terra. Le plancher près de la bordure sud semble être fortement érodé et on observe des affleurements clairs de roches sédimentaires.

MATERIALE CHIARO SUPERFICIALE NEL CRATERE ROEMER

Il cratere Roemer è un cratere di 120 km di diametro di forma irregolare in Noachis Terra. Il pavimento all'interno del bordo meridionale sembra essere pesantemente eroso e ci sono resti di materiale sedimentario dalle tonalità chiare.

ОБНАЖЕНИЯ СВЕТЛЫХ ТОНОВ В КРАТЕРЕ РЁМЕР

Кратер Рёмер – 120-километровый кратер неправильной формы на земле Ноя. Его дно с внутренней стороны южного вала, по-видимому, сильно эродировано. Имеются остатки осадочного материала с обнажениями светлых тонов.

9.15 LIGHT-TONED EXPOSURES IN ROEMER CRATER

ACQUISITION TIME: 2018-09-15T03:51:20.000
LONGITUDE: 8.3 E   LATITUDE: 28.7 S
REGION: NOACHIS

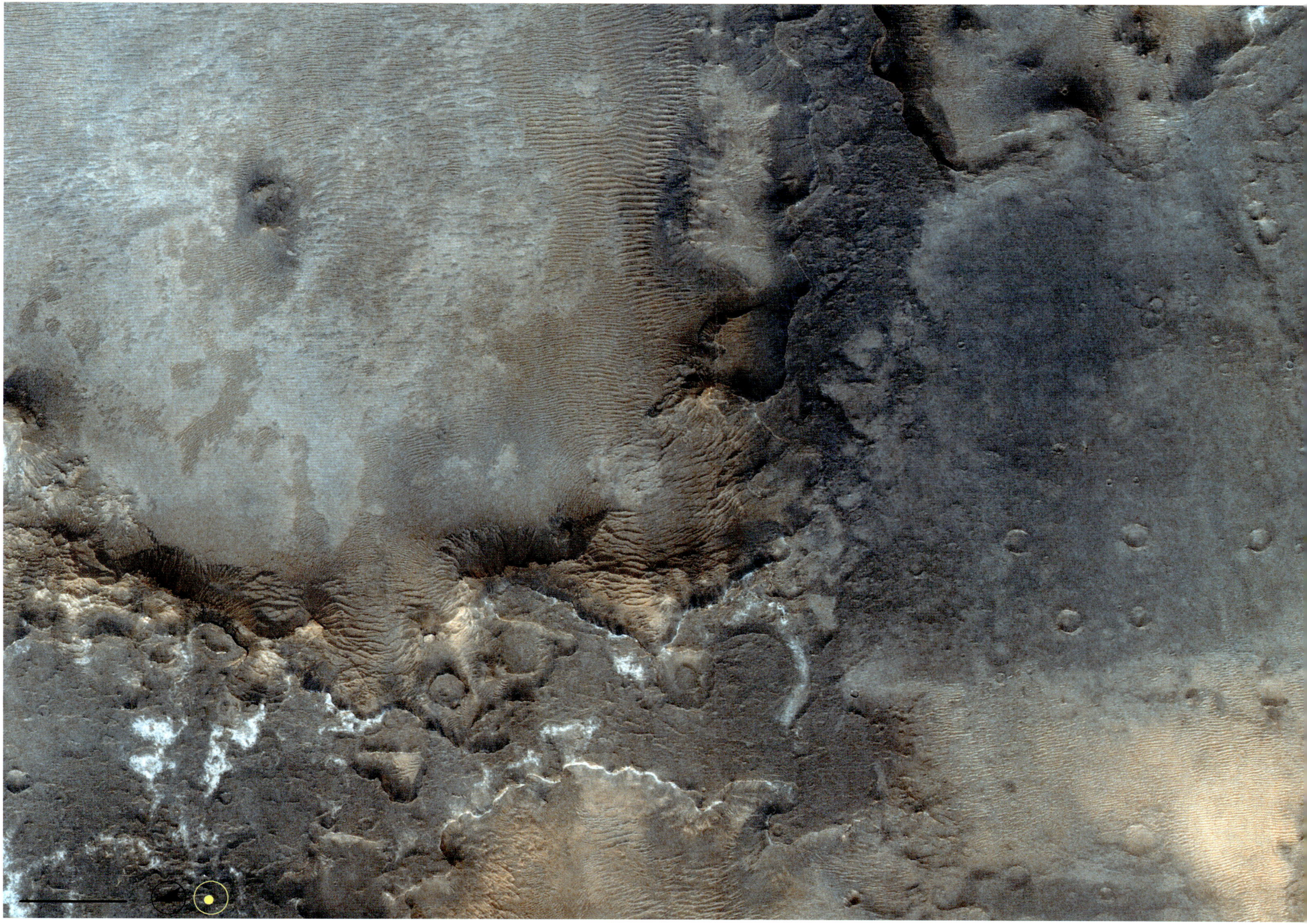

GESCHICHTETES GRUNDGESTEIN IN TERRA SIRENUM

Dieses optimierte synthetische RGB-Bild zeigt farbiges, geschichtetes Gestein in einem erodierten Krater auf Terra Sirenum nördlich des Cross-Kraters, der auch hell getöntes Material aufweist. Die hohe thermische Trägheit in diesem Gebiet deutet ausserdem auf dichtes Material hin.

LAYERED BEDROCK IN TERRA SIRENUM

This enhanced synthetic RGB image shows colourful layered bedrock in an eroded crater in Terra Sirenum north of Cross Crater which also shows light-toned material. The high thermal inertia seen in the area also indicates consolidated, bedrock material in this area.

SUBSTRAT ROCHEUX STRATIFIÉ DANS TERRA SIRENUM

Couches rocheuses colorées dans un cratère érodé de Terra Sirenum. Au nord, le cratère Cross présente également des affleurements de roche claire. La forte inertie thermique observée dans cette zone indique également un matériau consolidé.

STRATIFICAZIONI ROCCIOSE IN TERRA SIRENUM

Questa immagine composta nei filtri RGB mostra un colorato fondo roccioso stratificato con la presenza di materiale dai toni chiari in un cratere eroso in Terra Sirenum a nord del Cross Crater. L'alta inerzia termica misurata nell'area indica che il materiale è consolidato.

СЛОИСТАЯ ГОРНАЯ ПОРОДА В ЗЕМЛЕ СИРЕН

Данное усиленное синтетическое RGB-изображение показывает красочные стратифицированные коренные породы в эродированном кратере на земле Сирен к северу от кратера Кросс. Здесь также присутствует светлая порода. Высокая тепловая инерция, наблюдаемая в этой области, указывает на присутствие уплотнённых коренных пород.

9.16 LAYERED BEDROCK IN TERRA SIRENUM

ACQUISITION TIME: 2018-09-18T16:18:55.000
LONGITUDE: 203.1 E   LATITUDE: 28.2 S
REGION: SIRENUM

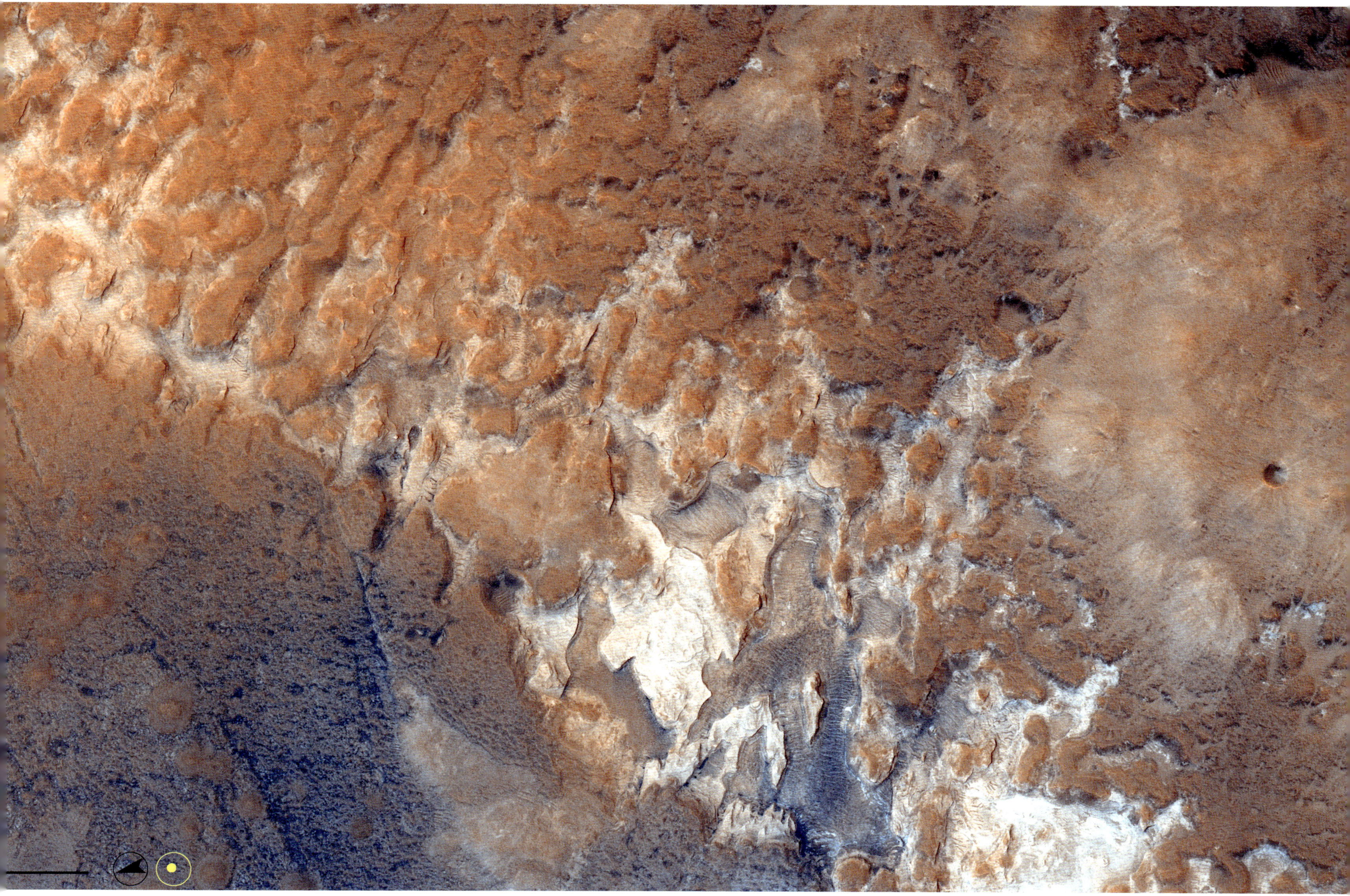

TONREICHES TERRAIN IN DER NÄHE DES OYAMA-KRATERS

Dies ist ein Bild von lehmreichem Gelände nordwestlich des Oyama-Kraters, der wiederum westlich von Mawrth Vallis liegt. Die gesamte Region ist reich an Schichtsilikaten und weist sowohl eine farbliche als auch eine morphologische Vielfalt auf. Dieses Bild ist ein gutes Beispiel dafür. Man beachte insbesondere die Wand des Kraters links und das hellere Terrain (oben, leicht links von der Mitte), das sich stratigrafisch unter dem dunkleren Material befindet.

CLAY-RICH TERRAIN NEAR OYAMA CRATER

This is an image of clay-rich terrains to the northwest of Oyama Crater which is itself west of Mawrth Vallis. The whole region is rich in phyllosilicates and shows both colour and morphological diversity. This image is a good example. Note in particular the wall of the crater to the left and the brighter terrain (top slightly left of centre) which is stratigraphically below the darker material.

TERRAIN RICHE EN ARGILE PRÈS DU CRATÈRE OYAMA

Cette image montre des terrains riches en argile au nord-ouest du cratère Oyama, lui-même à l'ouest de Mawrth Vallis. Toute la région est riche en phyllosilicates et présente une grande diversité de couleurs et de morphologies. Cette image en est un bon exemple. Notez en particulier la paroi du cratère à gauche et le terrain plus clair (en haut légèrement à gauche du centre) qui se trouve stratigraphiquement sous le matériau plus sombre.

TERRENO RICCO DI ARGILLA VICINO AL CRATERE OYAMA

Questa è un'immagine dei terreni ricchi di argilla a nord-ovest del cratere Oyama che è a sua volta a ovest della Mawrth Vallis. L'intera regione è ricca di fillosilicati e mostra una importante diversità sia nei colori che nella morfologia. Questa immagine ne è un buon esempio: si notino in particolare la parete del cratere a sinistra e il terreno più chiaro (in alto a sinistra del centro) che è stratigraficamente al di sotto del materiale più scuro.

МЕСТНОСТЬ, БОГАТАЯ ГЛИНИСТЫМИ МИНЕРАЛАМИ

Это снимок глинистой местности к северо-западу от кратера Ояма, находящегося к западу от долины Маврт. Весь регион богат филлосиликатами и демонстрирует как цветовое, так и морфологическое разнообразие. Данный снимок является хорошим примером. Обратите внимание, в частности, на стену кратера слева и более светлую поверхность (вверху чуть левее центра), которая стратиграфически (в последовательности слоёв) ниже более темного материала (то есть старше).

9.17 CLAY-RICH TERRAIN NEAR OYAMA CRATER

ACQUISITION TIME: 2020-01-06T01:53:34.771
LONGITUDE: 338.5 E   LATITUDE: 24.6 N
REGION: ARABIA

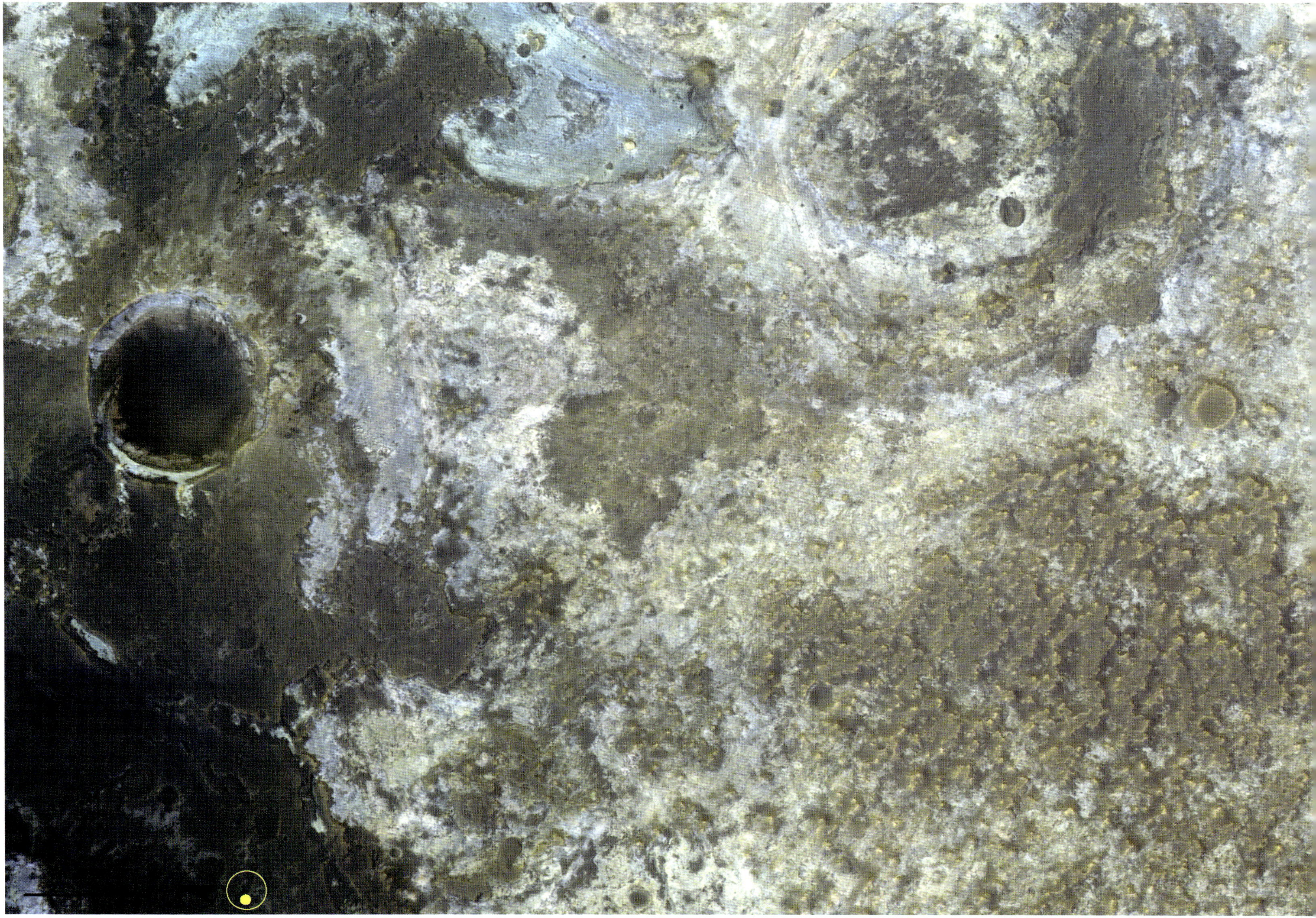

HELL GETÖNTES MATERIAL ÖSTLICH VON ARNUS VALLIS

Dies ist ein Bild eines schichtsilikatreichen Gebiets in Terra Sabaea, südlich der Nili Fossae und östlich von Arnus Vallis. Das hell getönte Material ist Teil eines komplexen Terrains, das durch mehrere Einschläge entstanden ist. Beachten Sie die ungleichmässige Färbung der Oberfläche um den 3 km grossen Einschlagkrater unten links.

LIGHT-TONED MATERIAL EAST OF ARNUS VALLIS

This is an image of a phyllosilicate rich area in Terra Sabaea, south of Nili Fossae and east of Arnus Vallis. The light-toned material is part of a complex patch of terrain produced by multiple impacts. Note the non-uniformity of the colour of the surface around the 3 km impact crater to the bottom left.

DÉPOTS DE COULEUR CLAIRE À L'EST D'ARNUS VALLIS

Image d'une zone riche en phyllosilicates dans la région de Terra Sabaea, au sud de Nili Fossae et à l'est d'Arnus Vallis. La formation claire fait partie d'une zone de terrain complexe, affectée par de multiples impacts. Notez l'hétérogénéité de la couleur de la surface autour du cratère d'impact de 3 km en bas à gauche.

MATERIALE CHIARO AD EST DI ARNUS VALLIS

Questa è un'immagine di un'area ricca di fillosilicati in Terra Sabaea, a sud di Nili Fossae e ad est di Arnus Vallis. Il materiale di colore chiaro fa parte di una complessa zona prodotta da impatti multipli. Si noti la disomogeneità del colore della superficie intorno al cratere da impatto di 3 km in basso a sinistra.

МАТЕРИАЛ СВЕТЛЫХ ТОНОВ К ВОСТОКУ ОТ ДОЛИНЫ АРНО

Это изображение богатой филлосиликатами области на Сабейской земле, к югу от борозд Нила и к востоку от долины Арно (ранее известной как уступ Арена). Материал светлых тонов является частью сложного участка, сформировавшегося в результате многочисленных метеоритных ударов. Обратите внимание на неоднородность цвета поверхности вокруг 3-километрового ударного кратера слева внизу.

9.18   LIGHT-TONED MATERIAL EAST OF ARNUS VALLIS

ACQUISITION TIME: 2020-01-10T21:53:20.000
LONGITUDE: 72.1 E   LATITUDE: 15.2 N
REGION: SABAEA

361

10 GRUNDGEBIRGE UND KOMPLEXE GELÄNDEFORMEN

BEDROCKS AND
COMPLEX TERRAINS

SUBSTRATS ROCHEUX ET
TERRAINS COMPLEXES

PAVIMENTI ROCCIOSI E
TERRENI COMPLESSI

КОРЕННЫЕ ПОРОДЫ И
СЛОЖНЫЕ РЕЛЬЕФЫ

GEBÄNDERTES TERRAIN IM HELLAS-BECKEN

Gebändertes Terrain findet sich nur am nordwestlichen Rand von Hellas Planitia innerhalb des Hellas-Beckens in der Nähe der tiefsten Punkte des Mars. Sie scheinen auf einen rezenten Materialfluss hinzuweisen. Auf der linken Seite ist zu erkennen, dass einige Bänder an den Seiten von höher gelegenen Hügeln zu entstehen scheinen. Die Oberflächenstrukturen der Bänder deuten darauf hin, dass sie eisreich sind.

BANDED TERRAIN IN HELLAS BASIN

Banded terrain is only found on the north-west edge of Hellas Planitia within Hellas Basin near the deepest points on Mars. It seems to indicate recent flow of material. Note on the left that some bands seem to originate on the sides of higher elevation mounds. There are indications from the surface structures of bands that they are ice-rich.

TERRAINS EN BANDES DANS LE BASSIN D'HELLAS

Ce type de terrain « en bandes » ne se trouve que sur le bord nord-ouest de Hellas Planitia, dans le bassin de Hellas, près des points les plus profonds de Mars. Il semble indiquer un écoulement récent de matériaux. Notez que certaines bandes sur la gauche de l'image semblent naître sur les parois de monticules plus élevés. Les structures de surface des bandes pourraient indiquer qu'elles sont riches en glace.

TERRENI A BANDE NEL BACINO DI HELLAS

I «terreni a bande» si trovano solo sul bordo nord-ovest di Hellas Planitia all'interno del bacino di Hellas vicino ai punti più profondi di Marte. Sembrano indicare un flusso recente di materiale. Nella immagine sulla sinistra alcune bande sembrano originarsi sui lati di cumuli di altezza superiore e dalle strutture superficiali delle bande si può dedurre che siano ricche di ghiaccio.

ПОЛОСЧАТЫЙ РЕЛЬЕФ В БАССЕЙНЕ ЭЛЛАДА

Так называемый «полосчатый рельеф» встречается только на северо-западном краю Равнины Эллада в пределах бассейна Эллада вблизи самых глубоких впадин на поверхности Марса. Он, по-видимому, свидетельствует о недавних потоках. Обратите внимание на левую часть снимка, где видны полосы, начинающиеся на сторонах наиболее высоких курганов. Поверхностная структура полос имеет признаки насыщенности льдом.

10.1 BANDED TERRAIN IN HELLAS BASIN

ACQUISITION TIME: 2021-01-06T20:31:21.305
LONGITUDE: 51.7 E   LATITUDE: 41.1 S
REGION: HELLAS

UNREGELMÄSSIGE DEPRESSION IN SYRTIS MAJOR

Dies ist ein Bild einer Muldengrube im Zentrum eines Kraters in den Nili Fossae auf der Nordostseite von Syrtis Major. Es ist bekannt, dass die Oberflächenschichten in diesem Gebiet aus Tonen bestehen. In der Stratigrafie sind verschiedene Farben zu erkennen, die auf Schichten aus unterschiedlichem Material hinweisen. Jarosit, ein wasserhaltiges Kalium- und Eisen(III)sulfat, wurde in der Nähe entdeckt, was auf eine Wechselwirkung von Wasser mit Gesteinsmaterial unter sauren Bedingungen hinweist.

IRREGULAR DEPRESSION IN SYRTIS MAJOR

This is an image of a trough pit at the centre of a crater in Nili Fossae on the north-east side of Syrtis Major. There are known to be clays in the surface layers in this area. Different colours can be seen in the stratigraphy indicating layers of different material. Jarosite, a hydrous sulphate of potassium and ferric iron, has been detected near here which indicates water interaction with rocky material under acidic conditions.

DÉPRESSION IRRÉGULIÈRE DANS SYRTIS MAJOR

Image d'un fossé au centre d'un cratère dans la région de Nili Fossae sur le bord nord-est de Syrtis Major. On sait qu'il y a de l'argile dans les couches de surface de cette zone. On peut voir différentes couleurs dans la stratigraphie, indiquant des couches de composition minéralogique différente. De la jarosite, un sulfate hydraté de potassium et de fer, a été détectée près d'ici, ce qui indique une interaction de l'eau avec les matériaux rocheux dans des conditions acides.

DEPRESSIONE IRREGOLARE IN SYRTIS MAJOR

Questa è un'immagine di una fossa trocleare al centro di un cratere a Nili Fossae sul lato nord-est di Syrtis Major. Si sa che sono presenti argille negli strati superficiali di quest'areai possono vedere diversi colori nella stratigrafia che indicano strati di materiale diverso. La giarosite, un solfato idrico di potassio e ferro(III), è stata rilevata qui vicino e indica una interazione acquosa con il materiale roccioso in condizioni acide.

ИРРЕГУЛЯРНАЯ ВПАДИНА В БОЛЬШОМ СИРТЕ

Это снимок желоба в центре кратера в бороздах Нила на северо-восточной стороне Большого Сирта. Известно, что в поверхностных слоях этой области присутствуют глины. Видны стратифицированные слои пород, различающиеся по цвету. Неподалеку был обнаружен ярозит – водный сульфат калия и трёхвалентного железа, что указывает на взаимодействие скальных пород с водой в условиях повышенной кислотности.

10.2 IRREGULAR DEPRESSION IN SYRTIS MAJOR

ACQUISITION TIME: 2018-09-17T01:19:13.000
LONGITUDE: 74.2 E   LATITUDE: 22.2 N
REGION: SYRTIS

EBENE MIT FREILIEGENDEN TONEN

Dieses Bild zeigt eine Ebene im Süden von Tyrrhena Terra südöstlich des Auce-Kraters. Es zeigt freiliegenden Ton als helles Material dort, wo die Oberflächenschicht erodiert ist. Auch die Oberflächenschicht ist nicht einheitlich gefärbt. Man kann zum Beispiel eine örtliche violette Färbung des älteren Geländes unten links erkennen. Tone sind normalerweise mit der Hydratation von Mineralien verbunden, und dies ist ein weiterer Beweis dafür, dass in der Vergangenheit flüssiges Wasser vorhanden war.

PLAIN WITH EXPOSED CLAYS

This image is of a plain in south Tyrrhena Terra southeast of Auce Crater. It shows exposed clays as light-toned material where the surface layer has been eroded. The surface layer is not uniform in colour either. One can see for example localised purplish colouring of the older terrain to the bottom left. Clays are normally associated with hydration of minerals and this is further evidence of liquid water having been present in the past.

PLAINE AVEC ARGILES EXPOSÉES

Cette image montre une plaine au sud de Tyrrhena Terra, au sud-est du cratère Auce. Elle montre des argiles exposées au sein d'une formation de couleur claire, là où la couche de surface a été érodée. Cette couche de surface n'est d'ailleurs pas non plus de couleur uniforme. On peut voir par exemple une coloration localement violacée du terrain ancien en bas à gauche. Les argiles sont normalement associées à l'altération aqueuse des minéraux et ceci est une preuve supplémentaire de la présence d'eau liquide dans le passé.

PIANURA CON ARGILLE ESPOSTE IN SUPERFICIE

Questa immagine ritrae una pianura nel sud di Tyrrhena Terra a sud-est del cratere Auce e mostra il materiale argilloso di colore chiaro, dove lo strato superficiale è stato eroso. Lo strato superficiale non ha un colore uniforme, e si può vedere per esempio una colorazione violacea localizzata del terreno più vecchio in basso a sinistra. Le argille sono normalmente associate all'idratazione dei minerali e questa è un'ulteriore prova della presenza di acqua liquida nel passato di questa regione.

РАВНИНА С ОБНАЖЕНИЯМИ ГЛИНИСТЫХ ПОРОД

Это снимок равнины на юге Тирренской Земли к юго-востоку от кратера Ауце. Слева, где поверхностный слой удален эрозией, на снимке видны светлые глины. Образование глин обычно связано с гидратацией минералов, и это еще одно доказательство того, что в прошлом здесь присутствовала жидкая вода. Сам поверхностный слой также не является однородным по цвету. Например, слева внизу можно видеть локальное слабо заметное фиолетовое окрашивание более древнего рельефа.

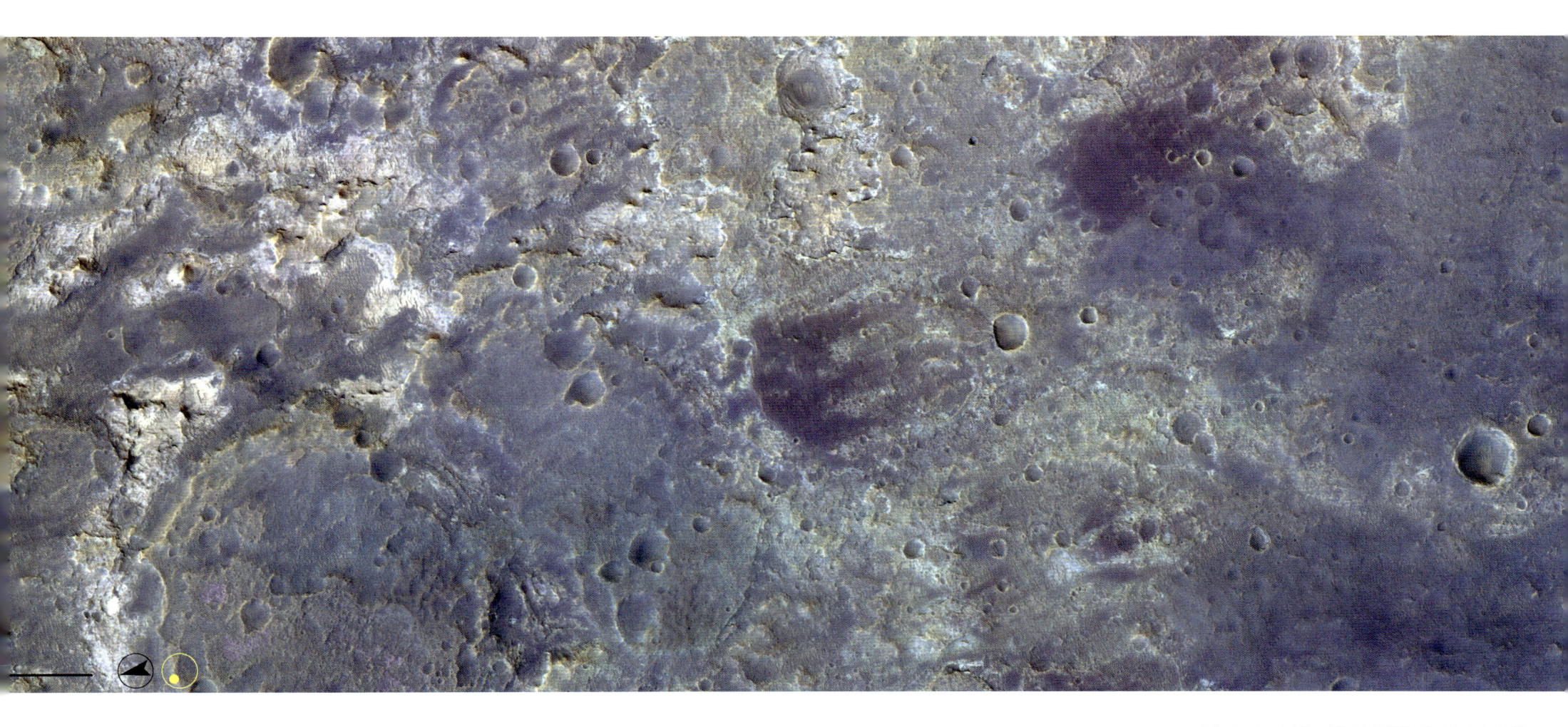

10.3   PLAIN WITH EXPOSED CLAYS

ACQUISITION TIME: 2018-05-31T02:48:55.000
LONGITUDE: 80.7 E   LATITUDE: 28.1 S
REGION: TYRRHENA

GESCHICHTETE GRUBENWAND IN SABAEA

Dies ist ein Bild des Randes einer Grube auf einem Kraterboden im Süden von Terra Sabaea, der freiliegende geschichtete Ablagerungen aufweist. Staub ist in diesen CaSSIS-Farbprodukten normalerweise hellblau, und an den Rändern der Klippen ist viel Blau zu sehen, was darauf hindeutet, dass dieses Material erodiert und daher recht brüchig ist.

LAYERED PIT WALL IN SABAEA

This is an image of the edge of a pit on a crater floor in south Terra Sabaea that has exposed layered deposits. Dust is normally bright blue in these CaSSIS colour products and a lot of blue is seen at the edges of the cliffs indicating that this material is eroding and therefore quite fragile.

PAROI DE FOSSÉ STRATIFIÉE DANS SABAEA TERRA

Image du bord d'un fossé sur le plancher d'un cratère dans le sud de Terra Sabaea qui a exposé des dépôts stratifiés. La poussière est normalement d'un bleu vif dans ces images CaSSIS et on voit beaucoup de bleu sur les bords des falaises, ce qui indique que ce matériau est en érosion et donc assez fragile.

PARETE STRATIFICATA DI UNA FOSSA IN SABAEA

Questa è un'immagine del bordo di una fossa sul pavimento di un cratere nella Terra Sabaea meridionale che mostra depositi stratificati. La polvere è normalmente blu brillante in queste immagini a colori di CaSSIS e si vede molto blu ai bordi delle scarpate, indicando che questo materiale è in erosione e quindi abbastanza fragile e non compattato.

МНОГОСЛОЙНЫЙ СКЛОН ВПАДИНЫ В САБЕЙСКОЙ ЗЕМЛЕ

Это снимок края депрессии на дне кратера на юге Сабейской Земли, которая вскрывает слоистые отложения. Как правило ярко-синие цвета на снимках CaSSIS связаны с присутствием пыли. Это относится и к синим пятнам у подножий обрывов, которые видны на этом снимке. В свою очередь, это указывает на процессы эрозии и достаточную хрупкость пород.

10.4 LAYERED PIT WALL IN SABAEA

ACQUISITION TIME: 2020-11-27T17:02:53.230
LONGITUDE: 50.2 E   LATITUDE: 28.1 S
REGION: SABAEA

ERODIERENDES TERRAIN IN TERRA CIMMERIA

Dies ist ein Gebiet mit freiliegendem Grundgestein in Terra Cimmeria nördlich von Cimmeria Tholi. Es wurde vermutet, dass sich in der Nähe Chloridsalze auf der Oberfläche befinden, und auf diesem Bild (Mitte links) sind einige kleine violette Flecken zu sehen, die in CaSSIS-Farbprodukten ein möglicher Indikator für Chloride sind. Der talartige Kanal im Süden (rechts) ist einer von vielen in diesem Teil von Cimmeria. Der Boden der Rinne ist reichlich mit Kratern versehen, was darauf hindeutet, dass es sich um ein recht altes Merkmal handelt.

ERODING TERRAIN IN TERRA CIMMERIA

This is an area of exposed bedrock in Terra Cimmeria north of Cimmeria Tholi. It has been suggested that there are chloride salts on the surface nearby and there are some small purple patches in this image (centre-left) which, in CaSSIS colour products, is a possible indicator of chlorides. The valley-like channel to the south (right) is one of many in this part of Cimmeria. The channel floor is heavily cratered suggesting this is quite an old feature.

TERRAIN EN ÉROSION DANS TERRA CIMMERIA

Image d'une zone de substrat rocheux exposé dans la région de Terra Cimmeria au nord de Cimmeria Tholi. Il a été suggéré que des sels chlorés soient présents sur la surface à proximité. Les quelques petites taches violettes dans cette image (centre gauche) sont, dans les images CaSSIS, un potentiel indicateur de ces chlorures. Le chenal en forme de vallée au sud (à droite) est l'un des nombreux chenaux de cette région. Le fond du chenal est fortement cratérisé, ce qui suggère qu'il s'agit d'une zone assez ancienne.

TERRENO IN EROSIONE IN TERRA CIMMERIA

Questa è un affioramento del pavimento roccioso in Terra Cimmeria a nord di Cimmeria Tholi. È stato ipotizzato che ci siano sali di cloruro sulle superfici nelle vicinanze e ci sono alcune piccole macchie viola in questa immagine (centro-sinistra) che, nei prodotti di colore di CaSSIS, sono un possibile indicatore di cloruri. Il canale simile a una valle a sud (a destra) è uno dei tanti in questa parte di Cimmeria. Il pavimento del canale è pesantemente craterizzato, il che suggerisce che si tratta di una conformazione piuttosto antica.

ЭРОДИРОВАННАЯ ПОВЕРХНОСТЬ В КИММЕРИЙСКОЙ ЗЕМЛЕ

Это участок выходов коренных пород на Киммерийской Земле к северу от Киммерийского купола. Предполагается, что на поверхности поблизости есть хлористые соли, и на этом снимке (в центре слева) есть несколько небольших пурпурных пятен, что на цветных снимках CaSSIS является возможным индикатором хлоридов. Долинообразный канал на юге (справа) – один из многих в этой части Киммерии. Дно канала густо покрыто кратерами, говорящими о существенной древности этого образования.

10.5 ERODING TERRAIN IN TERRA CIMMERIA

ACQUISITION TIME: 2020-12-17T18:30:52.270
LONGITUDE: 157.5 E   LATITUDE: 32.2 S
REGION: CIMMERIA

ERODIERENDES TERRAIN IN IANI CHAOS

Iani Chaos liegt am südlichen Ende von Ares Vallis und wird allgemein als Quellgebiet für den Ausflusskanal angenommen. Das chaotische Terrain könnte durch einen Kollaps nach dem Verschwinden des unterirdischen Eises entstanden sein. Dieses Bild ist jedoch faszinierend, weil es den Anschein hat, dass die Oberfläche von einer schichtartigen Ablagerung bedeckt war, die dann erodiert wurde und unregelmässige Mesas (isolierte flache Hügel) freilegte. Man beachte auch die freigelegten hellen Schichten und die sehr kleinen wabenförmigen Dünen links im Bild.

ERODING TERRAIN IN IANI CHAOS

Iani Chaos is at the southern end of Ares Vallis and is widely assumed to have been a source region for the outflow channel. Chaotic terrain may have been produced by collapse following removal of sub-surface ices. This image, however, is intriguing because it appears that there has been a layered deposit covering the surface that has then eroded exposing irregular mesas (isolated flat topped hills). Note also the exposed light-toned layers and the very small honeycomb dunes to the left of the image.

TERRAIN EN ÉROSION DANS IANI CHAOS

Iani Chaos se trouve à l'extrémité sud d'Ares Vallis et on suppose qu'il s'agit d'une région source du chenal d'écoulement. Le terrain chaotique peut avoir été produit par un effondrement consécutif au retrait des glaces du sous-sol. Cette image est cependant intrigante car il semble qu'il y ait eu un dépôt stratifié, couvrant la surface, qui a ensuite été érodé exposant des mésas irrégulières (collines isolées à sommet plat). Notez également les strates claires et les très petites dunes en nid d'abeille à gauche de l'image.

TERRENO IN EROSIONE IN IANI CHAOS

Iani Chaos si trova all'estremità meridionale di Ares Vallis e si presume che sia stata la regione di origine del canale di defflusso. Il terreno caotico può essere stato prodotto dal collasso della superficie in seguito alla rimozione dei ghiacci presenti al di sotto di essa. Questa immagine, tuttavia, è intrigante perché sembra che ci sia stato un deposito stratificato che copriva la superficie e che è stato successivamente eroso, esponendo mesa irregolari (colline isolate con la cima piatta). Si osservino anche gli strati chiari affioranti e le dune molto piccole con strutture a nido d'ape a sinistra dell'immagine.

ЭРОДИРОВАННАЯ ПОВЕРХНОСТЬ В ХАОСЕ ЯНУСА

Хаос Януса находится в южной части долины Ареса. По общему мнению, здесь располагались источники потока, сформировавшего эту долину. Хаотичный рельеф мог образоваться в результате обрушений, связанных с удалением подповерхностных льдов. Также на снимке можно различить признаки хаотично расположенных столовых гор (изолированных возвышенностей с плоскими вершинами), которые могли быть сформированы эрозией слоистых толщ. Обратите внимание на обнаженные слои светлых тонов и «сотовый» рельеф в левой части снимка, сформированный маленькими дюнами.

10.6 **ERODING TERRAIN IN IANI CHAOS**

ACQUISITION TIME: 2021-02-25T23:24:41.909
LONGITUDE: 342.0 E LATITUDE: 3.2 S
REGION: MARGARITIFER

REICHHALTIGE SCHICHTEN IN NILOSYRTIS MENSAE

Dieses Bild wurde in einer Region aufgenommen, die offiziell Terra Sabaea zugeordnet wird, sich aber eigentlich auf der Ostseite von Nilosyrtis Mensae befindet. Es zeigt bemerkenswert geschichtete Klippen als Teil eines grossen Tafelbergs (Bildmitte). Ein Grossteil von Nilosyrtis ist als abgenutztes Gelände mit Klippen, Tafelbergen und breiten, flachen Tälern klassifiziert. Es wird vermutet, dass sich das Gelände durch das Schmelzen oder Sublimieren von Gletschereis gebildet hat. In der Region wurde unterirdisches Eis entdeckt.

ABUNDANT LAYERS IN NILOSYRTIS MENSAE

This image was acquired in a region formally assigned to Terra Sabaea but is actually on the east side of Nilosyrtis Mensae. It shows remarkably layered cliffs as part of a large mesa (centre of the image). Much of Nilosyrtis is classified as fretted terrain which contains cliffs, mesas, and wide, flat, valleys. It is inferred that the terrain has evolved through melting or sublimation of glacial ice. Sub-surface ice has been detected in the region.

DES STRATES ABONDANTES DANS NILOSYRTIS MENSAE

Cette image a été acquise dans une région officiellement attribuée à Terra Sabaea mais qui se trouve en fait sur le côté est de Nilosyrtis Mensae. Elle montre des falaises finement stratifiées faisant partie d'une grande mésa (au centre de l'image). Une grande partie de Nilosyrtis est classée comme un terrain « fretté » qui contient des falaises, des mésas et de larges vallées plates. On en déduit que le terrain a évolué par la fonte ou la sublimation de glace qui a été détectée dans le sous-sol de la région.

ABBONDANTI STRATIFICAZIONI IN NILOSYRTIS MENSAE

Questa immagine è stata acquisita in una regione formalmente assegnata alla Terra Sabaea ma si trova in realtà sul lato est di Nilosyrtis Mensae. Mostra scarpate che fanno parte di una grande mesa (centro dell'immagine) notevolmente stratificate. Gran parte di Nilosyrtis è classificata come terreno eroso («fretted terrain») e contiene scarpate, mesa e ampie valli piatte. Se ne deduce che il terreno si è evoluto a causa della fusione o la sublimazione del ghiaccio glaciale. In questa regione è stato rivelato del ghiaccio sub-superficiale.

МНОГОЧИСЛЕННЫЕ СЛОИ В СТОЛОВЫХ ГОРАХ НИЛОСИРТ

Этот снимок был сделан в регионе, который формально относится к Сабейской земле, но фактически находится в восточной части региона столовых гор Нилосирт. Он показывает примечательные многослойные скалы, расположенные на склонах большой столовой горы (в центре снимка). Большая часть района Нилосирт классифицируется как «нарушенный рельеф» (fretted terrain), который состоит из утёсов, столовых гор и широких, плоских долин. Предполагается, что рельеф сформировался в результате таяния или сублимации ледников. В регионе обнаружен подповерхностный лёд.

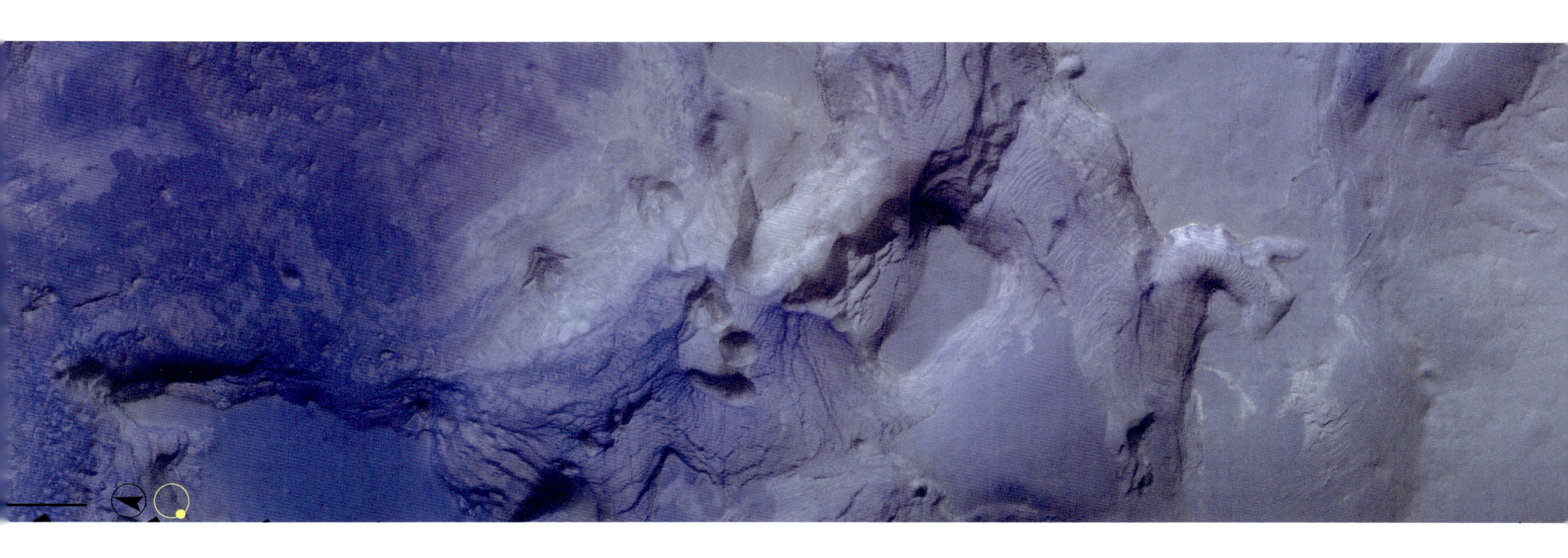

10.7 ABUNDANT LAYERS IN NILOSYRTIS MENSAE

ACQUISITION TIME: 2020-08-07T14:29:01.477
LONGITUDE: 75.9 E   LATITUDE: 28.3 N
REGION: NILOSYRTIS

ERODIERENDES POLARES TERRAIN

Dieses Bild gehört zu einer Reihe von Bildern, die vom Reynolds-Krater in Parva Planum aufgenommen wurden. Dieses Bild hier wurde Mitte des Herbstes auf der südlichen Hemisphäre des Mars aufgenommen. Die jahreszeitlich bedingte Eiskappe ist zu diesem Zeitpunkt des Marsjahres grösstenteils sublimiert (von der festen direkt in die gasförmige Phase übergegangen) und gibt den Blick auf die darunter liegende Oberfläche frei. Hier sehen wir erodierendes Terrain und scharfe Grenzübergänge in der Helligkeit und Farbe der Oberfläche.

ERODING POLAR TERRAIN

This image is from a sequence of images obtained from Reynolds Crater in Parva Planum. This particular image was obtained in mid-autumn in the southern hemisphere of Mars. The seasonal ice cap has mostly sublimed (turned directly from the solid phase to the gaseous phase) by this time in the Martian year revealing the surface below. Here we see eroding terrains and sharp boundaries in surface brightness and colour.

TERRAIN POLAIRE EN ÉROSION

Image du cratère Reynolds dans la région de Parva Planum prise au milieu de l'automne dans l'hémisphère sud de Mars. La calotte de glace saisonnière s'est en grande partie sublimée (passant directement de la phase solide à la phase gazeuse) à ce moment de l'année martienne, révélant la surface en dessous. Nous voyons ici des terrains activement érodés montrant des contrastes nets de couleur et d'albédo.

TERRENO POLARE IN EROSIONE

Questa immagine proviene da una sequenza di immagini ottenute dal cratere Reynolds in Parva Planum ed è stata ottenuta a metà autunno nell'emisfero meridionale di Marte. In questo periodo dell'anno marziano, la calotta di ghiaccio stagionale è per lo più sublimata (trasformata direttamente dalla fase solida alla fase gassosa) rivelando la superficie sottostante. Qui vediamo terreni in erosione e netti cambiamenti nella luminosità e nel colore della superficie.

ЭРОДИРОВАННЫЙ ПОЛЯРНЫЙ РЕЛЬЕФ

Это изображение относится к серии снимков кратера Рейнольдс на плато Парва. Оно было получено в середине осени в южном полушарии Марса. Большая часть сезонной ледяной шапки к этому времени марсианского года уже сублимировалась (перешла из твердой фазы в газообразную). На обнажившейся поверхности видны следы эрозии и участки различной яркости и цвета с резкими границами.

10.8 ERODING POLAR TERRAIN

ACQUISITION TIME: 2020-12-17T20:46:36.274
LONGITUDE: 198.9 E   LATITUDE: 74.0 S
REGION: PARVA

ERODIERENDE GRUBE INNERHALB EINES KRATERS

Dieses Bild zeigt eine grosse, stufenförmige Grube in der Mitte eines Kraters im Südwesten von Terra Sabaea in der Nähe des Mare Serpentis. Die Südwand (rechts) ist glatt und verfestigt (es gibt sogar Hinweise auf eine Schicht aus Schichtsilikatmaterial darin). Die Nordseite ist sehr rau und könnte stark erodiert und geschichtet sein.

ERODING PIT WITHIN A CRATER

This image is a large stepped pit in the centre of a crater in southwest Terra Sabaea near Mare Serpentis. While the south wall (to the right) is smooth and consolidated (there may even be some evidence for a layer of phyllosilicate material within it). The northern side is very rough and may be heavily eroded and layered.

FOSSE EN ÉROSION DANS UN CRATÈRE

Cette image représente une grande fosse avec plusieurs paliers au centre d'un cratère dans le sud-ouest de Terra Sabaea, près de Mare Serpentis. Alors que la paroi sud (à droite) est lisse et consolidée (il semble même y avoir des preuves d'une couche de phyllosilicates à l'intérieur), le côté nord est très rugueux et sans doute fortement érodé et stratifié.

FOSSA IN EROSIONE ALL'INTERNO DI UN CRATERE

Questa immagine è una grande fossa a gradini al centro di un cratere nella Terra Sabaea sudoccidentale vicino al Mare Serpentis. Mentre la parete sud (a destra) è liscia e consolidata (ci sono addirittura alcuni blandi indizi di uno strato di fillosilicati al suo interno), il lato nord è molto aspro ed è probabilmente pesantemente eroso e stratificato.

ЭРОДИРОВАННАЯ ЯМА ВНУТРИ КРАТЕРА

На этом снимке видна большая депрессия со ступенчатыми склонами, расположенная в центре кратера на юго-западе Сабейской земли вблизи Моря Змеи. Южная стена (справа) достаточно гладкая, что может свидетельствовать о наличии здесь прочных коренных пород силикатного состава. Территория в правой стороне снимка очень неровная, что, вероятно, связано с наличием слоистых толщ и сильной эрозией.

10.9   ERODING PIT WITHIN A CRATER

AACQUISITION TIME: 2018-10-30T09:22:11.919
LONGITUDE: 44.6 E   LATITUDE: 29.1 S
REGION: SABAEA

UNTERSCHIEDLICHE ZUSAMMENSETZUNG DES GELÄNDES IN DEN NILI FOSSAE

Der CaSSIS-Datensatz enthält viele farbenfrohe Bilder von den Nili Fossae, die die Vielfalt der Zusammensetzung zeigen. Das Bild hier ist ein Beispiel, das auch eine raue Oberfläche zeigt, die wahrscheinlich durch Erosion beeinflusst wurde. Dies hat dazu geführt, dass ursprünglich verdeckte Schichten durch die Erosion der oberen Schichten freigelegt wurden.

COMPOSITIONALLY DIVERSE TERRAIN IN NILI FOSSAE

Within the CaSSIS dataset there are many colourful images of Nili Fossae indicating compositional diversity. This is an example that also shows a rough surface that has probably been influenced by erosion. This has led to originally sub-surface layers being revealed through the erosion of the upper layers.

TERRAIN DE COMPOSITION DIVERSE DANS NILI FOSSAE

Dans le jeu de données de CaSSIS, de nombreuses images très colorées de Nili Fossae indiquent une grande diversité de composition minéralogique. Cet exemple montre également une surface rugueuse qui a probablement été influencée par l'érosion. L'érosion des couches supérieures a permis de révéler des couches plus profondes du sous-sol.

TERRENO COMPOSITIVAMENTE VARIO NELLE NILI FOSSAE

All'interno del set di dati di CaSSIS ci sono molte immagini colorate di Nili Fossae che indicano la sua diversità compositiva. Questo è un esempio che ne mostra anche la superficie aspra che è stata probabilmente modellata dall'erosione. Questo ha portato all'esposizione degli strati originariamente sub-superficiali attraverso l'erosione degli strati superiori.

РАЗНООБРАЗНАЯ ПО СОСТАВУ ПОВЕРХНОСТЬ В БОРОЗДАХ НИЛА

В наборе данных CaSSIS есть много красочных снимков борозд Нила, свидетельствующих о разнообразии слагающих их пород. Этот пример демонстрирует неровную поверхность, вероятно сформированную эрозией, частично уничтожившей верхние слои.

10.10 COMPOSITIONALLY DIVERSE TERRAIN IN NILI FOSSAE

ACQUISITION TIME: 2018-10-24T03:48:40.654
LONGITUDE: 73.4 E   LATITUDE: 21.1 N
REGION: SYRTIS

ERDHÜGEL IN EINER RAUEN EBENE

Diese quasi kreisförmigen Tafelberge befinden sich in der Nähe des südlichen Rands des Huygens-Kraters in Noachis Terra. Ihr Ursprung ist unklar, und vor den hochauflösenden Aufnahmen von HiRISE und später CaSSIS dachte man, dass es sich um Schlammvulkane handeln könnte. Als die hochauflösenden Aufnahmen gemacht wurden, gab es darin jedoch keinen Hinweis darauf. Die Strukturen sind wahrscheinlich Überbleibsel der Erosion der Schichten, die die rauen Ebenen überlagern.

MOUNDS ON A ROUGH PLAIN

These quasi circular mesas are close to the southern rim of Huygens Crater in Noachis Terra. Their origin is unclear and prior to the high resolution imaging afforded by HiRISE and subsequently CaSSIS, it was thought that they might be mud volcanoes. However, there was no real indication of this when high resolution observations were acquired. The structures are probably leftovers from the erosion of the layers that overlie the rough plains.

MONTICULES SUR UNE PLAINE ACCIDENTÉE

Ces mésas quasi circulaires sont proches de la bordure sud du cratère Huygens dans la région de Noachis Terra. Leur origine n'est pas claire et avant l'imagerie à haute résolution offerte par HiRISE et par la suite CaSSIS, on pensait qu'il pouvait s'agir de volcans de boue. Les observations ultérieures n'en ont cependant pas apporté la preuve. Ces structures sont probablement des restes de l'érosion des couches qui recouvrent les plaines rugueuses.

TUMULI SU UNA PIANURA SCABRA

Queste mesa quasi circolari sono vicine al bordo meridionale del cratere Huygens in Noachis Terra. La loro origine non è chiara e, prima delle immagini ad alta risoluzione offerte da HiRISE e successivamente da CaSSIS, si pensava che potessero essere vulcani di fango. Tuttavia, non c'era alcun indizio reale a supporto di questa ipotesi. Le strutture sono probabilmente resti dell'erosione degli strati che sovrastano le pianure aspre e irregolari.

СТОЛОВЫЕ ГОРЫ НА НЕРОВНОЙ РАВНИНЕ

Эти закруглённые столовые горы находятся недалеко от южного края кратера Гюйгенса на земле Ноя. Их происхождение неясно, и до появления изображений высокого разрешения считалось, что они могут являться грязевыми вулканами. Однако изучение данных высокого разрешения HiRISE, а затем и CaSSIS, не дало никаких подтверждений этой гипотезы. Вероятно, эти структуры являются останками эродированного слоя.

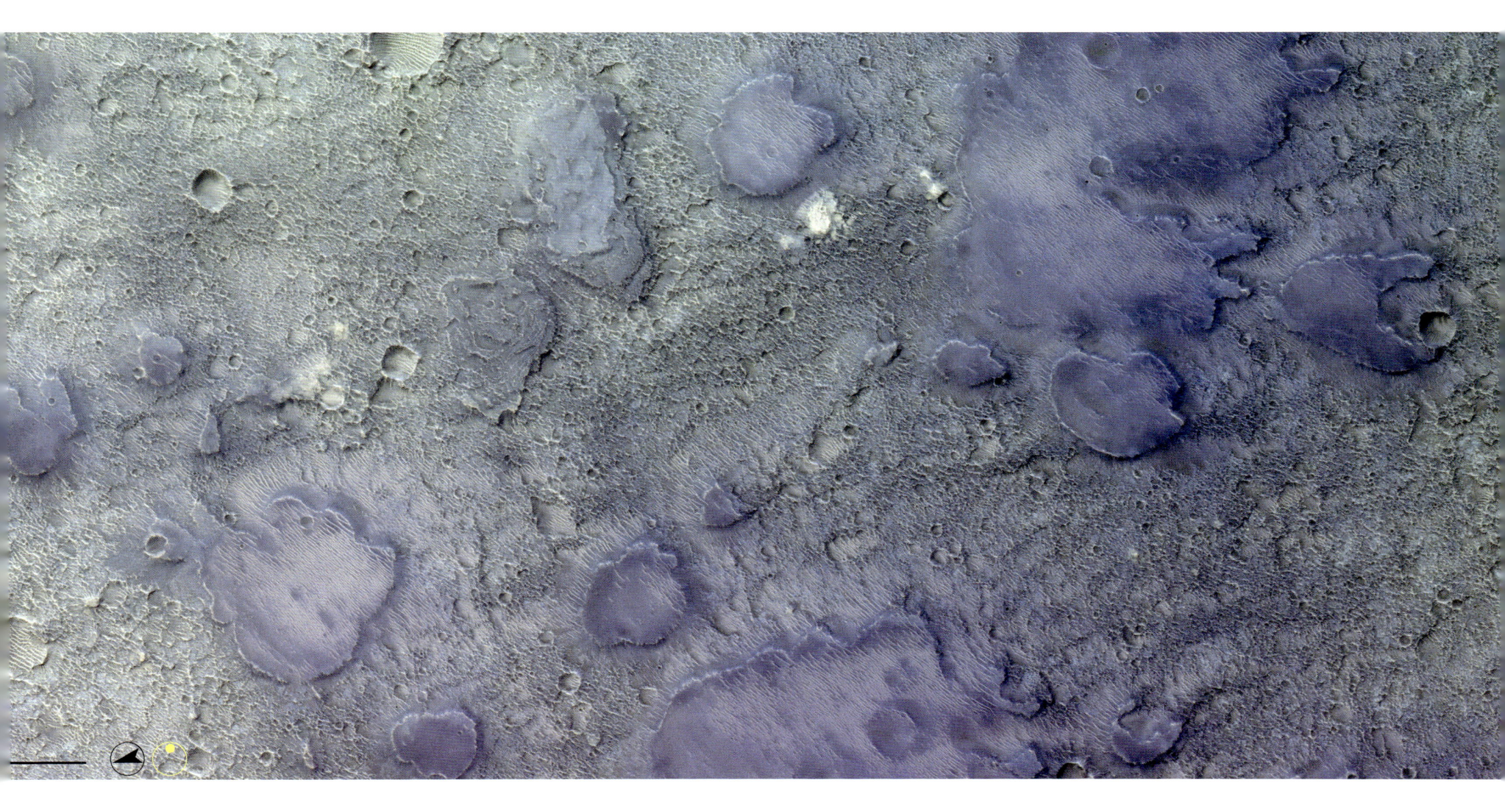

10.11 MOUNDS ON A ROUGH PLAIN

ACQUISITION TIME: 2018-09-30T07:17:40.642
LONGITUDE: 54.9 E   LATITUDE: 16.5 S
REGION: NOACHIS

FREIGELEGTES GRUNDGESTEIN IM KRATER BEI DEN HIMERA VALLES

Es handelt sich wahrscheinlich um freiliegendes Grundgestein in der Nähe des Zentrums eines Kraters südlich der Himera Valles. Der Krater ist unbenannt, liegt aber unmittelbar südlich eines (kleineren) benannten Kraters namens Yegros am Südrand von Margaritifer Terra. Der Kraterrand ist von Kanälen durchzogen, und der Kraterboden weist verschiedene fluviale Morphologien auf. Diese bemerkenswerten Aufschlüsse könnten mit der zentralen Hebung des ursprünglichen Kraters oder mit der nachfolgenden fluvialen Aktivität zusammenhängen.

EXPOSED BEDROCK IN CRATER NEAR HIMERA VALLES

This is probably exposed bedrock near the centre of a crater south of Himera Valles. The crater is unnamed but is immediately south of a (smaller) named crater called Yegros on the southern edge of Margaritifer Terra. The crater rim is degraded by channels and the crater floor has diverse fluvial morphologies. These remarkable exposures may be related to the central uplift of the original crater or to the subsequent fluvial activity.

SUBSTRAT ROCHEUX EXPOSÉ DANS UN CRATÈRE PRÈS DE HIMERA VALLES

Il s'agit probablement ici d'un substrat rocheux exposé près du centre d'un cratère au sud de Himera Valles. Le cratère n'a pas de nom mais se trouve immédiatement au sud d'un cratère (plus petit) appelé Yegros, à la limite sud de Margaritifer Terra. Le bord du cratère est dégradé par des chenaux et le fond du cratère présente diverses morphologies fluviales. Ces structures remarquables peuvent être liées au soulèvement central du cratère originel ou à une activité fluviale ultérieure.

FONDO ROCCIOSO IN UN CRATERE VICINO A HIMERA VALLES

Probabile affioramento vicino al centro di un cratere a sud di Himera Valles. Il cratere non ha nome, ma è immediatamente a sud di un cratere (più piccolo) chiamato Yegros sul bordo meridionale di Margaritifer Terra. Il bordo del cratere è degradato da alcuni canali e il pavimento presenta morfologie fluviali assortite. Questi notevoli affioramenti del pavimento roccioso possono essere legati al sollevamento centrale del cratere o alla successiva attività fluviale.

ОБНАЖЕНИЯ КОРЕННЫХ ПОРОД В КРАТЕРЕ ВБЛИЗИ ДОЛИНЫ ХИМЕРЫ

Данный снимок центральной части кратера к югу от долины Химеры показывает, вероятно, обнажения коренных пород. Кратер не имеет названия, однако находится непосредственно к югу от меньшего по размеру картера Егрос, что на южном краю Жемчужной Земли. Края кратера разрушены каналами, а дно имеет разнообразные морфологические формы, которые могут быть связаны с центральным поднятием первоначального кратера и/или с последующей флювиальной активностью.

10.12 EXPOSED BEDROCK IN CRATER NEAR HIMERA VALLES

ACQUISITION TIME: 2020-12-13T04:24:49.032
LONGITUDE: 336.3 E   LATITUDE: 22.8 S
REGION: MARGARITIFER

POLYGONAL ZERKLÜFTETES TERRAIN IN TERRA CIMMERIA

Dieser gemusterte Boden befindet sich in einem unbenannten Krater am westlichen Rand von Terra Cimmeria. Das weissliche Material ist in polygonaler Weise aufgerissen. Es gibt einige kleine dunkle Dünen, die das weisse Material durchziehen. Der Krater oben links ist verwittert und hat einen ungewöhnlich vielfarbigen Boden. Polygonale Muster können auf verschiedene Weise entstehen. Eine Möglichkeit ist die fortschreitende Austrocknung eines schlamm- oder tonhaltigen Sediments.

POLYGONALLY-FRACTURED TERRAIN IN TERRA CIMMERIA

This patterned ground is in an unnamed crater on the western edge of Terra Cimmeria. The whitish material is cracked in a polygonal manner. There are some small dark dunes crossing the white material. The crater to the top left is degraded with an unusual multi-coloured floor. Polygonal patterns can arise in one of several ways. Desiccation as a result of a mud or clay-bearing sediments drying out over time is one possibility here.

TERRAIN À FRACTURES POLYGONALES DANS TERRA CIMMERIA

Ces formations à la texture remarquable se trouvent dans un cratère sans nom à la limite ouest de Terra Cimmeria. La formation blanchâtre est craquelée de manière polygonale. Quelques petites dunes sombres la traversent également. Le cratère en haut à gauche est dégradé avec un sol inhabituel multicolore. Les motifs polygonaux peuvent apparaître de plusieurs façons. La dessiccation résultant de l'assèchement d'une boue ou de sédiments argileux au fil du temps est une possibilité ici.

TERRENO FRATTURATO POLIGONALMENTE IN TERRA CIMMERIA

Questo terreno modellato si trova in un cratere senza nome sul bordo occidentale di Terra Cimmeria. Il materiale biancastro è fratturato in strutture poligonali e ci sono alcune piccole dune scure che attraversano il materiale bianco. Il cratere in alto a sinistra è ormai deteriorato e mostra un insolito pavimento multicolore. Le strutture poligonali possono crearsi in molti modi diversi, in questo caso potrebbe essere il risultato dell'essiccazione di fango o di sedimenti argillosi che si sono asciugati con il passare del tempo.

ПОЛИГОНАЛЬНЫЙ РЕЛЬЕФ В КИММЕРИЙСКОЙ ЗЕМЛЕ

Эти «узорчатые» поверхности располагаются в безымянном кратере на западном краю Киммерийской земли. Гряда из небольших темных дюн пересекает светлую равнину, на которой хорошо виден полигональный узор, который мог быть сформирован при растрескивании грязевых или глинистых отложений при высыхании. Небольшой разрушенный кратер слева имеет необычное многоцветное дно.

10.13   POLYGONALLY-FRACTURED TERRAIN IN TERRA CIMMERIA

ACQUISITION TIME: 2020-12-12T18:36:19.011
LONGITUDE: 122.2 E   LATITUDE: 25.5 S
REGION: CIMMERIA

HELL GETÖNTE, ZERKLÜFTETE ERDHÜGEL

Dieses Bild zeigt helle Hügel im Zentrum von Terra Cimmeria. Die Hügel sind kilometergross und scheinen von Felsgraten oder Brüchen durchzogen zu sein. Ihr Ursprung ist nicht genau bekannt.

LIGHT-TONED FRACTURED MOUNDS

This image is of light-toned mounds in the centre of Terra Cimmeria. The mounds are kilometre scale and appear crossed by ridges or fractures. Their origin is not well understood.

MONTICULES FRACTURÉS DE COULEUR CLAIRE

Cette image présente des monticules aux tons clairs au centre de Terra Cimmeria. Les monticules sont de taille kilométrique et semblent traversés par des crêtes ou des fractures. Leur origine n'est pas bien comprise.

CUMULI FRATTURATI DI COLORE CHIARO

Questa immagine è costituita da tumuli di colore chiaro nel centro di Terra Cimmeria. I tumuli sono di dimensioni chilometriche e sembrano attraversati da creste o fratture. La loro origine è ancora oggetto di studio.

ТРЕЩИНОВАТЫЕ КУРГАНЫ СВЕТЛЫХ ТОНОВ

Это снимок светлых курганов в центре Киммерийской земли. Данные возвышения имеют примерно километровые размеры и, судя по снимку, пересечены хребтами или разломами. Их происхождение остается неясным.

10.14 **LIGHT-TONED FRACTURED MOUNDS**

ACQUISITION TIME: 2020-06-25T01:57:19.280
LONGITUDE: 162.4 E   LATITUDE: 36.6 S
REGION: CIMMERIA

FREIGELEGTES GRUNDGESTEIN AM HOHEN NÖRDLICHEN BREITENGRAD

Der Stokes-Krater liegt auf der nördlichen Hemisphäre. In seinem Zentrum ist helles Material freigelegt. Die grossen Gesteinsfragmente, die hier zu sehen sind, werden oft als Megabrekzien bezeichnet, aber der Entstehungsmechanismus ist durch diesen Begriff nicht definiert. Da sich dieses Material im Zentrum des Stokes-Kraters befindet, könnte es sogar durch den Einschlag selbst entstanden sein. Ungewöhnlich für die nördlichen Ebenen ist, dass das Material im Laufe der Zeit nicht mit Staub bedeckt wurde. In der Umgebung wurden auch Schichtsilikate nachgewiesen.

EXPOSED BEDROCK AT HIGH NORTHERN LATITUDE

Stokes Crater is in the northern hemisphere and its centre has exposures of bright material. The large rock fragments seen here are often referred to as megabreccia but the formation mechanism is not defined by this term. As this material is at the centre of Stokes Crater, it may have been produced by the impact event itself. Unusually for the northern plains, the material has not been covered with dust over time. Phyllosilicates have also been detected in the vicinity.

SUBSTRAT ROCHEUX EXPOSÉ À HAUTE LATITUDE NORD

Le cratère Stokes se trouve dans l'hémisphère nord et son centre présente des affleurements de roche claire. Les grands fragments de roches que l'on voit ici sont souvent appelés mégabreccia, mais le mécanisme de formation n'est pas défini par ce terme. Comme ce matériau se trouve au centre du cratère Stokes, il a pu être produit par l'impact lui-même. De façon inhabituelle pour les plaines du nord, le matériau n'a pas été recouvert de poussière au fil du temps. Des phyllosilicates ont également été détectés dans les environs.

STRATO ROCCIOSO SUPERFICIALE AD ALTA LATITUDINE NORD

Il cratere Stokes si trova nell'emisfero settentrionale e presenta un affioramento di materiale chiaro al suo centro. I grandi frammenti di roccia che si vedono qui sono spesso chiamati megabreccia, ma questo termine non specifica il loro meccanismo di formazione. Poiché questo materiale è al centro del cratere Stokes, potrebbe essere stato prodotto dall' impatto stesso. In modo del tutto insolito per gli oggetti presenti nelle pianure settentrionali, il materiale non si è coperto di polvere col passare del tempo. Nelle vicinanze sono anche stati rilevati fillosilicati.

ОБНАЖЕНИЯ КОРЕННЫХ ПОРОД В ВЫСОКИХ СЕВЕРНЫХ ШИРОТАХ

Кратер Стокса, в центре которого обнажился яркий материал, находится в северном полушарии Марса. Видимые здесь крупные фрагменты породы часто называют мегабрекчиями, но механизм их образования этот термин не определяет. Поскольку эти образования находятся в центре ударного кратера, можно предположить, что они сформировались в результате самого импактного события. Необычным для северных равнин является то, что выходы яркого материала по каким-то причинам не покрылись пылью, с течением времени, возможно они были обнажены недавно. В окрестностях также были обнаружены силикатные породы.

10.15 EXPOSED BEDROCK AT HIGH NORTHERN LATITUDE

ACQUISITION TIME: 2021-04-19T19:34:59.307
LONGITUDE: 171.3 E   LATITUDE: 55.9 N
REGION: VASTITAS BOREALIS FORMATION

MINERALAUFSCHLÜSSE IN DER ZENTRALEN SPITZE DES HALE-KRATERS

Der Hale-Krater wird recht häufig beobachtet, um nach Anzeichen für Bewegungen an den Hängen seines zentralen Gipfels zu suchen. Die CaSSIS-Aufnahmen zeigen nicht nur dunkleren, blauen Staub an den Hängen abseits der Erhebungen im Zentrum des Kraters, sondern auch die türkise Farbe, die typischerweise von mafischen Mineralien in CaSSIS-Farbprodukten herrührt. Das CRISM-Infrarotspektrometer auf dem Mars Reconnaissance Orbiter der NASA zeigt eine starke mafische Signatur in der zentralen Spitze von Hale.

MINERAL EXPOSURES IN THE CENTRAL PEAK OF HALE CRATER

Hale Crater is observed quite frequently to look for evidence of movement on the slopes of its central peak. CaSSIS imaging shows, not merely darker blue dust on the slopes away from the uplifts in the centre of the crater but also the turquoise colour that typically arises from mafic minerals in CaSSIS colour products. The CRISM infrared spectrometer on NASA's Mars Reconnaissance Orbiter shows a strong mafic signature in the central peak of Hale.

MINÉRAUX EXPOSÉS DANS LE PIC CENTRAL DU CRATÈRE HALE

Le cratère Hale est observé assez fréquemment pour rechercher des preuves de mouvement sur les pentes de son pic central. Cette image CaSSIS montre non seulement une poussière bleue plus foncée sur les pentes éloignées des soulèvements au centre du cratère, mais aussi la couleur turquoise qui provient généralement des minéraux mafiques dans les produits colorés CaSSIS. Le spectromètre infrarouge CRISM de la sonde Mars Reconnaissance Orbiter de la NASA montre une forte signature mafique dans le pic central de Hale.

MINERALI AFFIORANTI NEL PICCO CENTRALE DEL CRATERE HALE

I Il cratere Hale viene osservato abbastanza frequentemente per cercare prove di materiale in movimento sulle pendici del suo picco centrale. Le immagini di CaSSIS mostrano non solo polvere blu più scura sulle pendici lontane dai rilievi al centro del cratere, ma anche affioramenti di colore turchese che tipicamente sono associati ai minerali mafici. Lo spettrometro infrarosso CRISM del Mars Reconnaissance Orbiter della NASA rileva una forte firma mafica proveniente dal picco centrale di Hale.

МАФИЧЕСКИЕ МИНЕРАЛЫ В ЦЕНТРАЛЬНОМ ПОДНЯТИИ КРАТЕРА ХЭЙЛ

Кратер Хейл часто наблюдается в поисках признаков движения материала на склонах его центральной вершины. Съемка CaSSIS показывает не только темно-синюю пыль на склонах вдали от поднятий в центре кратера, но и бирюзовый цвет, который в фильтрах CaSSIS обычно связан с присутствием мафических минералов. Мафические минералы также диагностируются в пределах центрального поднятия кратера Хейл по данным инфракрасной спектрометрии CRISM на аппарате NASA Mars Reconnaissance Orbiter.

10.16 MINERAL EXPOSURES IN THE CENTRAL PEAK OF HALE CRATER

ACQUISITION TIME: 2021-05-11T07:32:42.453
LONGITUDE: 323.4 E   LATITUDE: 36.0 S
REGION: NOACHIS

FREIGELEGTES SCHICHTMATERIAL UND DÜNEN IN DEN NEREIDUM MONTES

Dieses Bild zeigt ein Gebiet in den Nereidum Montes, das sich westlich des Argyre-Beckens befindet. Oben in der Mitte des Bilds ist eine Farbenvielfalt zu erkennen, die durch unterschiedliche Lithologien entsteht. Es sind Schichten zu erkennen (z. B. unten in der Mitte), die sich weiter nach Süden (links) und über das Bild hinaus erstrecken. Oben links befindet sich ein komplexes Dünenfeld (dunkelblau), das häufig auf die Aktivität von Dünenrinnen untersucht wird.

EXPOSED LAYERED MATERIAL AND DUNES IN NEREIDUM MONTES

This image shows an area in Nereidum Montes which is to the west of the Argyre Basin. The centre top of the image shows variegation arising from different lithologies. There are layers evident (bottom centre, for example) which extend further to the south (left) and out of the frame. There is a complex dune field to the top left (dark blue) which is frequently monitored for dune gully activity.

DÉPOTS STRATIFIÉS EXPOSÉS ET DUNES DANS NEREIDUM MONTES

Cette image montre une zone de Nereidum Montes qui se trouve à l'ouest du bassin d'Argyre. La partie centrale supérieure de l'image présente une grande variabilité de couleurs liée aux différentes lithologies. Il y a des strates bien exposées (en bas au centre, par exemple) qui s'étendent plus loin au sud (à gauche) et hors du cadre. Un champ de dunes complexe en haut à gauche (bleu foncé) est fréquemment imagé pour étudier l'activité des ravines de dunes.

AFFIORAMENTO DI MATERIALE STRATIFICATO E DUNE IN NEREIDUM MONTES

Questa immagine mostra una zona di Nereidum Montes che si trova ad ovest del bacino di Argyre. Il centro in alto dell'immagine mostra evidenti sfumature di colore che appartengono a diverse litologie. Ci sono stratificazioni importanti (in basso al centro, per esempio) che si estendono più a sud (a sinistra) e fuori dall'inquadratura. Si può notare un complesso campo di dune in alto a sinistra (blu scuro) che è monitorato frequentemente per osservare l'attività dei calanchi dunali.

ОБНАЖЁННЫЙ СЛОИСТЫЙ МАТЕРИАЛ И ДЮНЫ В ГОРАХ НЕРЕИД

На этом снимке показана область в горах Нереид, которая находится к западу от равнины Аргир. В центре верхней части снимка наблюдается цветовое разнообразие, связанное с присутствием различных пород. Видны признаки слоёв осадочных пород (например, внизу в центре), простирающихся далее на юг (налево) за пределы кадра. Слева вверху находится область, занятая дюнами (темно-синий цвет), которая изучается на предмет образования оврагов.

10.17   EXPOSED LAYERED MATERIAL AND DUNES IN NEREIDUM MONTES

ACQUISITION TIME: 2021-04-13T10:54:06.947
LONGITUDE: 303.5 E   LATITUDE: 47.2 S
REGION: ARGYRE

VIELGESTALTIGE OBERFLÄCHE IM BECKENBODEN DER LADON VALLES

Die Ladon Valles scheinen in ein Becken zu münden, bei dem es sich wahrscheinlich um einen alten Einschlagkrater handelt. Dies ist ein Bild vom Boden des Beckens. Die Mündung des Ladon-Valles-Systems liegt etwa 120 km südwestlich. Das Bild zeigt hell getönte Ablagerungen, die für Tone in CaSSIS-Bildprodukten charakteristisch sind, und möglicherweise tektonische Vertiefungen.

VARIEGATED SURFACE IN LADON VALLES BASIN FLOOR

Ladon Valles appear to empty into a basin which is probably an ancient impact crater. This is an image from the floor of the basin. The mouth of the Ladon Valles system is about 120 km to the south-west. The image shows light-toned deposits that are characteristic of clays in CaSSIS image products, and possibly tectonic depressions.

SURFACE VARIÉE DANS LE FOND DU BASSIN DE LADON VALLES

Les vallées de Ladon semblent se déverser dans un bassin qui est probablement un ancien cratère d'impact. Voici une image du fond de ce bassin. L'embouchure du système des vallées de Ladon se trouve à environ 120 km au sud-ouest. L'image montre des dépôts de couleur claire qui sont caractéristiques des argiles dans les images CaSSIS, ainsi que de probables dépressions tectoniques.

SUPERFICIE VARIEGATA NEL FONDO DEL BACINO DI LADON VALLES

Le Ladon Valles sembrano svuotarsi in un bacino (il cui pavimento è ritratto in questa immagine) che è probabilmente un antico cratere da impatto. La bocca del sistema Ladon Valles si trova a circa 120 km a sud-ovest. L'immagine mostra depositi dai toni chiari che sono caratteristici dei materiali argillosi nelle immagini di CaSSIS, e probabilmente alcune depressioni tettoniche.

ПЕСТРАЯ ПОВЕРХНОСТЬ НА ДНЕ БАССЕЙНА ДОЛИН ЛАДОН

Долины Ладон впадают в котловину, которая, вероятно, является древним ударным кратером. Устье этой системы долин находится примерно в 120 км к юго-западу от самого кратера. На снимке дна кратера видны отложения светлых тонов, которые на снимках CaSSIS характерны для глин, и узкие вытянутые долины, возможно имеющие тектоническое происхождение.

10.18 VARIEGATED SURFACE IN LADON VALLES BASIN FLOOR

ACQUISITION TIME: 2018-11-14T20:32:04.576
LONGITUDE: 329.3 E   LATITUDE: 18.7 S
REGION: MARGARITIFER

OLIVINREICHE AUFSCHLÜSSE IM TAYTAY-KRATER

Der Taytay-Krater liegt am nordöstlichen Ufer des Ares Vallis. Er ist ein Krater in einem anderen Krater. Der Einschlag scheint olivinreiches Material aus dem Untergrund freigelegt zu haben. Olivin ist ein Magnesium-Eisen-Silikat, das sich in Gegenwart von Wasser leicht in ein spezielles Mineral namens Iddingsit umwandelt. Daher ist Olivin in Sedimentablagerungen selten zu finden. Dies spricht dafür, dass das Material im Taytay-Krater zu keiner Zeit in seiner Geschichte Wasser ausgesetzt war.

OLIVINE-RICH OUTCROPS IN TAYTAY CRATER

Taytay Crater is on the northeastern bank of Ares Vallis. It is a crater within another crater. The impact seems to have revealed olivine-rich material from below. Olivine is a magnesium-iron silicate and alters readily in the presence of water forming a specific mineral called iddingsite. Consequently, olivine is rarely found in sedimentary deposits. This argues that the material in Taytay Crater has not been exposed to water at any time in its history.

AFFLEUREMENTS RICHES EN OLIVINE DANS LE CRATÈRE TAYTAY

Le cratère Taytay se trouve sur la rive nord-est d'Ares Vallis, au sein d'un autre cratère plus grand. L'impact semble avoir révélé des roches riches en olivine du sous-sol. L'olivine est un silicate de magnésium et de fer qui s'altère facilement en présence d'eau en formant un minéral spécifique appelé iddingsite. Par conséquent, l'olivine est rarement présente dans les dépôts sédimentaires. Cela indique que cette formation du cratère Taytay n'a, à aucun moment de son histoire, été exposée à l'eau.

AFFIORAMENTI RICCHI DI OLIVINA NEL CRATERE TAYTAY

Il cratere Taytay si trova sulla riva nord-orientale di Ares Vallis e si tratta di un cratere all'interno di un altro cratere. L'impatto sembra aver rivelato il materiale ricco di olivina che si trovava al di sotto della superficie. L'olivina è un silicato di magnesio e ferro e si altera facilmente in presenza di acqua formando uno specifico minerale chiamato iddingsite. Di conseguenza, l'olivina non alterata si trova raramente nei depositi sedimentari. Questo indica che il materiale nel cratere Taytay non è stato esposto all'acqua in nessun momento della sua storia.

БОГАТЫЕ ОЛИВИНОМ ОБНАЖЕНИЯ В КРАТЕРЕ ТАЙТАЙ

Ударный кратер Тайтай находится у северо-восточного «берега» долины Ареса внутри другого более крупного кратера. Удар, вероятно, обнажил глубинные породы, богатые оливином. Оливин представляет собой магниево-железный силикат, который легко изменяется в присутствии воды и трансформируется в иддингсит – смесь смектита (Mg-содержащего глинистого минерала из группы монтмориллонита), хлорита, серпентина и гетита. Следовательно, оливин редко встречается в осадочных отложениях. Это говорит о том, что породы в кратере Тайтай не подвергались воздействию воды ни в один из периодов своей истории.

10.19 OLIVINE-RICH OUTCROPS IN TAYTAY CRATER

ACQUISITION TIME: 2018-11-03T05:35:45.082
LONGITUDE: 340.3 E LATITUDE: 7.8 N
REGION: MARGARITIFER

FREIGELEGTES GRUNDGESTEIN IM GANGES CHASMA

Der Boden des Ganges Chasma östlich von Ganges Mensa weist ein ausgedehntes Dünenfeld auf. Durch die Dünen ragen Hügel, die in den CaSSIS-Bildern bemerkenswert bunt sind. Dieses Bild hier ist ein schönes Beispiel. Wir haben nur wenige Spektralinformationen aus diesem Gebiet, aber in Analogie zu anderen CaSSIS-Bildern handelt es sich bei den türkisenen Farben wahrscheinlich um mafische (magnesium- und eisenreiche) Mineralien.

EXPOSED BEDROCK IN GANGES CHASMA

The floor of Ganges Chasma east of Ganges Mensa has an extensive dune field. Poking through the dunes are mounds that are remarkably colourful in CaSSIS imagery. Here is a nice example. We have little spectral information from this area but by analogy with other CaSSIS images, the turquoise colours are probably mafic (rich in magnesium and iron) minerals.

SUBSTRAT ROCHEUX EXPOSÉ DANS GANGES CHASMA

Le sol de Ganges Chasma, à l'est de Ganges Mensa, présente un vaste champ de dunes. Au travers des dunes, émergent des monticules qui apparaissent remarquablement colorés dans les images CaSSIS. En voici un bel exemple. Nous disposons de peu d'informations spectrales sur cette zone, mais par analogie avec d'autres images CaSSIS, les teintes turquoises indiquent probablement des minéraux mafiques (riches en magnésium et en fer).

PAVIMENTO ROCCIOSO NEL GANGES CHASMA

Il pavimento del Ganges Chasma a est di Ganges Mensa ha un esteso campo di dune tra cui spuntano dei tumuli che sono notevolmente colorati nelle immagini di CaSSIS, come l'impressionante esempio qui sopra. Si hanno poche informazioni spettrali da questa zona, ma per analogia con altre immagini di CaSSIS, i colori turchesi sono probabilmente minerali mafici (ricchi di magnesio e ferro).

ОБНАЖЕНИЯ КОРЕННЫХ ПОРОД В КАНЬОНЕ ГАНГ

На дне каньона Ганг к востоку от одноимённой столовой горы находится обширное дюнное поле. Сквозь дюны пробиваются курганы, которые на этом и других снимках CaSSIS выглядят удивительно красочно. По этому району имеется мало спектральной информации, но по аналогии с другими снимками CaSSIS, бирюзовые цвета, вероятно, создаются мафическими (богатыми магнием и железом) минералами.

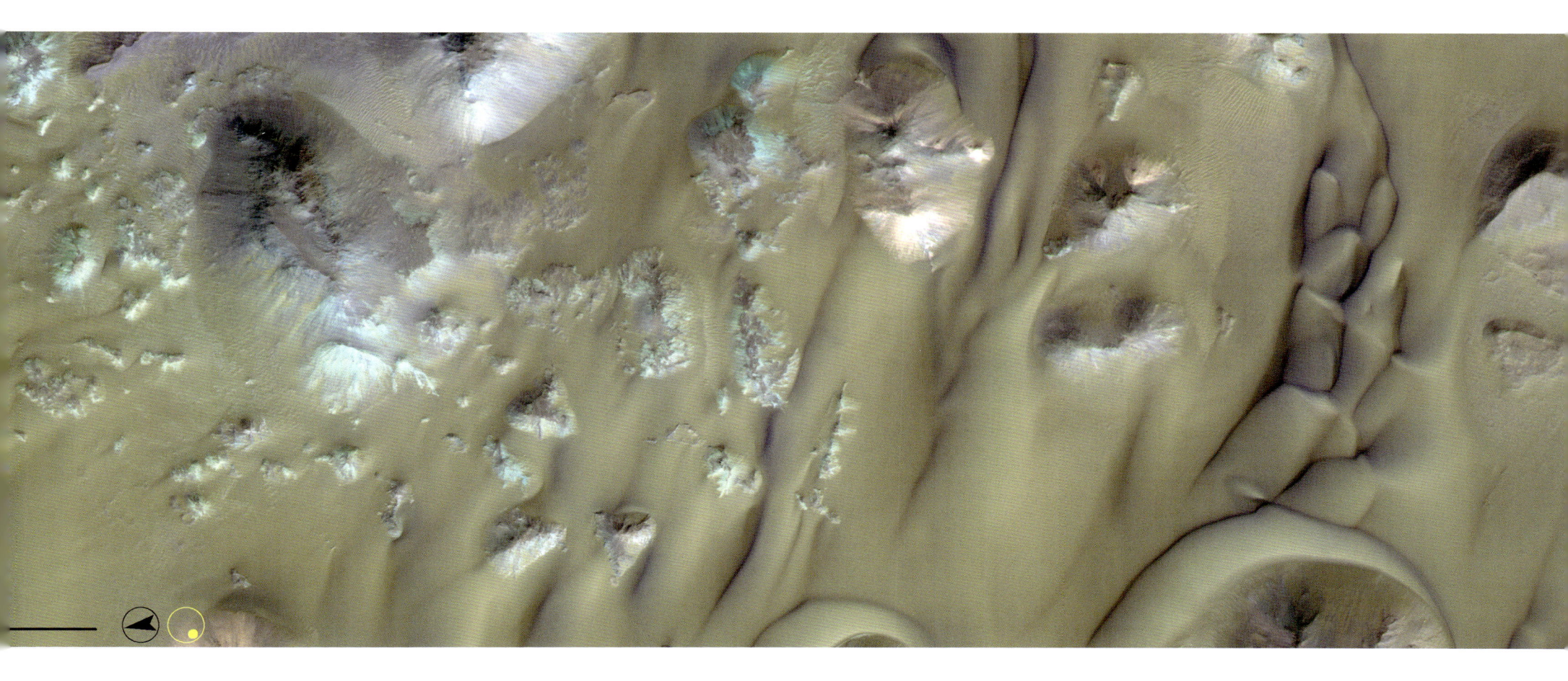

10.20 EXPOSED BEDROCK IN GANGES CHASMA

ACQUISITION TIME: 2020-10-03T11:21:16.381
LONGITUDE: 312.5 E   LATITUDE: 8.1 S
REGION: MARINERIS

KANÄLE, DÜNEN UND VERTIEFUNGEN

Dies ist ein Dünenfeld, das sich über den Rand einer zentralen Grube in einem Krater im Zentrum von Terra Cimmeria erstreckt. Auf dem Kraterboden im Südosten (oben rechts) befinden sich Kanäle. Bei näherer Betrachtung erkennt man, dass es möglicherweise Anzeichen für eine Sedimentschichtung gibt.

CHANNELS, DUNES AND DEPRESSIONS

This is a dune field which is across the edge of a central pit in a crater in central Terra Cimmeria. There are channels on the crater floor to the south-east (upper right). Close inspection reveals that there is possible evidence of sedimentary layering.

CHENAUX, DUNES ET DÉPRESSIONS

Champ de dunes recoupant la bordure d'un fossé central dans un cratère au centre de Terra Cimmeria. Il y a des chenaux sur le fond du cratère au sud-est (en haut à droite). Une inspection minutieuse révèle qu'il y a peut-être des traces de stratification sédimentaire.

CANALI, DUNE E DEPRESSIONI

Questo è un campo di dune che si trova al margine di una fossa centrale in un cratere della Terra Cimmeria centrale. Alcuni canali si estendono sul pavimento del cratere a sud-est (in alto a destra). Un'ispezione ravvicinata rivela che ci sono possibili indizi di stratificazione sedimentaria.

КАНАЛЫ, ДЮНЫ И ВПАДИНЫ

Это дюнное поле находится за краем центральной депрессии кратера в центральной части Киммерийской Земли. На юго-востоке (вверху справа) на дне кратера, присутствуют каналы. Также, при внимательном изучении видны вероятные признаки стратифицированных осадочных пород.

10.21 CHANNELS, DUNES AND DEPRESSIONS

ACQUISITION TIME: 2018-10-24T23:46:47.690
LONGITUDE: 158.1 E   LATITUDE: 36.1 S
REGION: CIMMERIA

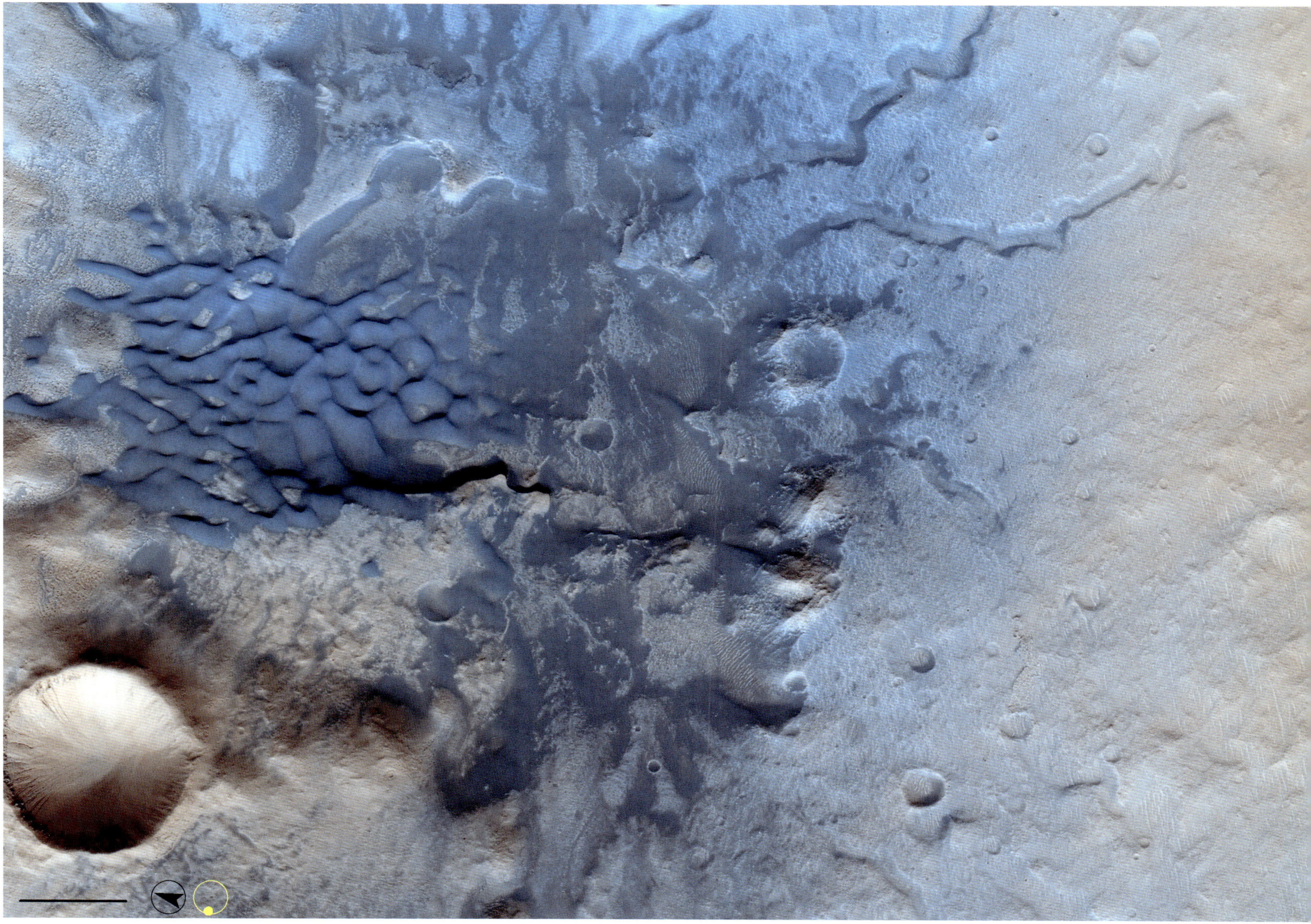

HÜGEL UND RINNEN IN DEN NEREIDUM MONTES

Dieses Bild zeigt ein Gebiet in den Nereidum Montes östlich von Argyre Planitia und westlich des Bozkir-Kraters. Der Hügel auf der linken Seite dieses Bilds grenzt an andere Hügel, die mit möglichen olivinreichen Aufschlüssen in Verbindung gebracht wurden. Die Farbvielfalt deutet darauf hin, dass hier verschiedene Lithologien vorhanden sind. Die Strukturen auf der rechten Seite (Süden) befinden sich in einem Kanal zwischen Hügeln und lassen vermuten, dass in der Vergangenheit Flüssigkeiten durch den Kanal geflossen sind. Beachten Sie auch die Rinnenaktivität auf der südlichen (rechten) Seite des Hügels.

MOUNDS AND CHANNELS IN NEREIDUM MONTES

This image is an area in Nereidum Montes east of Argyre Planitia and west of Bozkir Crater. The mound to the left of this image is adjacent to other mounds that have been associated with possible olivine-rich outcrops. The colour diversity here suggests different lithologies are present. The structures to the right (south) are in a channel between mounds and suggest that fluids have flowed through the channel in the past. Note also the gully activity on the south (right) side of the mound.

MONTICULES ET CHENAUX DANS NEREIDUM MONTES

Cette image montre une partie des monts Nereidum à l'est d'Argyre Planitia et à l'ouest du cratère Bozkir. Le monticule à gauche de cette image est adjacent à d'autres monticules qui ont été associés à de possibles affleurements riches en olivine. La diversité des couleurs ici suggère que différentes lithologies sont présentes. Les structures à droite (sud) sont dans un chenal entre les monticules et suggèrent que des fluides ont circulé dans ce chenal par le passé. Notez également l'activité de la ravine sur le côté sud (droit) du monticule.

TUMULI E CANALI IN NEREIDUM MONTES

Questa immagine è una zona di Nereidum Montes ad est di Argyre Planitia e ad ovest del cratere Bozkir. Il tumulo a sinistra di questa immagine è adiacente ad altri tumuli che sono stati associati a possibili affioramenti ricchi di olivina. La diversità di colore qui suggerisce la presenza di diverse litologie. Le strutture a destra (sud) si trovano in un canale tra tumuli e indicano che qualche liquido è fluito nel canale in passato. Si noti anche l'attività del calanco sul lato sud (destra) del tumulo.

КУРГАНЫ И КАНАЛЫ В ГОРАХ НЕРЕИД

На этом снимке видна область в Горах Нереид к востоку от равнины Аргир и к западу от кратера Бозкир. Курган в левой части снимка соседствует с другими возвышениями, которые были ассоциированы с возможными выходами пород, богатых оливином. Красочность цветов указывает на литологическое разнообразие. Образования неправильной формы, видимые справа (на юге), в канале между курганами, были, возможно, сформированы некими потоками. Обратите внимание также на оврагообразование на южном (правом) склоне кургана в центральной части снимка.

10.22 MOUNDS AND CHANNELS IN NEREIDUM MONTES

ACQUISITION TIME: 2020-09-06T12:09:06.993
LONGITUDE: 326.3 E   LATITUDE: 43.5 S
REGION: ARGYRE

FARBENVIELFALT IM WIRTZ-KRATER

Dieses Bild zeigt die nordöstliche Wand und den Boden des Wirtz-Kraters mit einem Durchmesser von 120 km. Der Krater liegt östlich von Argyre Planitia in Noachis Terra. Nur der südliche Rand des Wirtz-Kraters wurde mit dem hochauflösenden Bildspektrometer CRISM untersucht, und dieser Datensatz zeigte keine starken eindeutigen Mineralsignaturen. Dieses Bild zeigt jedoch eine spektakuläre Farbvielfalt. Man beachte auch, dass der Boden des Kraters ein polygonales Muster aufweist.

COLOUR DIVERSITY IN WIRTZ CRATER

This image shows the north-east wall and floor of the 120 km diameter Wirtz Crater. The crater is to the east of Argyre Planitia in Noachis Terra. Only the southern rim of Wirtz has been studied with the high resolution imaging spectrometer, CRISM, and that dataset did not show strong unambiguous mineral signatures. However, this image shows spectacular colour diversity. Note also that the floor of the crater exhibits polygonal patterned ground.

DIVERSITÉ DES COULEURS DANS LE CRATÈRE WIRTZ

Cette image montre la paroi nord-est et le plancher du cratère Wirtz de 120 km de diamètre. Le cratère se trouve à l'est de la plaine d'Argyre dans la région de Noachis Terra. Seul la bordure sud du cratère Wirtz a été étudiée par le spectromètre infrarouge CRISM, et cet ensemble de données n'a pas montré de fortes signatures spectrales minéralogiques. Cependant, l'image CaSSIS montre une diversité de couleurs spectaculaire. Notez également que le plancher du cratère présente une fracturation polygonale.

DIVERSITÀ DI COLORE NEL CRATERE WIRTZ

Questa immagine mostra la parete nord-est e il pavimento del cratere Wirtz di 120 km di diametro. Il cratere si trova ad est di Argyre Planitia in Noachis Terra. Solo il bordo meridionale di Wirtz è stato studiato con lo spettrometro ad alta risoluzione, CRISM, e quei dati non ha mostrato forti o inequivocabili firme minerali. Tuttavia, questa immagine mostra una spettacolare diversità di colori. Si noti anche che il pavimento del cratere mostra un terreno a forme poligonali.

РАЗНООБРАЗИЕ ЦВЕТОВ В КРАТЕРЕ ВИРТЦ

На этом снимке видны северо-восточная стена и часть дна 120-километрового кратера Виртц. Кратер находится к востоку от равнины Аргир на земле Ноя. Только южный край кратера Виртц был изучен с помощью спектрометра высокого разрешения CRISM, но полученные данные не выявили четко выраженных признаков конкретных минералов. Данное изображение примечательно цветовым разнообразием и полигональным узором дна кратера.

10.23 COLOUR DIVERSITY IN WIRTZ CRATER

ACQUISITION TIME: 2020-05-26T15:59:05.803
LONGITUDE: 334.9 E LATITUDE: 47.5 S
REGION: NOACHIS

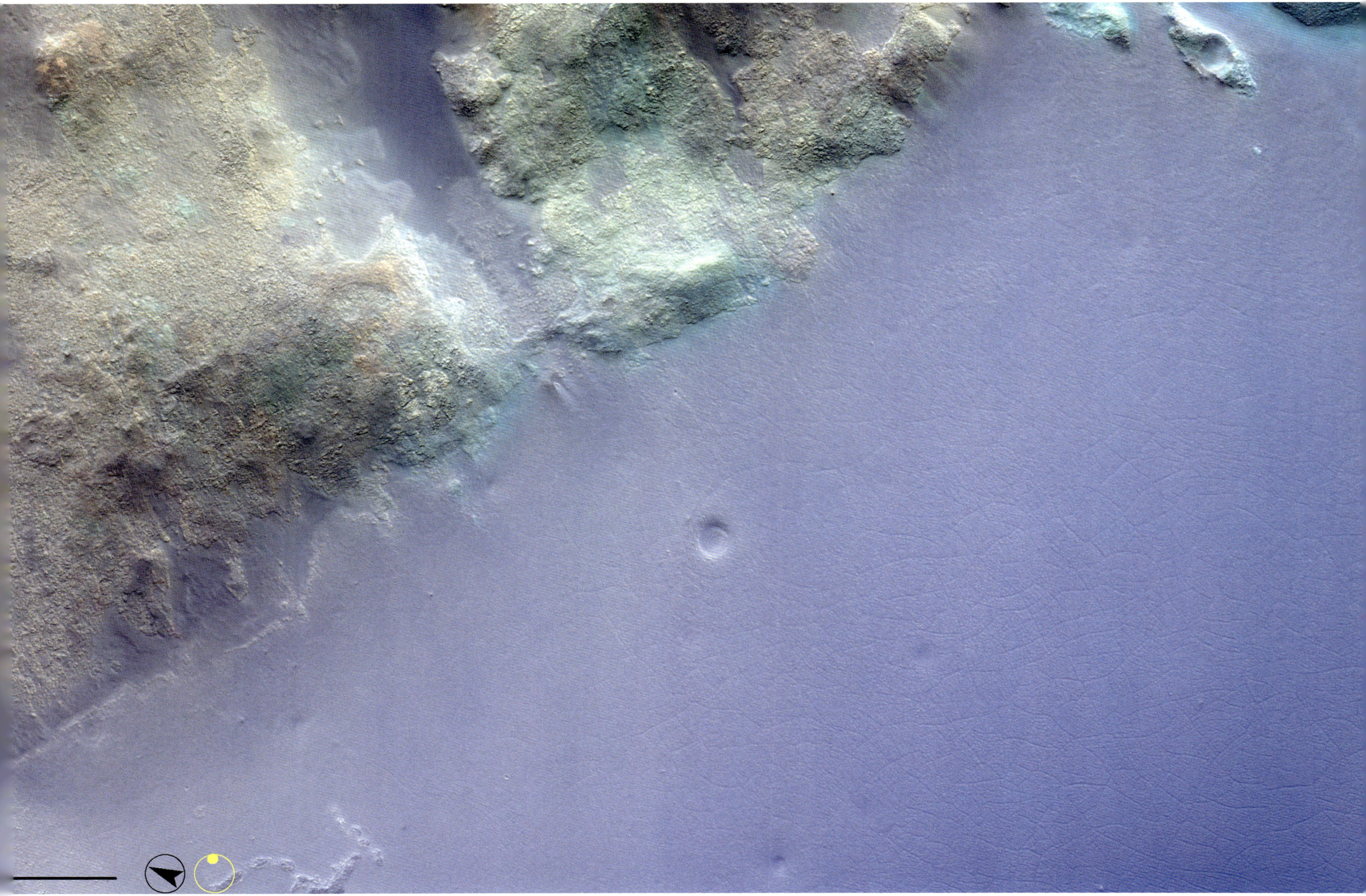

EINE VIELFARBIGE OBERFLÄCHE IM SAVICH-KRATER

Dies ist ein Bild des nordöstlichen Teils des Bodens des Savich-Kraters, eines Kraters mit 179 km Durchmesser in der südöstlichen Ecke von Tyrrhena Terra. Es gibt zahlreiche Fächer und Kanäle auf dem Kraterboden und durch den nordöstlichen Rand auch in den Krater hinein. Dieses Bild zeigt auch, dass es erhebliche Farbschwankungen gibt, die auf eine unterschiedliche Zusammensetzung hindeuten. Beachten Sie auch den Einschlagkrater in der Mitte des Bilds und das ihn umgebende, bläuliche (möglicherweise auf feines Material hinweisende) Auswurfmaterial.

A MULTI-COLOURED SURFACE IN SAVICH CRATER

This is an image of the north-east part of the floor of Savich Crater, a 179 km diameter crater in the south-east corner of Tyrrhena Terra. There are numerous fans and channels both on the crater floor and leading into the crater through the north-east rim. This image also shows that there is substantial colour variability suggesting variable composition. Note also the impact crater in the centre of the frame and the surrounding blue-ish (possibly indicating fine material) ejecta.

UNE SURFACE MULTICOLORE DANS LE CRATÈRE SAVICH

Image de la partie nord-est du plancher du cratère Savich, un cratère de 179 km de diamètre dans la partie sud-est de Tyrrhena Terra. De nombreux cônes de déjection et chenaux sont présents à la fois sur le plancher du cratère et menant au cratère par son bord nord-est. Cette image montre également une importante diversité de couleur suggérant une composition variable. Notez également le cratère d'impact au centre de l'image et les éjectas bleuâtres qui l'entourent (indiquant peut-être un matériau plus fin).

UNA SUPERFICIE MULTICOLORE NEL CRATERE SAVICH

Questa è un'immagine della parte nord-est del pavimento del Savich Crater, un cratere di 179 km di diametro nell'angolo sud-est di Tyrrhena Terra. Ci sono numerosi delta e canali sul fondo del cratere che si estendono nel cratere passando attraverso il bordo nord-est. Questa immagine mostra anche che c'è una sostanziale variabilità di colore che suggerisce una composizione variabile. Si noti anche il cratere d'impatto al centro dell'inquadratura e l'ejecta circostante di colore blu-azzurro (che potrebbe indicare polveri fini).

РАЗНОЦВЕТНАЯ ПОВЕРХНОСТЬ В КРАТЕРЕ САВИЧ

Это изображение северо-восточной части дна 179-километрового кратера Савич, расположенного на юго-востоке Тирренской Земли. На дне кратера имеются многочисленные каналы и аллювиальные вееры, часть из них, как это видно на снимке, пересекает вал кратера. Значительная вариация цветов на изображении указывает на разнообразный состав пород. Обратите внимание на небольшой ударный кратер в центре кадра и окружающее его светло-голубое кольцо, возможно, связанное с выбросом тонкого (мелко-зернистого) материала.

10.24   A MULTI-COLOURED SURFACE IN SAVICH CRATER

ACQUISITION TIME: 2018-10-25T03:39:10.697
LONGITUDE: 96.9 E   LATITUDE: 26.5 S
REGION: TYRRHENA

KRATER UND HELL GETÖNTE AUFSCHLÜSSE IN OXIA PLANUM

Dieses Bild stammt vom südöstlichen Ende der 200 km breiten, tonhaltigen Ebene Oxia Planum. Diese Region wurde als Landeplatz für den ExoMars-Rover der Europäischen Weltraumorganisation ausgewählt. Das Bild zeigt helle Aufschlüsse von Material, das in Analogie zu anderen CaSSIS-Bildern wahrscheinlich tonhaltig ist. Der kleine Einschlagkrater zeigt helles Material, das seine Wände umgibt, sowie helles Auswurfmaterial. Die Oberfläche ist morphologisch und in ihrer Zusammensetzung heterogen.

CRATER AND LIGHT-TONED EXPOSURES IN OXIA PLANUM

This image is from the southern-eastern end of the 200 km-wide clay-bearing plain, Oxia Planum. This region has been selected as the landing site of the European Space Agency's ExoMars rover. The image shows bright outcrops of material that, by analogy with other CaSSIS images, are likely to be clay-rich. The small impact crater shows a bright materials surrounding its walls and bright ejecta. The surface is morphologically and compositional heterogeneous.

CRATÈRE ET AFFLEUREMENTS CLAIRS DANS OXIA PLANUM

Cette image est prise à l'extrémité sud-est de la plaine argileuse Oxia Planum, de 200 km de large. Cette région a été choisie comme site d'atterrissage du rover ExoMars de l'Agence Spatiale Européenne. L'image montre des affleurements clairs d'unités géologiques qui, par analogie avec d'autres images CaSSIS, sont probablement riches en argile. Le petit cratère d'impact montre des matériaux clairs entourant ses parois et des éjectas clairs. La morphologie ainsi que la composition de la surface sont hétérogènes.

CRATERE E AFFIORAMENTI CHIARI IN OXIA PLANUM

Questa immagine proviene dall'estremità sud-orientale della pianura argillosa larga 200 km, Oxia Planum. Questa regione è stata scelta come sito di atterraggio del rover ExoMars dell'Agenzia Spaziale Europea. L'immagine mostra affioramenti di materiale chiaro che, per analogia con altre immagini di CaSSIS, sono probabilmente ricchi di argilla. Il piccolo cratere da impatto mostra un affioramento luminoso sulle sue pareti ed un ejecta chiaro. La superficie è morfologicamente e compositivamente eterogenea.

КРАТЕР И ОБНАЖЕНИЯ СВЕТЛЫХ ТОНОВ НА ПЛАТО ОКСИЯ

Это снимок юго-восточного края плато Оксия – глинистой равнины шириной 200 км. Данный регион был выбран в качестве места для посадки марсохода ExoMars Европейского космического агентства. На снимке видны яркие светлые пятна, которые, по аналогии с другими снимками CaSSIS, вероятно, фиксируют глинистые породы. Такие породы можно видеть также в стене и в выбросах небольшого ударного кратера. Поверхность неоднородна и по морфологическим особенностям, и по составу.

10.25 CRATER AND LIGHT-TONED EXPOSURES IN OXIA PLANUM

ACQUISITION TIME: 2020-12-09T01:55:27.819
LONGITUDE: 335.8 E   LATITUDE: 18.3 N
REGION: ARABIA

FARBENFROHES MASSIV IN DEN AUSONIA MONTES

Dieses Bild zeigt einen Ausschnitt eines Massivs (ein von tektonischen Prozessen wie Verwerfungen betroffenes Gebiet), das zu den Ausonia Montes im Westen von Hesperia Planum gehört. Die mineralogische Vielfalt, die sich hier in den Farbvariationen zeigt, ist sehr auffällig. Die nahe gelegenen fluvialen Landformen wurden wegen ihrer potenziellen biologischen Bedeutung erforscht, aber das Massiv wurde bisher nicht so detailliert untersucht. Infrarotspektrometerdaten zeigen starke Charakteristiken mafischer Mineralien in der Nähe.

COLOURFUL MASSIF IN AUSONIA MONTES

This is an image of a section of a massif (area affected by tectonic processes such as faults), which is part of Ausonia Montes in west Hesperia Planum. The mineralogical diversity evident here in the colour variation is quite striking. The nearby fluvial landforms have been the subject of research because of their potential biological significance but the massif has not been studied in such detail. Infrared spectrometer data show a strong signature of mafic minerals nearby.

MASSIF COLORÉ D'AUSONIA MONTES

Image d'une section d'un massif (zone affectée par des processus tectoniques tels que des failles), partie d'Ausonia Montes dans l'ouest d'Hesperia Planum. La diversité minéralogique, évidente ici dans la variabilité des couleurs, est assez frappante. Les reliefs fluviaux voisins ont fait l'objet de recherches en raison de leur importance biologique potentielle, mais le massif n'a pas été étudié avec le même niveau de détail. Les données des spectromètres infrarouge montrent de forte signatures de minéraux mafiques à proximité.

MASSICCIO COLORATO IN AUSONIA MONTES

Questa è un'immagine di una sezione di un massiccio (area interessata da processi tettonici come le faglie), che fa parte di Ausonia Montes nell'Hesperia Planum occidentale. La diversità mineralogica evidenziata dalle variazioni di colore è piuttosto sorprendente. Le vicine conformazioni fluviali sono state oggetto di ricerca a causa del loro potenziale significato biologico, ma il massiccio non è ancora stato studiato così in dettaglio. I dati dello spettrometro infrarosso mostrano una forte firma di minerali mafici nelle vicinanze.

КРАСОЧНЫЙ МАССИВ В ГОРАХ АВСОНИИ

Это изображение участка горного массива, в рельефе которого проявлены характерные признаки разломной тектоники. Массив является частью гор Авсонии на западе плато Гесперид. Минералогическое разнообразие, проявляющееся здесь в вариации цвета, поразительно. Близлежащие флювиальные формы рельефа были предметом исследования из-за их потенциальной биологической значимости, но сам массив пока не был детально изучен. Данные инфракрасной спектрометрии указывают на наличие в этом районе мафических минералов.

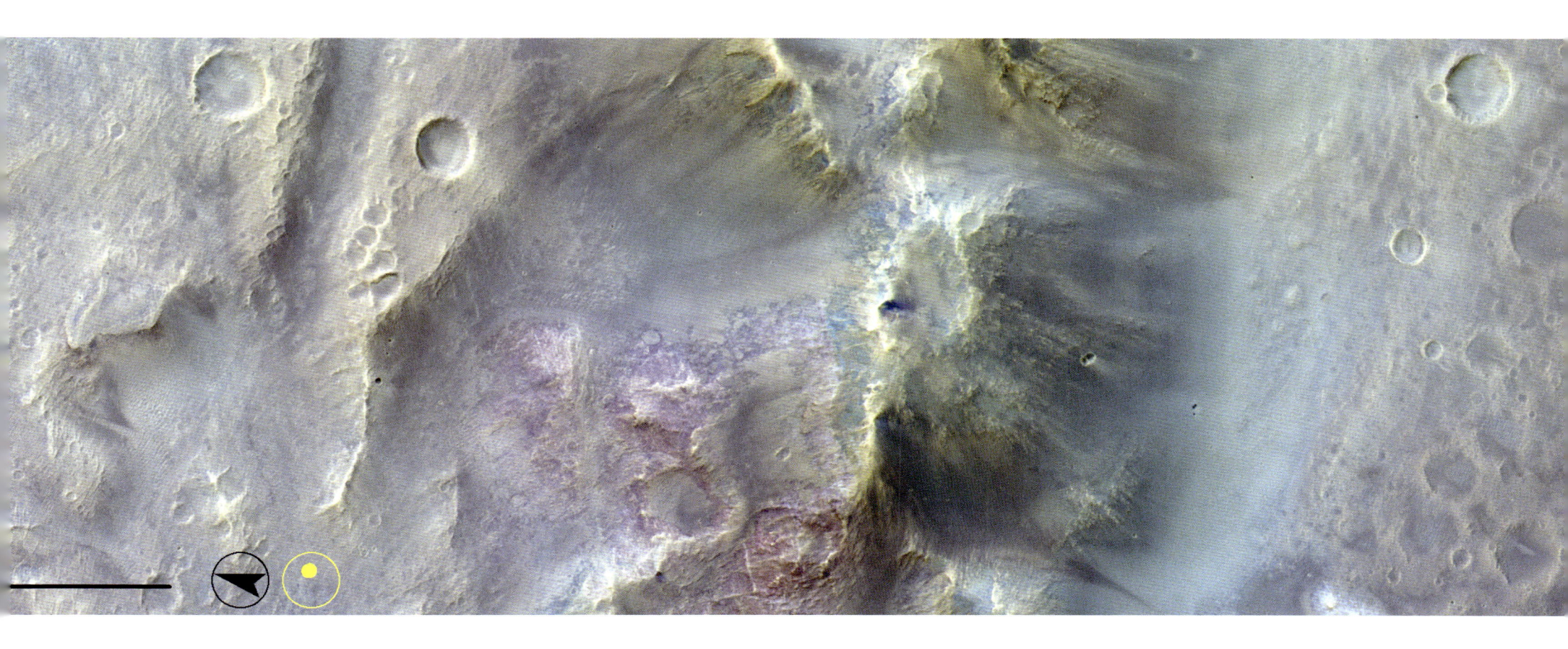

10.26 COLOURFUL MASSIF IN AUSONIA MONTES

ACQUISITION TIME: 2020-12-15T21:16:12.172
LONGITUDE: 101.3 E   LATITUDE: 24.1 S
REGION: HESPERIA

GEMUSTERTER BODEN IN ERYTHRAEUM CHAOS

Dieses synthetische RGB-Bild zeigt ein Gebiet in Erythraeum Chaos. Diese Region befindet sich am äussersten südöstlichen Rand von Margaritifer Terra und war wahrscheinlich eine Quelle für das Paraná-Valles-Talnetz im Osten und möglicherweise auch für das Loire-Valles-System im Nordwesten. Dieses Bild zeigt einen durchgängig gemusterten Boden in positivem Relief (auch bekannt als invertiertes Terrain).

PATTERNED GROUND IN ERYTHRAEUM CHAOS

This synthetic RGB image is of an area within Erythraeum Chaos. This region is on the extreme south-east edge of Margaritifer Terra and was probably a source for the Paraná Valles valley network to the east and possibly the Loire Valles system to the north-west. This image shows a pervasive patterned ground in positive relief (also known as inverted terrain).

TERRAINS TEXTURÉS DANS ERYTHRAEUM CHAOS

Cette image représente une zone de Erythraeum Chaos, au sud-est de Margaritifer Terra. Ce terrain chaotique était probablement une source pour le réseau de vallées du Paraná à l'est et peut-être pour le système des vallées de la Loire au nord-ouest. Cette image montre un sol à motifs omniprésents en relief positif (également connu sous le nom de terrain inversé).

TERRENO MODELLATO IN ERYTHRAEUM CHAOS

Questa immagine composta RGB è di un'area all'interno di Erythraeum Chaos. Questa regione si trova all'estremo margine sud-est di Margaritifer Terra ed è probabilmente l'origine della rete di valli Paraná Valles a est e del sistema di Loire Valles a nord-ovest. Questa immagine mostra un diffuso terreno modellato in rilievo positivo (noto anche come terreno invertito).

УЗОРЧАТЫЙ ГРУНТ В ЭРИТРЕЙСКОМ ХАОСЕ

Это синтезированное RGB изображение, показывает область, находящуюся в пределах Эритрейского хаоса на юго-восточном краю Жемчужной земли. Считается, что здесь располагался источник, обеспечивший формирование сети долин Параны на востоке и, возможно, долин Луары на северо-западе. На участках перевёрнутого рельефа можно заметить полигональные структуры.

10.27 PATTERNED GROUND IN ERYTHRAEUM CHAOS

ACQUISITION TIME: 2018-06-04T10:57:40.000
LONGITUDE: 347.8 E   LATITUDE: 21.7 S
REGION: MARGARITIFER

HELL GETÖNTE AUFSCHLÜSSE AUF EINEM KRATERBODEN BEI NIRGAL VALLIS

Dieses faszinierende Bild stammt vom Boden eines unbenannten Kraters von 43 km Durchmesser in Noachis Terra, nördlich von Argyre. Die Quellregion von Nirgal Vallis liegt etwa 180 km weiter nördlich (links). Links unten im Bild sind Dünen zu sehen, die sich in der dunkelblauen Farbe zeigen, die typisch ist für gestreckte CaSSIS-Bilder mit den Filtern PAN, BLU und NIR. Das dunklere, verfestigte Material in der Mitte zeigt Krater, was auf ein höheres Alter hinweist. Dieses Material wurde wahrscheinlich erodiert, wodurch das heller gefärbte Material zum Vorschein kommt. Die hellblauen Farben sind zerklüftet und könnten Megabrekzien sein, die von der zentralen Hebung des ursprünglichen Einschlags stammen, der den Krater gebildet hat.

LIGHT-TONED EXPOSURES ON A CRATER FLOOR NEAR NIRGAL VALLIS

This fascinating image is from the floor of an unnamed 43 km diameter crater in Noachis Terra, north of Argyre. The source region of Nirgal Vallis is around 180 km further to the north (left). There are dunes to the bottom left of the image that show up in the dark blue colour that is typical for CaSSIS image stretches using the PAN, BLU and NIR filters. The darker consolidated material in the centre is cratered indicating an older age. This material has probably been eroded and this reveals the lighter-toned material. The light blue colours are fractured and may be megabreccia from the central uplift of the original impact that formed the crater.

AFFLEUREMENTS DE COULEUR CLAIRE SUR LE PLANCHER D'UN CRATÈRE PRÈS DE NIRGAL VALLIS

Cette image fascinante provient du fond d'un cratère sans nom de 43 km de diamètre dans Noachis Terra, au nord d'Argyre. La région source de Nirgal Vallis est à environ 180 km plus au nord (à gauche). Des dunes apparaissent en bas à gauche de l'image avec leur couleur bleue foncée typique dans les images couleur CaSSIS. Le matériau consolidé plus sombre au centre est cratérisé, ce qui indique un âge plus avancé. Ce matériau a probablement été érodé, ce qui révèle la formation de couleur plus claire. Les roches de couleur bleue claire sont fracturées et pourraient être des mégabreccia provenant du soulèvement central de l'impact originel qui a formé le cratère.

AFFIORAMENTI CHIARI NEL PAVIMENTO DI UN CRATERE VICINO NIRGAL VALLIS

Questa affascinante immagine proviene dal pavimento di un cratere senza nome di 43 km di diametro in Noachis Terra, a nord di Argyre. La regione sorgente di Nirgal Vallis è circa 180 km più a nord (a sinistra). Ci sono delle dune in basso a sinistra dell'immagine evidenziate dal tipico colore blu scuro delle immagini di CaSSIS utilizzando i filtri PAN, BLU e NIR. Il materiale consolidato più scuro al centro è craterizzato e indica un'età più antica. Questo materiale è stato probabilmente eroso e rivela lo strato sub-superficiale di colore più chiaro. I colori azzurri appartengono a materiale fratturato e può trattarsi di megabreccia originata dal sollevamento centrale dell'impatto che ha formato in origine il cratere.

ОБНАЖЕНИЯ СВЕТЛЫХ ТОНОВ НА ДНЕ КРАТЕРА РЯДОМ С ДОЛИНОЙ НЕРГАЛ

Дно кратера диаметром 43 км на земле Ноя к северу от равнины Аргир. Примерно в 180 км к северу находится область вероятного расположения источников потоков, формировавших долину Нергал. В левой нижней части изображения видны дюны темно-синего цвета, характерного для фильтров CaSSIS. Более темная консолидированная поверхность в центре покрыта кратерами, что указывает на ее более древний возраст. Частичная эрозия этой поверхности обнажает материал светлых тонов. Разрозненные участки светло-голубого цвета могут быть остатками мегабрекчий центрального поднятия, созданного импактным событием, сформировавшим кратер.

10.28 LIGHT-TONED EXPOSURES ON A CRATER FLOOR NEAR NIRGAL VALLIS

ACQUISITION TIME: 2020-09-10T14:19:48.200
LONGITUDE: 313.2 E   LATITUDE: 30.0 S
REGION: NOACHIS

AUSGERICHTETE GRATE IM INNEREN DES TORUP-KRATERS

Dieses Bild stammt vom südöstlichen Rand von Tyrrhena Terra. Der südöstliche Rand des Savich-Kraters wurde von einem weiteren Einschlagkörper getroffen, der den Torup-Krater mit einem Durchmesser von 43 km erzeugt hat. Dieses Bild zeigt einen Ausschnitt des Bodens von Torup und zeigt ausgerichtetes, zerklüftetes, hell getöntes Material. Die drei braunen glatten Flecken auf der rechten Seite sind Staubrippel in Vertiefungen. Ein kleiner Kanal im Rand des Torup-Kraters deutet darauf hin, dass er in der Vergangenheit durchbrochen wurde und diese ungewöhnliche Struktur somit erodierte Sedimente darstellen könnte.

ALIGNED RIDGES IN THE INTERIOR OF TORUP CRATER

This image is from the south-east edge of Tyrrhena Terra. The south-east rim of Savich Crater has been hit by another impactor that has produced the 43 km diameter Torup Crater. This is an image of part of the floor of Torup and shows aligned fractured light-toned material. The three brown smooth patches to the right are dust ripples in depressions. There is a small channel in Torup Crater's rim indicating that it has been breached in the past and hence this unusual structure might represent eroded sediments.

CRÊTES ALIGNÉES À L'INTÉRIEUR DU CRATÈRE TORUP

Image de la marge sud-est de Tyrrhena Terra. La bordure sud-est du cratère Savich a été frappée par un autre impacteur qui a produit le cratère Torup de 43 km de diamètre. Cette partie du plancher du cratère montre une surface striée, fracturée et de couleur claire. Les trois unités marrons et lisses à droite sont des accumulations de poussière dans des dépressions. Un petit chenal dans la bordure du cratère Torup indique qu'il a été transpercé dans le passé et que la structure inhabituelle vue dans cette image pourrait être faite de sédiments érodés.

CRESTE ALLINEATE ALL'INTERNO DEL CRATERE TORUP

Questa immagine proviene dal bordo sud-est di Tyrrhena Terra. Il bordo sud-est del cratere Savich ha subito un altro impatto che ha prodotto il cratere Torup di 43 km di diametro. Questa è un'immagine di una parte del pavimento di Torup e mostra materiale chiaro fratturato linearmente. Le tre macchie marroni lisce sulla destra sono distese ondulate di polvere nelle depressioni. C'è un piccolo canale nel bordo del cratere Torup che indica che è stato scavato in passato da qualche liquido riversatosi nel cratere stesso, e quindi questa struttura insolita potrebbe rappresentare sedimenti erosi.

ВЫРОВНЕННЫЕ ХРЕБТЫ ВО ВНУТРЕННЕЙ ЧАСТИ КРАТЕРА ТОРУП

Юго-восточная часть вала кратера Савич, что на самом юго-востоке Тирренской земли, подверглась удару другого метеорита, и в результате образовался кратер Торуп диаметром 43 км. Это изображение части дна Торупа, на котором виден выровненный трещиноватый материал светлых тонов. Три гладких коричневых пятна справа — это впадины, заполненные пылью. Вал кратера Торуп прорезается небольшим каналом, что указывает на то, что в прошлом кратер представлял собой водосборный бассейн. Следовательно, видимые на снимке породы с необычной морфологией поверхности могут представлять собой эродированные осадочные отложения.

10.29   ALIGNED RIDGES IN THE INTERIOR OF TORUP CRATER

ACQUISITION TIME: 2021-05-09T22:10:56.379
LONGITUDE: 97.5 E   LATITUDE: 28.2 S
REGION: TYRRHENA

DER BODEN DES KASHIRA-KRATERS

Dies ist ein Bild des westlichen Teils des Bodens des Kashira-Kraters mit einem Durchmesser von 66 km. Dieser Krater war einst ein möglicher Landeplatz für den Mars Rover 2020. Die hellen Erhebungen auf dem Kraterboden zeigen die Signatur von Aluminiumphyllosilikaten oder hydroxyliertem Siliziumdioxid in Infrarotspektrometermessungen (CRISM). Es sind Felsgrat-Netzwerke zu erkennen. Dabei könnte es sich um beständigere Brüche handeln, die z. B. durch die Zementierung durch Flüssigkeiten entstanden sind.

THE FLOOR OF KASHIRA CRATER

This is an image of the western part of the floor of the 66 km diameter Kashira Crater. This crater was once a candidate landing site for the Mars 2020 rover. Light-toned mounds on the crater floor exhibit the signature of aluminium phyllosilicates or hydroxylated silica in infrared spectrometer data (CRISM). Ridge networks can be seen. These could be more resistant fractures arising from cementation by fluids, for example.

LE PLANCHER DU CRATÈRE KASHIRA

Cette image montre la partie ouest du plancher du cratère Kashira, d'un diamètre de 66 km. Ce cratère était un site d'atterrissage potentiel pour le rover de la mission Mars 2020. Les monticules de couleur claire sur le plancher du cratère présentent la signature de phyllosilicates d'aluminium ou de silice hydroxylée dans les données du spectromètre infrarouge CRISM. Des réseaux de crêtes sont visibles. Il pourrait s'agir d'anciennes fractures rendues plus résistantes à l'érosion par une cimentation résultant d'une circulation de fluides.

IL PAVIMENTO DEL CRATERE KASHIRA

Questa è un'immagine della parte occidentale del pavimento del cratere Kashira di 66 km di diametro. Questo cratere fu un candidato per il sito di atterraggio per il rover Mars 2020. I cumuli di colore chiaro sul pavimento del cratere sembrano essere fillosilicati di alluminio o silice idrossilata, identificabili nei dati dello spettrometro infrarosso (CRISM). Si possono vedere fitte reti di creste che potrebbero derivare dalla cementazione di materiale resistente all'erosione trasportato da liquidi all'interno di fratture preesistenti.

ДНО КРАТЕРА КАШИРА

Это снимок западной части дна 66-километрового кратера Кашира, который когда-то был кандидатом на место посадки марсохода «Марс-2020». Светлые возвышенности на дне кратера, по данным инфракрасного спектрометра CRISM, демонстрируют признаки филлосиликатов алюминия и гидроксилированного кремнезема. Видна сеть гряд, которые могут представлять собой бывшие трещины, сцементированные эрозионно-устойчивыми застывшими флюидами, например.

10.30 THE FLOOR OF KASHIRA CRATER

ACQUISITION TIME: 2018-10-25T11:30:42.711
LONGITUDE: 341.2 E   LATITUDE: 26.9 S
REGION: NOACHIS

ZERKLÜFTETES GELÄNDE BEI OKAVANGO VALLIS

Dies ist ein Bild des Bodens eines unbenannten Kraters in Arabia Terra. Der Krater befindet sich am nordöstlichen Ende der Okavango Valles. Der Kraterboden zeigt dieses bemerkenswerte, konzentrisch zerklüftete Terrain. Der Ursprung dieser Morphologie ist nicht geklärt, aber das Material könnte eisreich sein, und es könnten Analogien zu Séracs (Gletschereisblöcken) in Betracht gezogen werden, da mehrere strömungsähnliche Strukturen in der Nähe als mit Schutt bedeckte Eisströme gedeutet wurden.

FRACTURED TERRAIN NEAR OKAVANGO VALLIS

This is an image of the floor of an unnamed crater in Arabia Terra. The crater is at the northeastern end of Okavango Valles. The crater floor shows this remarkable concentric fractured terrain. The origin of this morphology has not been established but the material might be ice-rich and analogies with seracs (blocks of glacial ice) could be considered given that several flow-like structures nearby have been interpreted as debris-covered ice flows.

TERRAIN FRACTURÉ PRÈS D'OKAVANGO VALLIS

Image du sol d'un cratère sans nom dans la région d'Arabia Terra. Le cratère se trouve à l'extrémité nord-est d'Okavango Valles. Le plancher du cratère présente ce remarquable terrain fracturé concentrique. L'origine de cette morphologie n'a pas été établie mais le matériau pourrait être riche en glace et des analogies avec les séracs (blocs de glace fracturés) pourraient être envisagées étant donné que plusieurs structures semblables à proximité ont été interprétées comme des écoulements de glace couverts de débris rocheux.

TERRENO FRATTURATO VICINO OKAVANO VALLIS

Questa è un'immagine del pavimento di un cratere senza nome in Arabia Terra. Il cratere si trova all'estremità nord-orientale delle Okavango Valles e il suo pavimento mostra questo notevole terreno fratturato concentricamente. L'origine di questa morfologia non è stata stabilita, ma il materiale superficiale potrebbe essere ricco di ghiaccio e si potrebbero considerare una formazione analoga ai seracchi (blocchi di ghiaccio glaciale), dato che diverse strutture morfologiche nelle vicinanze sono state interpretate come colate di ghiaccio coperte da detriti.

ПОКРЫТАЯ ТРЕЩИНАМИ МЕСТНОСТЬ В РАЙОНЕ ДОЛИН ОКАВАНГО

Это снимок дна безымянного кратера в земле Аравия, который находится в северо-восточной части долин Окаванго. В видимых на дне кратера породах, разбитых впечатляющими концентрическими трещинами, может присутствовать лёд. Кроме того, несколько структур поблизости были интерпретированы как ледяные потоки, покрытые обломками породы. Таким образом, можно предположить, что снимок показывает радиальное сползание вещества к центру кратера, формирующее аналог земных сераков (частично отколовшихся фрагментов ледника).

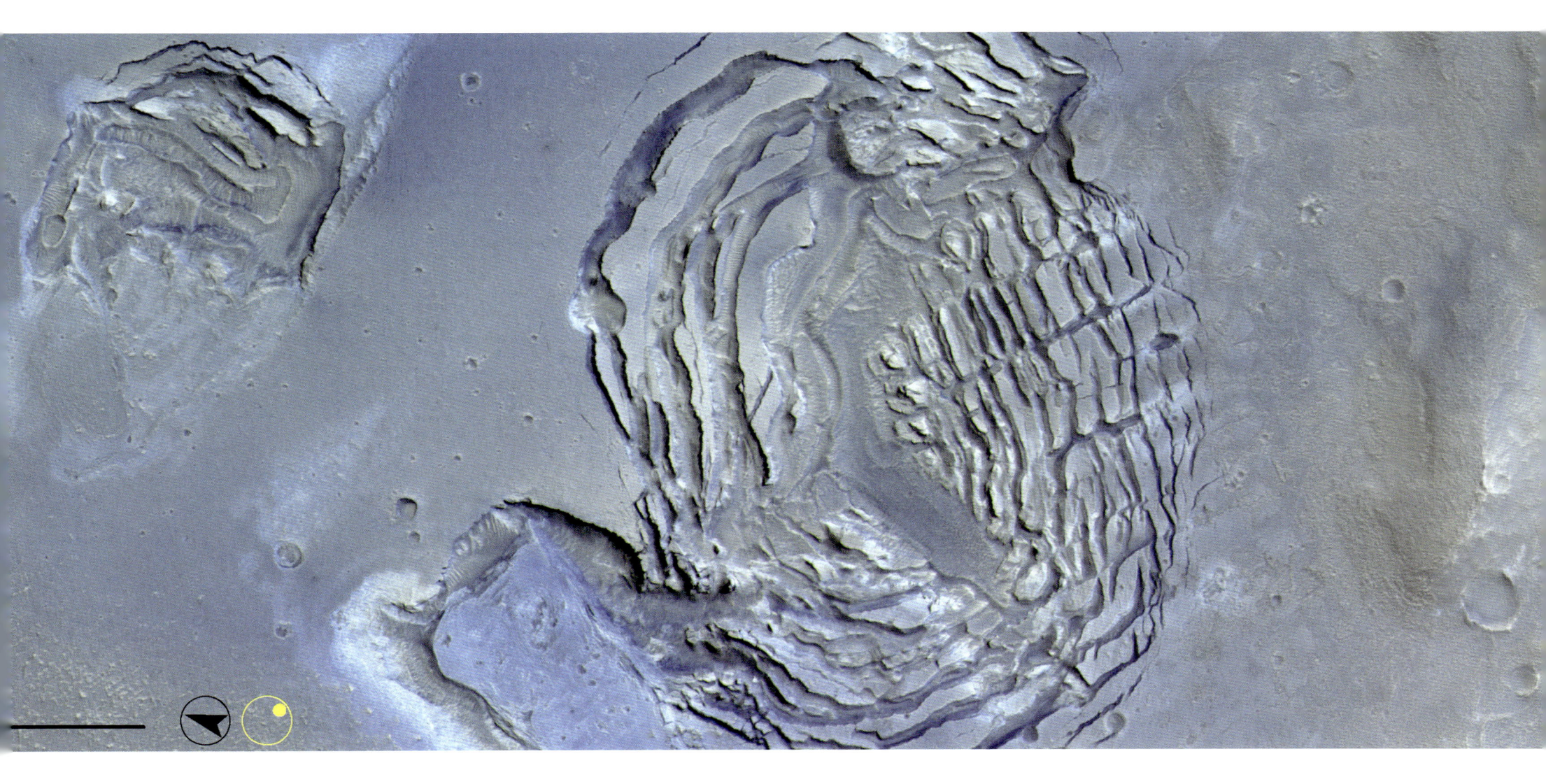

10.31 FRACTURED TERRAIN NEAR OKAVANGO VALLIS

ACQUISITION TIME: 2021-04-06T01:32:32.472
LONGITUDE: 6.6 E   LATITUDE: 39.6 N
REGION: ARABIA

VIELFARBIGES TERRAIN IN TERRA CIMMERIA

Dies ist ein Bild von Terra Cimmeria unmittelbar nördlich des Herschel-Kraters. Es zeigt mehrere interessante Merkmale. Die violette Farbe (unten in der Mitte) wird in CaSSIS-Daten häufig mit Chloridsalzablagerungen in Verbindung gebracht. Bei den kleinen Kanälen in der Nähe könnte es sich um alte zerfallene Flusskanäle handeln, die durch Wasser entstanden sind. Das braune Material auf der linken Seite ist unbekannten Ursprungs, aber ungewöhnlich, da es so isoliert ist. In der Mitte des Kraters ganz rechts befindet sich ein Dünenfeld.

MULTI-COLOURED TERRAIN IN TERRA CIMMERIA

This is an image of Terra Cimmeria just north of Herschel Crater. It shows several interesting features. The purple colour (bottom centre) is often associated with chloride salt deposits in CaSSIS data. The small channels nearby could be old degraded fluvial channels produced by water. The brown material to the left is of unknown origin but unusual in being so isolated. The centre of the crater to the extreme right contains a dune field.

TERRAIN MULTICOLORE DANS TERRA CIMMERIA

Image de Terra Cimmeria, juste au nord du cratère Herschel, qui présente de nombreuses structures intéressantes. La couleur violette (en bas au centre) est souvent associée à des dépôts de chlorures dans les données CaSSIS. Les petits chenaux à proximité pourraient être d'anciens chenaux fluviaux. L'unité brune à gauche de l'image est d'origine inconnue mais inhabituelle car complètement isolée. Le cratère tout à droite contient un champ de dunes en son centre.

TERRENO MULTICOLORE IN TERRA CIMMERIA

Questa è un'immagine della Terra Cimmeria appena a nord del cratere Herschel. Mostra diverse caratteristiche interessanti. Il colore viola (in basso al centro) è spesso associato a depositi di sale cloruro nei dati di CaSSIS. I piccoli canali vicini potrebbero essere vecchi canali fluviali prodotti dall'acqua liquida e degradati dal tempo. Il materiale marrone a sinistra è di origine sconosciuta, ma è piuttosto inusuale poichè molto isolato. Il centro del cratere all'estrema destra contiene un campo di dune.

РАЗНОЦВЕТНЫЙ РЕЛЬЕФ В КИММЕРИЙСКОЙ ЗЕМЛЕ

Это снимок Киммерийской Земли сразу к северу от кратера Гершель. На нем видно несколько интересных особенностей. Пурпурный цвет (внизу в центре) в данных CaSSIS часто ассоциируется с отложениями хлористых солей. Небольшие каналы рядом могут быть старыми деградировавшими руслами, образованными водой. Природа коричневого материала, видимого слева, неизвестна, но он необычен своей изолированностью. Центр небольшого кратера (справа на снимке) заполнен дюнами.

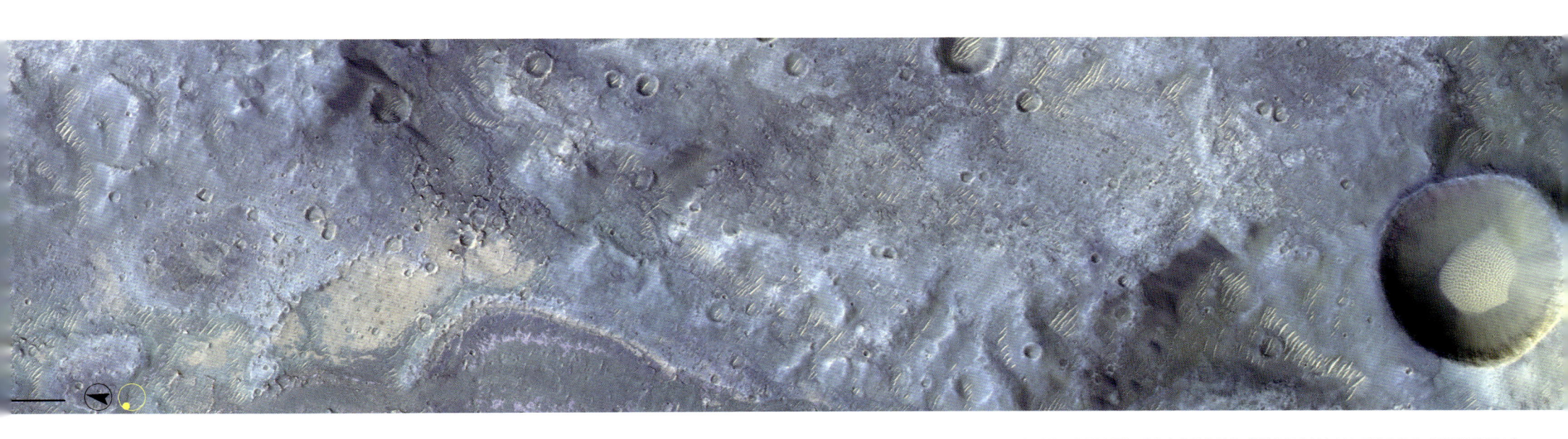

10.32 MULTI-COLOURED TERRAIN IN TERRA CIMMERIA

ACQUISITION TIME: 2021-03-17T10:12:28.933
LONGITUDE: 127.8 E   LATITUDE: 9.4 S
REGION: CIMMERIA

VERZWEIGTES NETZWERK IM ANTONIADI-BECKEN

Das Antoniadi-Becken ist eine noachische Einschlagstruktur von 330 km Durchmesser in der Region Terra Sabaea. Das Bild zeigt einen Teil des Beckenbodens, der ein bemerkenswertes, verzweigtes Felsgrat-Netzwerk aufweist. Über den Ursprung dieser Strukturen wurde viel diskutiert. Der Konsens tendiert zur Idee, dass diese Strukturen das Ergebnis zähflüssiger, wahrscheinlich vulkanischer Ströme sind.

BRANCHED NETWORK IN ANTONIADI BASIN

Antoniadi Basin is a 330 km diameter Noachian impact structure in the Terra Sabaea region. The image shows part of the floor of the basin which has a remarkable branched ridge network. There has been considerable debate as to the origin of these structures. The consensus is moving towards the idea that they are the result of viscous flows, probably volcanic.

RÉSEAU RAMIFIÉ DANS LE BASSIN D'ANTONIADI

Le bassin d'Antoniadi est une structure d'impact noachienne de 330 km de diamètre dans la région de Terra Sabaea. L'image montre une partie du fond du bassin qui présente un remarquable réseau de crêtes ramifiées. L'origine de ces structures a fait l'objet de nombreux débats. Le consensus s'oriente vers l'idée qu'elles sont le résultat d'écoulements visqueux, probablement volcaniques.

RETE DI RAMIFICAZIONI NEL BACINO ANTONIADI

Il bacino di Antoniadi è una struttura da impatto noachiana di 330 km di diametro nella regione della Terra Sabaea. L'immagine mostra parte del pavimento del bacino, che presenta una notevole rete di creste ramificate. C'è stato un considerevole dibattito sull'origine di queste strutture e sta prendendo sempre più piede l'ipotesi che siano il risultato di colate di fluidi viscosi, probabilmente vulcanici.

РАЗВЕТВЛЕННАЯ СЕТЬ ХРЕБТОВ В БАССЕЙНЕ АНТОНИАДИ

Бассейн Антониади – это ударная структура нойского периода диаметром 330 км в Сабейской земле. На снимке показана часть дна котловины, которая имеет замечательную разветвленную сеть хребтов. Происхождение этих структур вызывает много споров, но постепенно консенсус складывается в пользу того, что они являются следами вязких потоков, вероятно, вулканического происхождения.

10.33 BRANCHED NETWORK IN ANTONIADI BASIN

ACQUISITION TIME: 2021-04-02T21:00:23.306
LONGITUDE: 61.2 E   LATITUDE: 21.8 N
REGION: SABAEA

DIE GEOLOGISCHEN PERIODEN DES MARS

Geologen haben Zeitalter in der Erdgeschichte definiert, und die Namen einiger dieser Zeitalter sind inzwischen allgemein bekannt. Dank Hollywood ist die Jurazeit wahrscheinlich die bekannteste. Die historischen Zeitalter, die den Zeitraum von über 2,5 Milliarden Jahren vor heute abdecken, werden Archaikum und Hadäikum genannt. Es wurde auch versucht, Zeitalter in der Geschichte des Mars zu definieren. Diese sind in der Tabelle aufgeführt, und es wird sofort deutlich, dass sie im Vergleich zur Erde viel weniger detailliert sind. Die Zeitpunkte der Übergänge zwischen diesen Zeitaltern sind mit erheblichen Unsicherheiten behaftet. Dennoch werden wir in unseren Bildbeschreibungen gelegentlich auf diese Zeitalter Bezug nehmen. Es sei darauf hingewiesen, dass es Versuche gegeben hat, eine andere Nomenklatur zu definieren, die jedoch noch nicht allgemein anerkannt sind.

NOACHISCHES ZEITALTER

Früheste Periode der Marsgeschichte, die auf 3,7 bis 4,1 Milliarden Jahre geschätzt wird, als die Einschlagkraterrate sehr hoch war.

HESPERIANISCHES ZEITALTER

Zeitraum in der Marsgeschichte vor etwa 3,0 bis 3,7 Milliarden Jahren, als vulkanische Aktivitäten und katastrophale Überflutungen die Oberfläche beeinflussten.

AMAZONISCHES ZEITALTER

Die jüngste geologische Periode von vor 3,0 Milliarden Jahren bis heute, als die Bedingungen auf der Marsoberfläche trocken und kalt wurden.

THE GEOLOGICAL PERIODS OF MARS

Geologists have defined periods in Earth's history and the names of some of these periods have become common knowledge. Thanks to Hollywood, the Jurassic period is probably the best known. The ancient periods covering the time beyond 2.5 billion years are called the Archean and the Hadean. There has also been an attempt to define periods in Martian history. These are listed in the table and it will be immediately obvious that there is much less detail when compared to Earth. The times of the changes between these periods have significant uncertainty. Nonetheless, we shall refer to these periods occasionally in our descriptions of the images. It should be noted that there have been attempts to define another nomenclature but these have not been universally accepted at this time.

NOACHIAN

Earliest period in Mars history estimated to be 3.7–4.1 billion years ago when impact cratering rates were very high.

HESPERIAN

Period in Mars history between about 3.0–3.7 billion years ago when volcanic activity and catastrophic flooding were influencing the surface.

AMAZONIAN

The most recent geological period from 3.0 billion ago to the present when Martian surface conditions became dry and cold.

TEMPS GÉOLOGIQUES MARTIENS

Les géologues ont défini divers intervalles pour décrire l'histoire de la Terre dont certains noms sont passés dans le langage courant. Ainsi, le jurassique est certainement le nom de période le plus connu grâce à Hollywood. Les éons les plus anciens, plus vieux que 2.5 milliards d'années, sont appelés Archéen et Hadéen. On tente également de définir de tels intervalles dans l'histoire martienne. Les éons sont répertoriés dans le tableau et il est immédiatement évident qu'il y a beaucoup moins de détails que pour la Terre. Les limites temporelles entres ces éons sont également entachées d'incertitudes importantes. Néanmoins, nous ferons occasionnellement référence à ces périodes dans nos descriptions des images. Il convient de noter que d'autres nomenclatures ont parfois été proposées, mais celles-ci ne sont pas universellement acceptées à l'heure actuelle.

NOACHIEN

Période la plus ancienne de l'histoire de Mars estimée à 3,7–4,1 milliards d'années, lorsque les taux de cratérisation étaient très élevés.

HESPÉRIEN

Période de l'histoire de Mars entre environ 3,0 et 3,7 milliards d'années, lorsque l'activité volcanique et les inondations catastrophiques influençaient la surface.

AMAZONIEN

La période géologique la plus récente, débutant il y a 3,0 milliards d'années, lorsque les conditions de surface martiennes sont devenues sèches et froides.

I PERIODI GEOLOGICI DI MARTE

I geologi hanno definito dei periodi nella storia della Terra, e i nomi di alcuni di questi periodi sono diventati di uso comune. Grazie a Hollywood, il periodo giurassico è probabilmente il più conosciuto. I periodi che coprono il tempo prima di 2,5 miliardi di anni fa sono chiamati Archeano e Adeano. Sono state anche proposte alcune definizioni dei periodi storici di Marte, che sono elencati nella tabella e sarà subito ovvio al lettore che c'è molto meno dettaglio rispetto alla Terra. Il momento preciso di distinzione tra questi periodi ha un' incertezza significativa. Tuttavia, faremo occasionalmente riferimento a questi periodi nelle nostre descrizioni delle immagini. Va notato che ci sono stati tentativi di definire un'altra nomenclatura, ma questi non sono ancora universalmente accettati.

NOACHIANO

Il periodo della storia di Marte più antico, stimato tra i 3.7–4.1 miliardi di ani fa, quando la craterizzazione era molto frequente.

HESPERIANO

Il periodo della storia di Marte tra circa 3.0–3.7 miliardi di anni fa, quando l'attività vulcanica e alluvioni catastrofiche influenzavano la morfologia di superficie.

AMAZZONIANO

Il period geologico più recente, da 3.0 miliardi di anni fa al presente, quando le condizioni superficiali di Marte sono diventate asciutte e fredde.

ГЕОЛОГИЧЕСКИЕ ПЕРИОДЫ МАРСА (ГЛОССАРИЙ)

Геологи определили периоды в истории Земли, и названия некоторых из них стали общеизвестными. Благодаря Голливуду, Юрский период, вероятно, является самым известным. Древние периоды, охватывающие время до 2,5 миллиардов лет назад, называются Археем и Катархеем. Также была предпринята попытка выделить периоды в марсианской истории. Они перечислены в таблице, и сразу очевидно, что по сравнению с Землей здесь гораздо меньше деталей. Время переходов между этими периодами имеет значительную неопределенность. Тем не менее, мы будем периодически ссылаться на данные периоды в наших описаниях изображений. Следует отметить, что были попытки определить другую номенклатуру, но они не получили всеобщего признания в настоящее время.

НОЙСКИЙ (НОАХИЙСКИЙ)

Самый ранний период истории Марса, по оценкам охватывающий промежуток от 4,1 до 3,7 миллиардов лет назад, когда интенсивность образования ударных кратеров была очень высока.

ГЕСПЕРИЙСКИЙ

Период в истории Марса примерно между 3,7–3,0 миллиардами лет назад, когда на поверхность оказывали влияние вулканическая активность и катастрофические наводнения.

АМАЗОНИЙСКИЙ

Текущий геологический период с 3,0 миллиардов лет назад по настоящее время, в котором условия на марсианской поверхности стали сухими и холодными.

REGIONS OF MARS

Brief descriptions of specific regions on Mars. The longitude (east longitude is used) and latitude of the approximate centres of the regions are given. Note that some of the regions are very large. Please also refer to the map on the inside front and back cover. Check the website for translations into other languages.

REGION	SHORT DESCRIPTION	EAST LONG.	LAT.
ACIDALIA PLANITIA	Plain between Tharsis and Arabia Terra with evidence of flows although the origins of the flows are uncertain (volcanic, glacial, or fluvial).	339°	50°
AONIA TERRA	Upland region in the southern hemisphere with large-scale cratering.	263°	−60°
AMAZONIS PLANITIA	Smooth northern plain between Tharsis and Elysium containing Lycus Sulci and the 1000 km long Medusae Fossae.	197°	26°
ARABIA TERRA	Heavily cratered and eroded upland, one of the oldest terrains with many craters and canyons.	6°	21°
ARCADIA PLANITIA	Smooth lowland plain dominated by Amazonian-aged volcanic flows.	188°	49°
ARGENTEA PLANUM	High latitude plain consisting of Hesperian aged polar units.	298°	−72°
ARGYRE PLANITIA	Second deepest impact basin on Mars and approximately 1800 km wide.	317°	−50°
CHRYSE PLANITIA	Smooth circular plain in the northern equatorial region and sink for many outflow channels.	320°	28°
TERRA CIMMERIA	Ancient heavily cratered highlands in the southern hemisphere.	148°	−33°
CLARITAS FOSSAE	Highland terrain south of Tharsis, densely-dissected by many graben.	256°	−28°
DEUTERONILUS COLLES	Part of Deuteronilus Mensae located at the center of a circular area with mesas and hummocky small knobs.	22°	42°
ELYSIUM PLANITIA	Broad plain that straddles the equator showing evidence of lava flows and graben.	155°	3°
HELLAS BASIN / PLANITIA	The largest impact structure on Mars.	70°	−42°
HELLESPONTUS MONTES	Rocky peaks at the western edge of Hellas Basin.	43°	−44°
HESPERIA PLANUM	Lava plain in the southern highlands with notable impact craters, wrinkle ridges and home to the ancient volcano Tyrrhena Mons.	110°	−21°
ISIDIS PLANITIA	A plain with the third largest impact basin on Mars.	88°	14°

MALEA PLANUM	High volcanic plateau south-west of Hellas Basin in the Circum-Hellas Volcanic Province.	63°	—66°
MARGARITIFER TERRA	Heavily cratered region close to the equator characterised by dense valley networks and evidence for past lakes.	335°	—2°
MERIDIANI PLANUM	Equatorial plain hosting crystalline hematite. Landing site of the Opportunity rover.	357°	0°
NILOSYRTIS MENSAE	Area of fretted terrain along the dichotomy boundary.	68°	35°
NOACHIS TERRA	Highlands west of Hellas Basin highly cratered with evidences of fluvial processes.	355°	—50°
PARVA PLANUM	High latitude plain in Aonia Terra.	265°	—74°
PROMETHEI TERRA	Vast highlands region east of Hellas Basin made of rugged terrain with high-standing massifs and deep crater forms.	97°	—64°
PROTONILUS MENSAE	Region at the dichotomy boundary between Deuteronilus Mensae and Nilosyrtis Mensae made of fretted terrains.	49°	44°
TERRA SABAEA	Large region adjacent to Arabia Terra with irregular topography and a high density of craters.	51°	3°
SINAI PLANUM	High plateau south of Valles Marineris capped by small volcanic shields and lava-flow materials.	272°	—14°

TERRA SIRENUM	Highlands massively cratered and faulted with marginal basins and tectonic structures.	206°	—39°
SISYPHI PLANUM	Vast elevated plain at the southern circum-polar area of Noachis Terra.	6°	—70°
SOLIS PLANUM	Large lava plain near Tharsis.	270°	—26°
SYRTIS MAJOR PLANUM	Dark albedo feature in Terra Sabaea defined as a low-relief shield volcano hosting active dune fields.	67°	9°
TEMPE TERRA	Highland at the transition zone between old and young terrains (NE of Tharsis) with evidence of crustal fracturing and deformation.	289°	39°
THARSIS	Volcanic plateau containing Tharsis Montes (Arsia, Ascraeus and Pavonis Mons) and Olympus and Alba Mons on its north-western edge.	260°	8°
THAUMASIA PLANUM	Sloping volcanic plain associated with Tharsis.	295°	—22°
TYRRHENA TERRA	Typical southern terrae, massively cratered with very old volcanoes.	89°	—12°
UTOPIA PLANITIA	Broad lava plain where Viking 2 and Tianwen-1 landed.	118°	47°
VALLES MARINERIS	4000 km long system of canyons in the equatorial region of Mars.	301°	—14°
XANTHE TERRA	Large area with numerous ancient river valleys and deltas.	312°	3°

TERMS IN PLANETARY GEOLOGY

WELL-ESTABLISHED DEFINITIONS

CALDERA	A large, quasi-circular, volcanic depression created by collapse into a magma reservoir.
CATENA	A term describing a chain of craters that maybe be of any origin.
CAVUS (CAVI)	A term describing an irregular steep-sided depression often found in clusters.
CHAOS	A term describing areas of broken terrain where features such as ridges, cracks, mesas, and plains appear jumbled incoherently.
CHASMA (CHASMATA)	A descriptive term for a deep, elongated, steep-sided depression.
COLLIS (COLLES)	A term describing a smaller hill or knob or in the plural a collection thereof.
CRATER	A term referring to the result of a high velocity impact onto the surface of Mars.
DORSUM (DORSA)	A term for a ridge.
FOSSA (FOSSAE)	A descriptive term for a long, narrow depression or trough. Can be the result of a number of geological processes but most on Mars are probably graben.
GRABEN	A depressed block of the crust of a planet such as bordered by parallel faults producing escarpments on each side. Indicative of tensional forces.
HORSTS	Parallel blocks between graben.
LABYRINTHUS (LABYRINTHI)	A term for complex intersecting valleys or ridges.
MENSA (MENSAE)	A term describing a flat-topped prominence with cliff-like edges. The term mesa is sometimes used as a synonym.
MONS (MONTES)	A term describing a mountain or in the plural a chain of mountains on Mars. These are typically (but not always) larger than Tholi. The term is generic and does not specify any origin.
PATERA	An irregular or complex crater with sharp rims. On Mars these are usually of volcanic origin although the term does not specify origin.
PLANUM	A term describing a plateau or a high plain.
RUPES	A term describing one or more escarpments of significant length. Their origin is uncertain.
SCOPULUS	A term describing a lobate or irregular escarpment.
SULCUS (SULCI)	A descriptive term for an area of complex parallel or subparallel ridges and furrows on Mars.
THOLUS (THOLI)	A small dome-shaped hill or mountain. It is often associated on Mars with volcanic domes but the term is actually generic and does not imply any geological origin.
VALLIS (VALLES)	A term meaning valley. Although often associated with the flow of water, the term is more generic (e.g. rift valleys).
VASTITAS	A descriptive term for an extensive plain.

LESS WELL DEFINED TERMS

BANDED TERRAIN	Dissected smooth material found almost exclusively in the north-west part of Hellas Basin.
BARCHAN (BARCHANOID) DUNES	Crescent-shaped dunes produced by wind from a preferred direction.
CENTRAL PIT CRATER	A complex crater with a pit at its centre.
CENTRAL UPLIFT	The centres of complex craters may have a central peak that is often referred to as a central uplift.
DICHOTOMY BOUNDARY	One of the most noticeable features on Mars. It separated the northern lowlands from the southern highlands with a difference in elevation across the boundary of 6–7 km.
DIKE	A sheet of rock, either magmatic or sedimentary in origin, formed in a pre-existing fracture.
EJECTA	Material excavated and deposited on the surface after a high velocity impact.
EXTENSIONAL FAULTING	A fault caused by stretching of the crust or surface layer.
FRETTED TERRAIN	A complicated mix of mesas, buttes, and small-scale canyons found mostly at the dichotomy boundary.
GULLY	An erosional feature produced by downward flow of material (wet or dry) incising into a steep slope (on Mars this is commonly a crater wall).
INTERIOR LAYERED DEPOSIT	A set of sedimentary deposits in discrete layers found within craters or depressions.
INVERTED CHANNEL	A feature that has the appearance of a channel but is topographically inverted (i.e. the channel is above the surroundings). Often seen where sediment has solidified in the channel and the surroundings have eroded.

LAHAR	A fast, destructive mudflow.
LINEATED VALLEY FILL	Material within a valley showing aligned ridges and/or grooves possibly indicating flow.
MESA	An isolated, flat-topped, steep-sided, hill.
MEGABRECCIA	Collection of very large rock fragments often found in the central uplifts of impact craters.
OROGRAPHIC CLOUDS	Clouds produced when an air mass is forced upwards as it moves over rising terrain.
PERIGLACIAL PROCESSES/ MORPHOLOGIES	Processes (morphologies) resulting from seasonal or long timescale freeze/thaw cycles.
POLYGONAL TERRAIN	Terrain with cracks and channels that appears in irregular polygonal patterns. Thought to be associated with ice (periglacial) processes in many cases.
ROOTLESS CONES	A feature resembling a volcanic crater but with no evidence of magma. On Earth, they are formed by steam explosions when hot lava travels over water-rich surfaces.
SCALLOPS	An arcuate depression thought to be the result of periglacial processes.
SLOPE STREAKS	Evidence of downslope motion of dusts and/or sands that is active today on Mars.
WRINKLE RIDGES	Low, sinuous ridges produced when lava cools and contracts.

A more comprehensive list and description can be found in the Encyclopedia of Planetary Landforms by Hargitai and Kereszturi (https://doi.org/10.1007/978-1-4614-9213-9). Check the website for translations into other languages

ACKNOWLEDGEMENTS

I should like to thank Gabriele Cremonese for his forthright support of the CaSSIS instrument from the moment the idea of using a spare detector from the BepiColombo/SIMBIOSYS instrument came about. My thanks also to Alfred McEwen for organizing NASA support for our operations software. It has contributed enormously to the output from CaSSIS and to quality of the images in this book.

The support of the CaSSIS Science Team, our guest scientists and our students is gratefully acknowledged. They are Antoine Pommerol, Anton Ivanov, Candice Hansen, Chris Okubo, Vania da Deppo, Ernst Hauber, John Bridges, James Wray, Laszlo Keszthelyi, Livio Tornabene, Lucia Marinangeli, Marek Banaskiewicz, Matteo Massironi, Alfred McEwen, Nicolas Mangold, Ramy El Maarry, Piotr Orleanski, Randolph Kirk, Shane Byrne, Stefano Debei, Pawel Wajer, Susan Conway, Daniel Mege, Alice Lucchetti, Cristina Re, Eric Pilles, Ganna Portyankina, Jon Bapst, Klaus Gwinner, Marcello Coradini, Maurizio Pajola, Peter Grindrod, Adomas Valantinas, Camila Cesar, Peter Fawdon, Riccardo Pozzobon, Francesco Salese, Emanuele Simioni, Sarah Sutton, Thomas Roatsch, Veronica Bray, James Burley, Tatiana Drozhzhova, Stepan Tulyakov, and Patricio Becerra.

The CaSSIS Operations team continues to do a great job keeping the experiment operating and improving our radiometric and geometric pipeline processing. Members of the team include Miguel Almeida, Matthew Read, Charlotte Marriner, Chris Schaller, Rod Heyd, Guy McArthur, Audrie Fennema, Jeannie Backer, Austin Stevens, and Kristin Berry as well as specific members of the Science Team. I particularly want to thank Jason Perry for the algorithm development to maximize the colour contrast that you see in the images herein.

I also wish to thank the spacecraft and instrument engineering teams for the successful completion of the instrument and specifically Ruth Ziethe, Michael Gerber, Matthias Brändli, Giordano Bruno, Marc Erismann, Lisa Gambicorti, Thomas Gerber, Kaustav Ghose, Mario Gruber,

Pascal Gubler, Harry Mischler, Jürg Jost, Daniele Piazza, Martin Rieder, Vicky Roloff, Anthony Servonet, Werner Trottmann, T. Uthaicharoenpong, Claudio Zimmermann, Denis Vernani, Michael Johnson, Elena Pelò, Thomas Weigel, Jacques Viertl, Nico De Roux, Patrick Lochmatter, Guido Sutter, Antonio Casciello, Thomas Hausner, Iacopo Ficai Veltroni, W. Nowosielski, Tom Zawistowski, Sandor Szalai, Balint Sodor, Gabor Troznai, Don McCoy, Albert Haldemann, and Duncan Goulty.

We also gratefully acknowledge the support of the ESA operations staff in Darmstadt and Madrid. In particular, Pia Mitschsdorfer, Peter Schmitz, Mark Sweeney, Johannes Bauer, Robert Guilanya, Leo Metcalfe, David Frew, Bernhard Geiger, Marc Costa, and Tanya Lim.

My thanks as well to the TGO Science Working Team members (Hakan Svedhem, Ann-Carine Vandaele, Oleg Korablev, Igor Mitrofanov, Alexei Malakhov, Franck Montmessin, Francois Forget, and Colin Wilson) for their collegial and constructive approach to maximizing the science from TGO.

CaSSIS is a project of the University of Bern and funded through the Swiss Space Office/SBFI via the Mesure d'Accompaignment programme and ESA's PRODEX programme. The instrument hardware development was also supported by the Italian Space Agency (ASI) (ASI-INAF agreement no.I/018/12/0), INAF/Astronomical Observatory of Padova, and the Space Research Center (CBK) in Warsaw. Support from SGF (Budapest), the University of Arizona (Lunar and Planetary Lab.) and NASA are also gratefully acknowledged. Operations support from the UK Space Agency under grant ST/R003025/1 is also acknowledged.

The support from our funding agencies has been critical. I would personally like to thank Daniel Neuenschwander, Renato Krpoun, Oliver Botta, Andreas Werthmueller, Kamlesh Brocard, Urs Frei, Valerie Koller, Barbara Negri, David Parker, Sue Horne, and Luigi Colangeli.

The image descriptions were checked by Alfred McEwen, Susan Conway, James Wray, Ernst Hauber, Ramy El Maarry, Laz Kestay, Pete Grindrod, Maurizio Pajola, and Adam Valantinas but please note that any remaining errors are entirely mine! The translations were a lot of work and I thank Tina Rothenbuehler, Antoine Pommerol, Camila Cesar, Stefano Spadaccia, Nico Haslebacher, Caroline Haslebacher, Kelly Pasquon, Axel Noblet, Daniel Gurevich and Tatiana Drozhzhova for their help.

Finally, I would like to thank Madeleine Hadorn, our editor at Weber Verlag, for advice and support and Sonja Berger for the layout.

All images are Credit: ESA/Roscosmos/CaSSIS and are based upon data delivered to the European Space Agency's Planetary Science Archive by the CaSSIS project.

Nick Thomas
Bern, February 2022

AUTORENPORTRÄT

Nicolas Thomas (*1960) ist Professor für Experimentalphysik an der Universität Bern und leitender Forscher des Colour and Stereo Surface Imaging System (CaSSIS). Er wurde in Shrewsbury, Grossbritannien, geboren und schloss 1986 sein Studium an der Universität York ab. Im Jahr 2003 kam er nach Bern. Er hat sich auf die Entwicklung von Fernerkundungsinstrumenten für die detaillierte Untersuchung von Objekten in unserem Sonnensystem spezialisiert. Zu seinen Hauptinteressen gehören Kometen, die Monde des Jupiters und der Mars. Er hat dieses Buch mit umfassender Unterstützung des CaSSIS-Teams erstellt.

AUTHOR PORTRAIT

Nicolas Thomas (*1960) is Professor of Experimental Physics at the University of Bern and the Principal Investigator of the Colour and Stereo Surface Imaging System (CaSSIS). He was born in Shrewsbury, Great Britain and graduated from the University of York in 1986. He came to Bern in 2003. He has specialised in developing remote sensing instruments for the detailed study of objects in our Solar System. His main interests include comets, the moons of Jupiter, and Mars. He has produced this book with extensive help from the CaSSIS team.

PORTRAIT DE L'AUTEUR

Nicolas Thomas (*1960) est professeur de physique expérimentale à l'Université de Berne et chercheur principal du Colour and Stereo Surface Imaging System (CaSSIS). Il est né à Shrewsbury, en Grande-Bretagne, et a obtenu son diplôme à l'Université de York en 1986. Il est arrivé à Berne en 2003. Il s'est spécialisé dans le développement d'instruments de télédétection pour l'étude détaillée des objets de notre système solaire. Ses principaux centres d'intérêt sont les comètes, les lunes de Jupiter et Mars. Il a réalisé ce livre avec l'aide de l'équipe CaSSIS.

RITRATTO D'AUTORE

Nicolas Thomas (*1960) è professore di fisica sperimentale all'Università di Berna e ricercatore principale del Colour and Stereo Surface Imaging System (CaSSIS). È nato a Shrewsbury, Gran Bretagna, e si è laureato all'Università di York nel 1986. È arrivato a Berna nel 2003. Si è specializzato nello sviluppo di strumenti di telerilevamento per lo studio dettagliato degli corpi celesti del nostro sistema solare. I suoi interessi principali includono le comete, le lune di Giove e Marte. Ha prodotto questo libro con l'importante contributo da parte di tutti i membri del team CaSSIS.

ОБ АВТОРЕ

Николас Томас (*1960) – профессор экспериментальной физики Бернского университета, руководитель исследовательской группы CaSSIS. Он родился в Шрусбери, Великобритания, и окончил Йоркский университет в 1986 году. В 2003 году Николас приехал в Берн. Профессор Томас специализируется на разработке приборов дистанционного зондирования для детального изучения объектов нашей Солнечной системы. Его основные интересы включают кометы, луны Юпитера и Марса. Он подготовил эту книгу при значительной помощи команды CaSSIS.

IMPRESSUM

Für die grosszügige Unterstützung geht unser Dank an das Swiss Space Office (SSO), eine Abteilung des Staatssekretariats für Bildung, Forschung und Innovation SBFI.

Alle Angaben in diesem Buch wurden vom Autor nach bestem Wissen und Gewissen erstellt und von ihm und dem Verlag mit Sorgfalt geprüft. Inhaltliche Fehler sind dennoch nicht auszuschliessen. Daher erfolgen alle Angaben ohne Gewähr. Weder Autor noch Verlag übernehmen Verantwortung für etwaige Unstimmigkeiten.

Alle Rechte vorbehalten, einschliesslich derjenigen des auszugsweisen Abdrucks und der elektronischen Wiedergabe.

© 2022 Weber Verlag AG, CH-3645 Thun/Gwatt

IDEE UND TEXTE
CaSSIS-Team

FOTOS
Credit: ESA / Roscosmos / CaSSIS

WEBER VERLAG:

GESTALTUNG
Sonja Berger

SATZ
Sonja Berger, Celine Lanz, Nina Ruosch, Cornelia Wyssen

LEKTORAT
Alain Diezig, Madeleine Hadorn

KORREKTORAT
David Heinen

ISBN 978-3-03922-151-6

www.weberverlag.ch

Der Weber Verlag wird vom Bundesamt für Kultur mit einem Strukturbeitrag für die Jahre 2021–2024 unterstützt.

CHRYSE PLANITIA

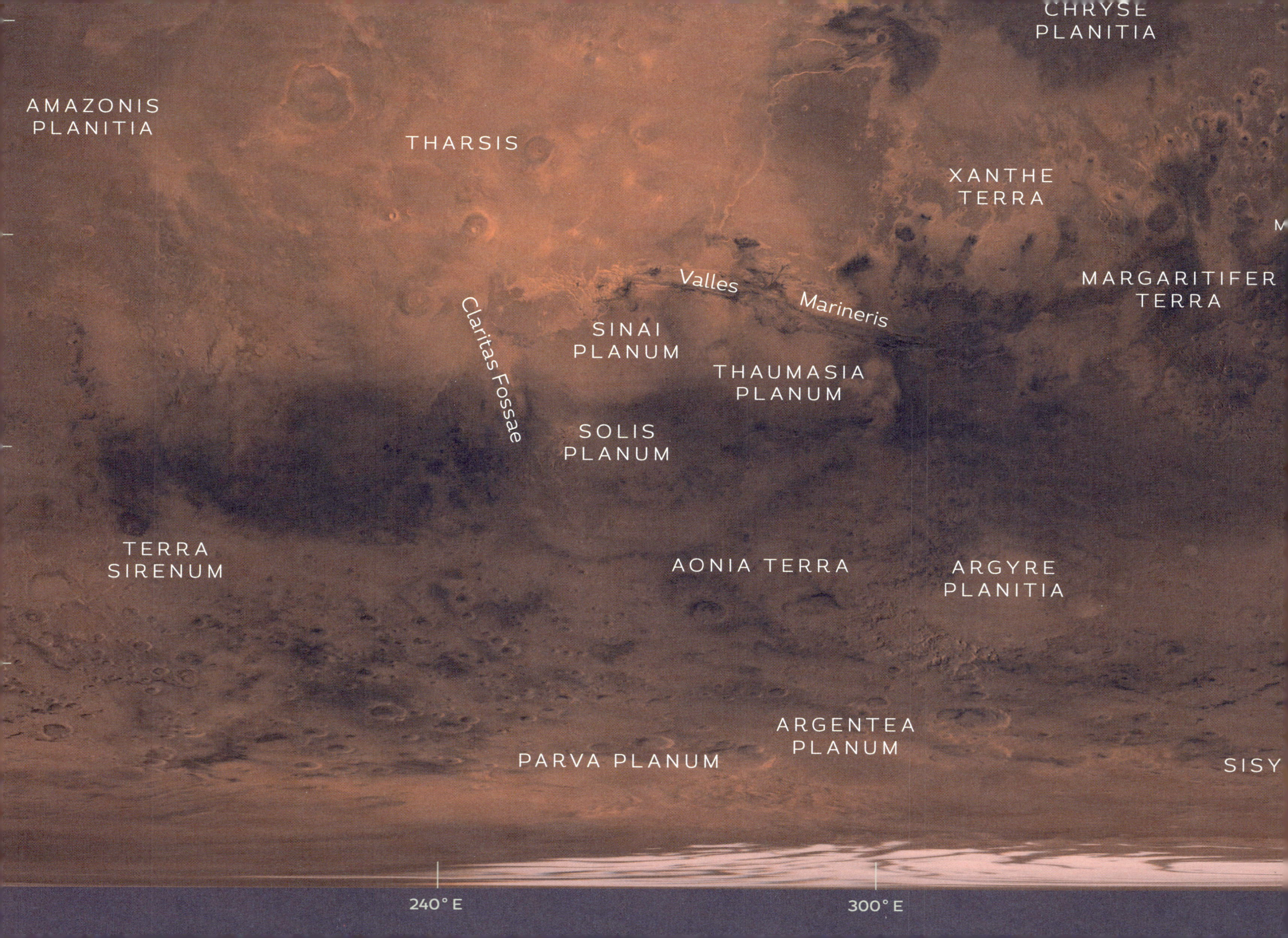

AMAZONIS PLANITIA

THARSIS

XANTHE TERRA

MARGARITIFER TERRA

Valles Marineris

Claritas Fossae

SINAI PLANUM

THAUMASIA PLANUM

SOLIS PLANUM

TERRA SIRENUM

AONIA TERRA

ARGYRE PLANITIA

ARGENTEA PLANUM

PARVA PLANUM

SISY

240° E   300° E